INTERNATIONAL UNION OF CRYSTALLOGRAPHY
BOOK SERIES

IUCr BOOK SERIES COMMITTEE

E. N. Baker, *New Zealand*
J. Bernstein, *Israel*
G. R. Desiraju, *India*
A. M. Glazer, *UK*
J. R. Helliwell, *UK*
P. Paufler, *Germany*
H. Schenk (*Chairman*), *The Netherlands*

IUCr Monographs on Crystallography
1. *Accurate molecular structures*
   A. Domenicano, I. Hargittai, editors
2. *P. P. Ewald and his dynamical theory of X-ray diffraction*
   D. W. J. Cruickshank, H. J. Juretschke, N. Kato, editors
3. *Electron diffraction techniques, Vol. 1*
   J. M. Cowley, editor
4. *Electron diffraction techniques, Vol. 2*
   J. M. Cowley, editor
5. *The Rietveld method*
   R. A. Young, editor
6. *Introduction to crystallographic statistics*
   U. Shmueli, G. H. Weiss
7. *Crystallographic instrumentation*
   L. A. Aslanov, G. V. Fetisov, G. A. K. Howard
8. *Direct phasing in crystallography*
   C. Giacovazzo
9. *The weak hydrogen bond*
   G. R. Desiraju, T. Steiner
10. *Defect and microstructure analysis by diffraction*
    R. L. Snyder, J. Fiala and H. J. Bunge
11. *Dynamical theory of X-ray diffraction*
    A. Authier
12. *The chemical bond in inorganic chemistry*
    I. D. Brown
13. *Structure determination from powder diffration data*
    W. I. F. David, K. Shankland, L. B. McCusker, Ch. Baerlocher, editors

14 *Polymorphism in molecular crystals*
    J. Bernstein
15 *Crystallography of modular materials*
    G. Ferraris, E. Makovicky, S. Merlino
16 *Diffuse X-ray scattering and models of disorder*
    T. R. Welberry
17 *Crystallogrphy of the polymethylene chain: an inquiry into the structure of waxes*
    D. L. Dorset
18 *Crystalline molecular complexes and compounds: structure and principles*
    F. H. Herbstein
19 *Molecular aggregation: structure analysis and molecular simulation of crystals and liquids*
    A. Gavezzotti

IUCr Texts on Crystallography
1 *The solid state*
    A. Guinier, R. Julien
4 *X-ray charge densities and chemical bonding*
    P. Coppens
5 *The basics of crystallography and diffraction, second edition*
    C. Hammond
6 *Crystal structure analysis: principles and practice*
    W. Clegg, editor
7 *Fundamentals of crystallography, second edition*
    C. Giacovazzo, editor
8 *Crystal structure refinement: a crystallographer's guide to SHELXL*
    P. Müller, editor

Molecular Aggregation

Structure Analysis and Molecular Simulation of Crystals and Liquids

ANGELO GAVEZZOTTI

University of Milano

OXFORD
UNIVERSITY PRESS

OXFORD
UNIVERSITY PRESS

Great Clarendon Street, Oxford OX2 6DP

Oxford University Press is a department of the University of Oxford.
It furthers the University's objective of excellence in research, scholarship,
and education by publishing worldwide in

Oxford New York

Auckland Cape Town Dar es Salaam Hong Kong Karachi
Kuala Lumpur Madrid Melbourne Mexico City Nairobi
New Delhi Shanghai Taipei Toronto

With offices in

Argentina Austria Brazil Chile Czech Republic France Greece
Guatemala Hungary Italy Japan Poland Portugal Singapore
South Korea Switzerland Thailand Turkey Ukraine Vietnam

Oxford is a registered trade mark of Oxford University Press
in the UK and in certain other countries

Published in the United States
by Oxford University Press Inc., New York

© Angelo Gavezzotti, 2007

The moral rights of the author have been asserted
Database right Oxford University Press (maker)

First published 2007

All rights reserved. No part of this publication may be reproduced,
stored in a retrieval system, or transmitted, in any form or by any means,
without the prior permission in writing of Oxford University Press,
or as expressly permitted by law, or under terms agreed with the appropriate
reprographics rights organization. Enquiries concerning reproduction
outside the scope of the above should be sent to the Rights Department,
Oxford University Press, at the address above

You must not circulate this book in any other binding or cover
and you must impose the same condition on any acquirer

British Library Cataloguing in Publication Data
Data available

Library of Congress Cataloging in Publication Data
Data available

Typeset by Newgen Imaging Systems (P) Ltd., Chennai, India
Printed in Great Britain
on acid-free paper by
Biddles Ltd., King's Lynn Norfolk

ISBN 0–19–857080–5   978–0–19–857080–6

1 3 5 7 9 10 8 6 4 2

Preface

In the late afternoon of a fine September day I was visiting the Elettra Synchrotron facility on the top of the Basovizza hill, near Trieste, among a group of attendees of the Annual School of the Italian Crystallographic Association. Our guide, a physicist, looked upon us chemists with mild sympathy, the attitude one would take in a kindergarten or in an obedience school for puppy dogs. He explained that, deep under our feet, at that very instant, electrons were generated at an astonishing rate, then accelerated to an astonishing speed by a monster machine called a Linac. They were then forced by giant magnets to circle in a toroidal tunnel. Suddenly, as I stood, a warm and thin breeze gently buffeting my shirt, the sea a faraway glow under the slanted rays of the setting sun, I expected to hear a rumble in my ears, to feel a trembling of the earth under my feet or, at the very least, to see my hair rise under some mysterious electric power seeping out of the violated mountain.

None of that, of course. A distant hum was all that could be heard, but that was the air-conditioning plant. Our guide explained that we were protected by many metres of solid rock. Protected from what? I mused, "what indeed *is* an electron?" We humans – and other animals maybe, for all we know – tend to appropriate objects and objective phenomena by seeing, feeling, hearing, smelling, and tasting, usually in that order of preference. None of these senses works with an electron. Scientists of natural philosophy use a higher (or just different?) level of categorization: models, mental constructions usually involving a fairly large amount of mathematics. As we walked towards the main synchrotron building, I realized that, being a chemist, the best model of an electron I could come up with was a spread of bits and pieces of negative electric charge. I immediately realized that the physicist guiding us would have been horrified by such a crude picture. And that "negative", "electric" and "charge" were also things not to be seen or heard or felt. Fortunately, at that moment we entered the large circular room where the beamlines are located. My jaw dropped, and I was for the moment relieved from the burden of further philosophizing.

In a synchrotron, the circling electrons release a shower of frequency-tunable, extremely intense electromagnetic radiation in beams that run tangentially to the main loop. These beams are used to irradiate samples of matter of many kinds, and the scientists in charge of each experiment extract information about the constitution of these samples by recording their response to that irradiation: absorption, reflection, diffraction. The amount and variety of apparatus needed for this sort of abstruse sorcery are incredible for a newcomer. A typical beamline consists of tons of heavy hardware, concrete blocks, stainless steel, because most experiments must run under high vacuum. Aluminum foil covers many parts of the machinery, giving them the silvery glitter and the splendor of a pharaoh's treasure. The "wiring" consists of kilometers of brightly colored wires and tubes of all sizes, folded in thick loops, or just bunched on the floor, each carrying an indispensable fluid or the magic pulse of an experimental signal. There are computer screens, galore; literally piled one upon the other. The humans, of whom there were very few, looked exhausted and bored – as they indeed were, because they couldn't waste the precious allocated beam time with such trifling occupations as eating or sleeping. A bicycle, intended for quick commuting between beamlines, leant against the wall, a silent demonstration of the vastness of the premises.

Strangely enough, especially for a layman, is to see that all that care, all that display of mass and ingenuity are devoted to samples that usually are no bigger than a pencil's tip. A chemist knows, however, that they are made of an incredibly large number of elementary objects, the molecules within them. Molecules are made up of atoms, and hence of nuclei and electrons. So I summarized, what the scientists are doing here ultimately is analyze matter by shooting electrons at other electrons. I looked at my finger. A finger does not evaporate (or at least, not very fast) because electrons and nuclei in atoms stick together by intermolecular interactions, that is, electric forces. In all human activities, we use and appreciate bodies for their properties: we need objects to be hard, brittle or soft, viscous or elastic, red or blue, conducting or insulating, opaque, reflecting or transparent, and all this depends on how the electrons inside matter arrange themselves at a minimum potential energy, compatible with the kinetic energy of the nuclei.

Intermolecular interactions stem from the electric properties of atoms. Being the cause of molecular aggregation, intermolecular forces are at the roots of chemistry and are the fabric of the world. They are responsible for the structure and properties of all condensed bodies – our bodies, the food we eat, the clothes we wear, the drugs we take, the paper on which this book is printed. In the last forty years or so, theoretical and experimental research in this area has struggled to establish correlations between the structure of the constituent molecules, the structure of the resulting condensed phase, and the observable properties of any material. As in all scientific enterprise, the steps to follow are analysis, classification, and prediction, while the final goal is control, which in this case means the deliberate design of materials with specified properties. This last step requires a synthesis and substantial command of the three preceding steps.

This book aims to describe the concepts and techniques for the study of the structures, energies and dynamic evolution of organic condensed matter at a molecular level. It is mainly oriented to theoretical concepts and their computer simulations, experimental techniques being introduced only to a limited extent. The presentation of historical developments provides some aid to understanding the developments: the reader will forgive me for using, in some places, stories of my own life, but to my partial justification I can say that although it is not a particularly interesting life, it is one that I know very well; and my wanderings through the still sparsely inhabited meadows of organic structural chemistry in the early 1960s are indeed representative, and do give a perspective that is, in my opinion, worth considering.

Throughout the book, the description of methods and techniques is mostly on an intuitive level, in an attempt to provide a grasp of the physicochemical foundations of the subject rather than replicating extensive mathematical derivations that can be found in more specialized books. Specialists in the various sectors may argue with this approach, but in this way no theory, no law, no method, no experimental technique is mentioned without an explanation, and the subject index is designed to do away with the frustration a reader often feels when an author mentions a concept or a method without further comment, in the impertinent assumption that "everybody must know that", where "that" may mean such diverse topics as inertial matrix, molecular shape, Slater determinant, density functional theory, distributed multipole, Pauli exchange repulsion, Bragg peak, $P2_12_12_1$, lattice phonon, Bloch orbital, melting entropy, principal component analysis, smart Monte Carlo, and radial distribution function. The language has been kept as light and colloquial as possible, to accommodate the widely different backgrounds of the readers, and keeping in mind that a scientific book does not necessarily have to make dull reading. There is a survey of significant results, drawing from my own experiences, or from selected literature examples, a complete review being obviously impossible and probably useless; those who have contributed important work should not be offended if they are not mentioned here. Besides, a book should be a convenient vehicle for the expression of the author's current ideas on a given subject, rather than a complete collection of literature citations.

The potential readership is as broad as the community of intellectually curious persons with backgrounds in chemistry, physics, and biology – presumably in that order of descending appeal. Nearly any educational level, from a PhD down to a high school degree, should enable a potential reader to fruitfully approach most of this book. The ideal reader, one who will be able to exploit all the material given here in full, is a university graduate, working either in academia or in an industrial environment, with a strong interest in the investigation of the behaviour of matter at a molecular level, mainly by X-ray diffraction and molecular simulation techniques.

Some book prefaces include long lists of people who are thanked for their individual help. I would simply say that I thank Claudia for being there, Giuseppe Filippini for a lifelong collaboration, and all the colleagues with whom I have shared so many years of intellectual discipline and of good times; indeed, every scientist I have talked to in the last twenty years or so. Some of them have also become close personal friends; like Lia Addadi, Larry Bartell, Joel Bernstein, Hans-Beat Bürgi, Jack Dunitz, Carlo Maria Gramaccioli, Meir Lahav, Leslie Leiserowitz, Mike McBride, Massimo Simonetta, Pierluigi Bellon, and Vladimiro Scatturin, the stepfathers of the Dipartimento di Chimica Strutturale. Also, Egbert Keller, University of Freiburg, generously gave his permission to use his program SchaKal for the preparation of many pictures in this book. Finally, the Oxford University Press staff did a wonderful job typesetting and editing a very complex manuscript.

Angelo Gavezzotti
Milano, Spring 2006

Contents

PART I FUNDAMENTALS

1 The molecule: structure, size and shape ... 3
   1.1 Atoms and bonds ... 3
   1.2 Classification concepts in many particle systems ... 5
      1.2.1 Structure ... 5
      1.2.2 Symmetry ... 8
      1.2.3 Order ... 9
      1.2.4 Enthalpy and entropy ... 9
   1.3 Must a molecule have a size? ... 10
      1.3.1 Mass and length dimensions ... 10
      1.3.2 Atomic radii ... 13
      1.3.3 Molecular volume and surface ... 15
      1.3.4 Size in terms of electron density ... 18
   1.4 Must a molecule have a shape? ... 20
   1.5 Historical portraits: a chemistry course in the early 1960's ... 24

2 Molecular vibrations and molecular force fields ... 30
   2.1 Vibrational modes and force constants ... 30
   2.2 Molecular mechanics ... 35
   2.3 Evolution of molecular force fields ... 39
   2.4 Appendix: an example of coordinate transformation ... 43
   2.5 Historical portraits: Got a force constant? ... 45

3 Quantum chemistry ... 53
   3.1 Some fundamentals of quantum mechanics ... 53
      3.1.1 Dynamic variables, wavefunctions, operators ... 54
      3.1.2 The Schrödinger equation and stationary states ... 54
      3.1.3 Linear momentum ... 56
      3.1.4 Angular momentum ... 57
      3.1.5 Spherical harmonics ... 59
      3.1.6 The harmonic oscillator ... 60
   3.2 The hydrogen atom and atomic orbitals ... 61
   3.3 Spin ... 63
   3.4 Many-electron systems ... 65
   3.5 Molecular orbitals: The Fock and Roothaan equations ... 67
      3.5.1 The variational principle ... 69
      3.5.2 Why do a molecular orbital calculation? ... 72

	3.6	Approximate quantum chemical methods: NDO and EHT	74
	3.7	Evolution of quantum chemical calculations: Beyond Hartree–Fock	76
		3.7.1 Configuration interaction and Møller–Plesset perturbation methods	77
		3.7.2 Density functional theory (DFT) methods	78
		3.7.3 Other beyond-Hartree–Fock methods	80
		3.7.4 Ethylene in perspective	81
	3.8	Dimerization energies and basis set superposition error	81
	3.9	Historical portraits: early experiences in quantum chemistry	82

4 **The physical nature and the computer simulation of the intermolecular potential**   87
- 4.1 Experimental facts and conceptual framework   87
- 4.2 The representation of the molecular charge distribution and of the electric potential   89
  - 4.2.1 Full electron density   89
  - 4.2.2 Central multipoles   90
  - 4.2.3 Distributed multipoles   92
  - 4.2.4 Point charges   92
- 4.3 Coulombic potential energy   94
- 4.4 Polarization (electrostatic induction) energy   96
- 4.5 Dispersion energy   99
- 4.6 Pauli (exchange) repulsion energy   101
- 4.7 Total energies versus partitioned energies   103
  - 4.7.1 Pairwise additivity   103
  - 4.7.2 Interpretation   104
- 4.8 Intermolecular hydrogen bonding   105
- 4.9 Simulation methods   106
  - 4.9.1 The intermolecular atom–atom model for organic crystals   106
  - 4.9.2 Distributed multipole methods   110
  - 4.9.3 Other density-based methods   111
- 4.10 Ad hoc or transferable? Force field fitting from ab initio calculations   112

5 **Crystal symmetry and X-ray diffraction**   120
- 5.1 A structural view of crystal symmetry: bottom-up crystallography   120
- 5.2 Space group symmetry and its mathematical representation   127
- 5.3 von Laue's idea, 1912   130
- 5.4 The structure factor   131
  - 5.4.1 Scattering by one or two charge points   131
  - 5.4.2 The atomic scattering factor   133

		5.4.3	The molecular structure factor	134
		5.4.4	The structure factor for infinite periodic systems	135
	5.5	Miller indices and Bragg's law		137
	5.6	The electron density in a crystal		139
	5.7	The atomic prejudice		139
	5.8	Structure and X-ray diffraction: Some examples		140
	5.9	Historical portraits: Training of a crystallographer in the 1960s		144
6	Periodic systems: Crystal orbitals and lattice dynamics			153
	6.1	The mathematical description of crystal periodicity		153
		6.1.1	Equivalent positions and systematic absences in diffraction patterns	153
		6.1.2	Reciprocal space, wave vector, Brillouin zone	155
		6.1.3	Bloch functions	155
	6.2	The electronic structure of solids		157
		6.2.1	The crystal orbital approach	157
		6.2.2	Band structures: Complicated but not difficult	158
		6.2.3	Comparison with experiment; electronic density of states	162
	6.3	Lattice dynamics and lattice vibrations		163
		6.3.1	Periodic vibrations in infinite crystals	163
		6.3.2	Comparison with experiment; measuring lattice-vibration frequencies	167
7	Molecular structure and macroscopic properties: Calorimetry and thermodynamics			172
	7.1	Molecules and macroscopic bodies		172
	7.2	Energy		174
		7.2.1	The partition function: Molecules	174
		7.2.2	The partition function: Macroscopic systems	176
		7.2.3	Internal energy I: From statistics and quantum mechanics	177
		7.2.4	Internal energy II: From thermal and mechanical experiments	179
	7.3	Heat capacity		179
	7.4	Entropy		180
		7.4.1	Classical entropy	181
		7.4.2	Statistical entropy	182
		7.4.3	The calculation of entropy for chemical systems	182
	7.5	Free energy and chemical equilibrium		183
		7.5.1	Chemical potential	183
		7.5.2	Free energy	184
		7.5.3	Chemical potentials in practice	184
	7.6	Thermodynamic measurements		186

		7.6.1	Heat capacity	186
		7.6.2	Melting enthalpies	188
		7.6.3	Sublimation enthalpies	190
	7.7	Derivatives		193

8 Correlation studies in organic solids — 196
- 8.1 The Cambridge Structural Database (CSD) of organic crystals — 196
- 8.2 Structure correlation — 198
- 8.3 Retrieval of molecular and crystal structures from the CSD — 199
- 8.4 The SubHeat database — 201
- 8.5 The geometrical categorization of intermolecular bonding — 202
- 8.6 Space analysis of molecular packing modes — 203
  - 8.6.1 Empty space versus filled space — 203
  - 8.6.2 Close packing in crystals — 204
- 8.7 The calculation of intermolecular energies in crystals — 207
  - 8.7.1 Lattice energies: Some basic concepts — 207
  - 8.7.2 Convergence problems in lattice sums — 212
  - 8.7.3 Sublimation entropies and vapor pressures of crystals — 213
- 8.8 General-purpose force fields for organic crystals — 214
- 8.9 Accuracy and reproducibility — 217
- 8.10 Correlation between molecular and crystal properties: Fact or fiction? — 220
  - 8.10.1 Bivariate analysis — 220
  - 8.10.2 Principal component analysis — 223
- 8.11 Acceptable crystal structures — 225
- 8.12 Historical portraits: Lattice energies and the phase problem in the old days — 225

9 The liquid state — 230
- 9.1 Proper liquids — 230
- 9.2 Molecular dynamics (MD) — 230
  - 9.2.1 Equations of motion — 231
  - 9.2.2 Temperature — 232
  - 9.2.3 Pressure — 233
  - 9.2.4 NPT and NVT simulations — 234
  - 9.2.5 Performance and constraints in molecular dynamics simulations — 235
- 9.3 The Monte Carlo (MC) method — 236
- 9.4 Structural and dynamic descriptors for liquids — 237
  - 9.4.1 Radial distribution functions — 238
  - 9.4.2 Correlation functions — 241
- 9.5 Physicochemical properties of liquids from MD or MC simulations — 243
  - 9.5.1 Enthalpy, heat capacity and density — 243

	9.5.2	The Jorgensen school	244
	9.5.3	Crystal and liquid equations of state	245
9.6	Polarizability and dielectric constants		246
9.7	Free energy simulations		247
9.8	A theme with variations		249
9.9	Water		249

10 Computers — 254
- 10.1 Bits and pieces — 254
- 10.2 Operating systems — 256
- 10.3 Computer programming — 258
- 10.4 Bugs and program checking and validation — 260
- 10.5 Reproducibility — 261
- 10.6 "Because it's there"? — 262

PART II   THE FRONTIER

11 Structure-property and structure–activity relationships — 269
- 11.1 Fundamental research and applied technology — 269
- 11.2 The structure–activity dogma — 270
- 11.3 Crystal dissolution — 273
- 11.4 Thermal properties — 275
  - 11.4.1 Thermal expansion coefficients — 275
  - 11.4.2 Heat capacity and heat transport — 277
- 11.5 Strain and stress, elastic and viscous properties — 278
- 11.6 Optical, electric and magnetic properties — 284
  - 11.6.1 Color — 284
  - 11.6.2 Optical properties and the polar axis — 289
  - 11.6.3 Electric and magnetic properties — 292

12 Intermolecular bonding — 296
- 12.1 The decline of the intermolecular atom–atom bond — 296
  - 12.1.1 The Feynman–Ehrenfest chemical bond — 296
  - 12.1.2 More familiar models: distance–energy analyses — 298
- 12.2 Full density models: the SCDS–Pixel method — 304
  - 12.2.1 Theory — 304
  - 12.2.2 Coulombic energy — 305
  - 12.2.3 Polarization energy — 306
  - 12.2.4 Dispersion energy — 307
  - 12.2.5 Repulsion energy — 308
  - 12.2.6 Total energies and parameters — 308
  - 12.2.7 The generation of crystal coordinates — 309
  - 12.2.8 Pixel calculations: General features — 310

	12.2.9	Pixel theory: Pros and cons	314
12.3	Systematic application of the Pixel theory to intermolecular bonding		315
	12.3.1	A glossary of intermolecular recognition modes	315
		12.3.1.1 Interactions not involving hydrogen	316
		12.3.1.2 C–H$\cdots$X interactions	318
		12.3.1.3 The O–H$\cdots$X and N–H$\cdots$X hydrogen bond	319
		12.3.1.4 π-interactions	323
	12.3.2	Crystal energies	325
12.4	Directed bonds versus diffuse bonding		326

13 Phase equilibria, phase changes, and mesophases: Analysis and simulation — 330

- 13.1 Things and molecules — 330
- 13.2 Basic thermodynamic functions — 331
- 13.3 Melting — 332
- 13.4 Solid–liquid equilibrium and nucleation from the melt — 338
- 13.5 Vapor–liquid and vapor–solid equilibrium — 341
- 13.6 Glasses — 342
- 13.7 Liquid crystals — 345
- 13.8 Nucleation and growth from solution: Experiments — 347
  - 13.8.1 Overview — 347
  - 13.8.2 Light scattering, calorimetry — 348
  - 13.8.3 Chemical spectroscopy — 349
  - 13.8.4 X-ray scattering and diffraction — 350
- 13.9 Crystal growth and morphology — 351
  - 13.9.1 Crystal faces, attachments energies, and morphology prediction — 351
  - 13.9.2 Electron micrography and atomic force microscopy (AFM) — 355
- 13.10 Evolutionary molecular simulation — 356

14 Crystal polymorphism and crystal structure prediction — 367

- 14.1 A fundamental fact — 367
- 14.2 What are crystal polymorphs? — 368
  - 14.2.1 The taxonomy of organic crystals — 368
  - 14.2.2 Phenomenology of crystal polymorphism — 369
  - 14.2.3 Crystal structure fingerprints: Detecting real polymorphism — 375
  - 14.2.4 Analysis of crystal polymorphism by Pixel and quantum chemical calculations — 379
- 14.3 The construction of crystal structures by computer — 383
  - 14.3.1 "Brute force" approach — 384

		14.3.2	The "Prom" sequential approach	385
		14.3.3	Molecular clusters with one symmetry operator	385
		14.3.4	Combination of two or three symmetry operators	387
		14.3.5	The translation search: Sorting and ranking	389
		14.3.6	The Prom algorithm: Pros and cons	389
		14.3.7	Some examples of crystal structure generation	390
	14.4	Crystal structure prediction by computer		395
		14.4.1	The aims	397
		14.4.2	The tools	398
		14.4.3	Are crystal structures predictable?	398
15	Epilogue: A theory of crystallization?			405
	15.1	Laws and theories		405
	15.2	Aggregation stages		406
		15.2.1	Oligomers	407
		15.2.2	Nanoparticles	407
		15.2.3	Mesoparticles	410
	15.3	Macroscopic crystals		410
	15.4	Thermodynamics, kinetics, and symmetry		411
	15.5	The language of the theory		416

Index   419

Supplementary material

OPiX, an open-code computer program package for crystal packing analysis, polymorph generation and prediction, and Pixel-SCDS calculation; manuals and source codes

Other open-source program codes for the calculations described in the book

The SubHeat database (see in Section 8.4)

The polymorph database (see in Section 14.2)

Details of calculations (lists of atomic coordinates, etc.)

This material is available at the author's website: http://users.unimi.it/gavezzot

I

Fundamentals

1

The molecule: structure, size and shape

... the effective force acting on a nucleus in a molecule can be calculated by simple electrostatics as the sum of the coulombic forces exerted by the other nuclei and by a hypothetical electron cloud whose charge density ... is found by solving the electronic Schrödinger equation.

Levine, I.N. Quantum Chemistry, 5th edn, 2000, Prentice Hall, p. 474.

We shall say that there is a chemical bond between two atoms or two groups of atoms in case that the forces acting between them are such as to lead to the formation of an aggregate with sufficient stability to make it convenient for the chemist to consider it as an independent chemical species.

Pauling, L. The Nature of the Chemical Bond,
Cornell University Press, 2nd edn, 1939, Chapter 1.

1.1 Atoms and bonds

The idea that matter is composed of a very large number of extremely small particles that we cannot see is very old, and most likely stems from the simple observation that any piece of matter can be divided into smaller and smaller parts without any apparent limit to the number of times this operation can be carried out. Traditionally, Lucretius is credited with the first sketch of a theory on the existence and properties of atoms, literally, things that cannot be cut into smaller pieces, although what he and the Greek philosophers before him meant by an atom has not even the vaguest affinity with our present concept.

While geometry and mathematics have been for millennia considered as disciplines worth cultivating by something as noble as the human mind, chemistry has been for a very long time relegated to the lesser enterprises of higher intellect, something a philosopher – somebody who cultivates knowledge, in the Greek sense of the word – would hardly consider worthy of attention. Two thousand years ago mankind possessed an almost perfect geometry, a substantial theory of numbers and a good algebra, a decent theory of planetary motions [1], but knowledge of the inner texture of matter reached not far beyond Aristotle's entirely wrong and incredibly naive – at least for a philosopher of his caliber – theory of the four elements. For many centuries, chemistry was developed for grocery, food manufacturing, and jewelry, whilst alchemic manipulations with transmutation between different aspects of matter no doubt must have looked like sorcery to the cultivated, as well as to the general public.

Strange as it may seem, the key to progress towards a reliable atomic conception of matter was a very humble instrument, the balance. It was in fact through weight laws – Boyle and Lavoisier – that modern science arrived at a proper understanding of the atomic and molecular structure of matter. The discovery of the electron and quantum mechanics completed the job at the beginning of the last century.

The relationships between atoms and molecules are mediated by the theory of chemical bonding. Atoms were forged in the extreme conditions that occurred at the cores of stars in the early stages of the development of our universe, but they did not survive as isolated systems in the much milder conditions of mature planets, where more stable structures are formed by highly stabilizing interactions between electrons of different atoms, in terms of lowering of the potential energy and of spin pairing.

The first epigraph at the beginning of this chapter is a statement of the Hellman–Feynman electrostatic theorem [2]. It says that in a molecule nuclei see a static, smeared-out electron cloud and the forces acting at nuclei are just coulombic forces exerted by other nuclei and by the electron cloud. That these forces are electrostatic in nature reflects the fact that the only potential energy term appearing in a molecular Hamiltonian is a q/R term; at least in this aspect, quantum mechanics does not stray very far from the classical picture, and there are no mysterious quantum mechanical forces acting in atoms and molecules. Equilibrium structures result from zero net forces at nuclei; hence, atomic nuclei are taken as reference points for structure, although one should not forget that the electronic cloud also adjusts itself under the action of the same electrostatic forces. Chemistry is the science of changing one set of nuclear positions with its surrounding electron cloud into a new set of nuclear positions with a new electron distribution: bond breaking and bond making.

Although there is little doubt that a molecule is a structure into which atoms are bound, there have been, and there still are today, some doubts as to what exactly is meant by a bond between two atoms in a polyatomic molecule. The definition of a bond requires the identification of the bound objects. If a molecule is a sea of electrons with a few, very small nuclei swimming in it, what are the bound objects? Even in this diffuse model, for many practical purposes it is useful to preserve to some extent the atomic identities in the molecule and to identify pairs of atoms that we consider as joined by a chemical bond. This is called the molecular connectivity, and then of course molecular structure is commonly described in terms of some geometrical parameters depending on nuclear positions after chemical bonding has occurred. The emphasis is on the position of the nuclei because the particle model of an electron is not applicable in chemical problems, where electrons are much more conveniently treated as a continuous distribution of negative electrical charge in the space around the nuclei, and it is meaningless to speak about the "position" of each single electron. But it should never be forgotten that much if not all of chemistry depends on the rather subtle detail of the electron distribution; chemistry is made by electrons, rather than by nuclei [3].

Recourse to the details of the electron distribution for the explanation of chemical phenomena is rigorous but sometimes awkward, and chemistry has always striven to develop more practical models of chemical bonding. In this perspective, a very simple but not entirely wrong picture of a chemical bond is that of an accumulation of negative electrical charge in between positively charged nuclei, so that attraction exactly balances repulsion; this is the root of the popular "two dot" picture introduced by Lewis to describe ordinary intramolecular chemical bonds.

The second epigraph to this chapter, a statement of charming positivistic flavor, is taken from Linus Pauling's famous book [4]. Very well; but then, when does one absolutely need to, and when does one stop finding it convenient to draw a chemical bond? Pauling's insight is wiser than it seems at first sight. His definition hints at the fact that while the bonding within an organic molecule can be adequately and convincingly described in terms of electron-pair accumulation between atoms, other bonding situations may exist, so that the definition astutely includes the possibility of the formation of bonds between groups of atoms, rather than between individual atoms. Since this book is mainly concerned with intermolecular structure and properties, the expression "chemical bonding" is here preferred over the term "chemical bond" in order to avoid any atomic prejudice from the very beginning. Bonding is here taken to signify net cohesion even when it is impossible to identify two atoms as bond partners, and the attractive force and the stabilizing potential are spread over a considerable part of, or even over the entire electron density of the molecule. This is usually the case in intermolecular bonding between organic molecules.

The matter of atoms in molecules and bonds between them is taken up again in Chapter 12.

1.2 Classification concepts in many particle systems

The word "structure" comes from a Latin root (*construere*) signifying organization, or the act of building ("construction" and "instruction" stem from the same root). The concept of structure apparently hints at a process by which a number of objects evolves from a more random state to a more ordered state. However, structure should not to be confused with order, which is a qualitative and in addition a very often subjective concept. Order sometimes goes hand in hand with symmetry, at least in common thinking, but again the two concepts are very different in the exact sciences – one could even argue that order is not a scientific concept. Symmetry is instead a powerful tool in the scientific description of a structure. The following definitions and considerations may be proposed.

1.2.1 *Structure*

The structure of a chemical system is the result of restrictions on the possible positions of its atomic nuclei and on the distributions of its electron density in space, due to potentials or other quantum mechanical restraints acting among the nuclei and the electrons: the more restrictions there are, the more complex is the structure. The structure of a chemical system is described by a set of nuclear coordinates (x, y, z), together with an electron distribution $\rho(x, y, z)$. Many quantitative structural descriptors can be prepared on the basis of nuclear positions alone.

Chemistry is so much concerned with structure because one of the basic assumptions of modern chemical thinking is that inter- and intramolecular structures determine stability, physical properties and reactivity of chemical systems. Structural chemistry rests upon the definition and use of structural descriptors, which come in large numbers and in general depend on the nature of the compound and on the state of aggregation. According to the above definitions, the only chemical system that has no structure at all is the ideal gas; since the ideal gas does not exist, it follows that all chemical systems have a certain type and amount of structure. The structural descriptor in a sample of argon gas is the diameter of the argon atom, which poses a restriction on the possible positions of other atoms because of Pauli repulsion between outer electrons. Hydrogen gas needs two structural descriptors, the equilibrium distance between the two hydrogen atoms within the molecule and a molecular diameter – a better model may include more descriptors, such as, for instance, the bond distance and the principal axes of an ellipsoid. And so on and on; a molecular crystal apparently needs an infinite number of structure descriptors, but a consideration of symmetry and periodicity brings the number of independent ones down to a few independent atomic positions, plus lattice parameters and symmetry operators.

When attempting a general definition of structure in chemical terms, unexpected difficulties arise, especially when large and complex systems are considered. One the one hand, it might be said that the concept of chemical structure is a uniquely defined and a fundamentally quantitative one: structure is the electron distribution and the set of nuclear coordinates for which Feynman forces vanish. On the other hand, this definition is general but very difficult to visualize and to use, and human consideration of structure very often resorts to qualitative and even plainly subjective ideas. In particular, chemical thinking pays due regard to the peculiar human attitude that, among means of perception of the objective world, considers seeing with one's eyes as the highest form of mental appropriation. This habit is characteristic of an animal species that possesses highly efficient and specialized visual hardware and software: the eyes and the pattern recognition ability in the human brain. There is no structural chemistry without pictures of structures, and, as a sign of extreme devotion to the powers of sight, it is not uncommon to read in structural papers such expressions as a "beautiful" structure, a "neat" structure, and so on.

For molecular modeling purposes, one has to move from pictures to numbers. The location of atomic nuclei must be specified in terms of some geometrical coordinates, preferably expressed in a molecular reference frame. The terms "interatomic" and "internuclear" are often used interchangeably, it being traditionally understood that the location of a given atom in space is identified by the location of its nucleus. When the locations of all the N nuclei of a molecule are known, for example in terms of $3N$ atomic cartesian coordinates, one may calculate: (1) distances between atoms joined by chemical bonds, called bond lengths; (2) distances between atoms not joined by chemical bonds, or non-bonded distances; (3) angles between bond vectors, called bond angles; and (4) torsion angles. Bond lengths, bond angles, non-bonded distances and torsion angles are internal coordinates. The number of independent internal coordinates is also equal to $3N$. Internal coordinates give a more immediate structural picture and convey more immediate structural information than cartesian nuclear coordinates (see Sections 2.1 and 2.2).

Interatomic distances and angles are indeed related to, and actually caused by, the electronic bonding structure, but to infer the electron distribution from a perusal of molecular geometry is not a trivial task. Nevertheless, in the 1970s and the 1980s the ever increasing amount of distance and angle data accumulating from an exponentially rising number of determinations of molecular structures in crystals gave rise to a vast literature on the correlation between molecular geometry and electronic structure. Whether this exercise has been fruitful or not is anybody's judgment. In this perspective, the crystal was considered as a convenient container in which molecules could be kept standing (and perhaps saying "cheese") for an X-ray photograph of them to be taken. These years of crystal structure determination for molecular structure determination have produced a very important result: extremely detailed tables of standard bond lengths between atomic species, including different hybridization states, have been prepared [5]. The same has not been done for bond angles, because there are too many different types of X–Y–Z triads, and even considering only the ten or so atomic species of interest to organic chemistry, the number of different combinations would be impossibly large.

The overall external aspect of a molecule does not change much for small variations of interatomic distances and angles, but may change substantially by rotations about single bonds, also called rotations along torsion angles. The barrier to such torsional rotations is of the same order of magnitude as RT, and therefore quite often a gaseous or liquid sample of a given substance whose molecules have torsional freedom may consist of a thermally equilibrated mixture of different structures. The basic concepts of modern stereochemistry are exposed in a number of excellent textbooks and will not be repeated here. We will recall only the notions of molecular constitution (the way in which atoms are connected), molecular configuration (spatial arrangement not including rotation around single bonds, e.g. enantiomery), and molecular conformation (spatial arrangement produced by rotation around single bonds, e.g. staggered-eclipsed, *trans-gauche*, etc.).

Intermolecular structures are nowadays considered more important than intramolecular ones. Intermolecular forces determine the internal texture of materials – a term which in a general sense encompasses all forms of matter whose physical properties are technologically important; a molecular crystal becomes a material as soon as the chemist's concerns shift towards the relationships between inner structure and macroscopic mechanical, thermal, optical or magnetic properties. Today the main task of X-ray crystallography, aside from biological macromolecules, is the determination of the structure of the crystal, rather than the structure of the constituting molecule.

Descriptors of intermolecular structure are less easy to construct than descriptors for isolated molecules. When the crystal structure is known in full detail (see Chapter 5), including cell dimensions, space group and atomic coordinates for the symmetry-independent part of the molecular constitution of the material, the position of all molecules in the crystal is known. The whole crystal is a sort of giant supermolecule whose individual components do not go beyond small amplitude oscillations around an equilibrium position of the center of mass. For lack of a better approach many simple structure descriptors based on nuclear positions have been taken over from intramolecular chemistry, and one may speak of intermolecular interatomic bonds, the hydrogen bond being a good example. When only weaker interactions are present in a crystal or in a liquid, however, this atom–atom approach to structure description runs into difficulties, because bonding is delocalized. It is quite often the case that descriptors including groups of atoms, sometimes even the entire molecule, are better suited than just a few interatomic distances: typical examples are aromatic ring planes and their stacking, sometimes an important driving force in the formation of organic crystal structures. Besides, in periodic systems it is sometimes very useful to also recognize partial periodic patters, such as ribbons or layers. The degree of arbitrariness in the choice and use of such descriptors obviously increases with the increasing complexity of the descriptor itself; any medium with translational periodic symmetry perforce includes an infinite number of layers and ribbons. Subjective sorting out of particular motifs, especially if they are expected on the basis of one's own theory of molecular packing, is very likely to seep in.

1.2.2 *Symmetry*

Symmetry is the invariance of an object (in chemistry, an object made of nuclei and electrons) with respect to a geometrical operation that transforms the object into itself without distortion. A particular aspect of symmetry is periodicity; a system is said to have a structural periodicity when certain structural motifs or descriptors repeat themselves in space by translation with a given period. Symmetry and periodicity concern nuclear positions and electron density as well, and are extremely useful in describing the structure of crystalline systems.

A molecule is said to be symmetrical if its whole structure can be constructed by repeating part of it through space according to certain geometrical rules. Symmetry elements of interest here are the inversion center, the rotation axes and the mirror plane; symmetry operations are inversion through a center, rotation about a rotation axis, and reflection through a mirror. If two symmetry operations are performed one after another, it can be shown pictorially (and demonstrated mathematically) that the result is another symmetry operation that could have been performed independently. In mathematical language one says that symmetry operations in a symmetric molecule form a symmetry group, which in this case is called a point group. Space groups of crystalline systems will be described in Chapter 5.

In common thinking, symmetry is something related to the space arrangement of macroscopic objects, and by the usual mental prejudice, it is appreciated by seeing. The fleeting nature of this conception is demonstrated by the fact that through the centuries symmetrical decoration has been considered either as the highest achievement of an inspired artist (think of the khalifs who were the masters of the Alhambra in Granada), or as the cheap trick of an insipid dauber (in more recent times). The importance of symmetry in chemistry is underscored by the fact that whenever a structure is symmetrical, so must be the potential that induced it, and this is a great simplification in the description and understanding of chemical phenomena.

Chirality is a manifestation of asymmetry having to do with the truly mysterious fact that the space we live in has (or better, is ordinarily best described in) three dimensions. Switching between two enantiomeric forms of a compound without breaking any chemical bonds requires a brief excursion into the fourth dimension. Chirality is not a tool for simplification, but a tough experimental reality with which all practicing and theoretical chemists have to come to grips.

A beginner's experiment in chirality is as follows. Take two pieces of steel wire about 10 cm long, and bend both of them at 90° at about one third of the length. The two objects are still superimposable. Then, bend them again at 90° at about two thirds of the length, but one to the right and the other to the left: the two objects are no longer superimposable, and are each other's mirror image. Although the force field that produced them (your fingers) had exactly the same intensity, they are two different objects in terms of the orientation in space of some of their parts, and have different properties and serve different purposes against substrates which suffer from the same spatial pathology.

1.2.3 *Order*

Order is a subjective concept that should be carefully defined in each case, it being very difficult, if not altogether impossible, to provide a unique and general definition of order in a given system. The descriptors of structural periodicity within a chemical system are sometimes called order parameters. Since there is no general definition of order, any order parameter defines itself and also defines the kind of order it is trying to describe. Many different order parameters may be proposed, on the basis of geometrical or physical properties of the system. They should perhaps be more properly called "structuring parameters" or "periodicity parameters". For example, for a nematic liquid crystal a geometrical order-parameter could be constructed using the ratio of the number of molecules aligned along the z direction to the average number of molecules aligned along z in an isotropic system (see Section 13.7). Order parameters can also be constructed on the basis of some physical property, for example spin alignment in magnetic materials.

1.2.4 *Enthalpy and entropy*

Enthalpy is the prime quantitative descriptor of chemical energetic stability; enthalpy variations reflect the energetic gain cashed in, or the energetic price paid, when the nuclei and electrons of a chemical system break apart from one structure and coalesce into another structure. Enthalpy differences can be measured by calorimetric experiments, either directly or through specific heat measurements, or calculated by theoretical methods. It is usually true that periodic symmetric arrangements of objects of arbitrary shape that attract one another have more stabilizing cohesive energies than non-periodic ones – hence, the positive value of the enthalpy of melting.

Entropy is related to the number of indistinguishable ways in which a given distribution of objects in space, or of energies among available quantum states, can be attained within a given chemical system. (Entropy cannot be described in terms of order, although many sloppy definitions do just that.) Entropy is influenced by structure and symmetry, and together with enthalpy, it determines the relative thermodynamic stability of chemical systems. Entropy differences can be measured through measurements of equilibrium enthalpy differences. The direct theoretical calculation of the entropy of a chemical system is generally more difficult than the calculation of enthalpies.

1.3 Must a molecule have a size?

1.3.1 *Mass and length dimensions*

X-ray diffraction and spectroscopy have provided the modern chemist with an amazing wealth of structural information on organic molecules. Molecules long ago ceased to be just lists of symbols and numbers on a sheet of paper, at best with a few dots and dashes here and there, and nowadays spring into the third dimension with their stereochemical characterization. Modern chemistry is stereochemistry. As a consequence, encouraged by the beautiful models built from balls and sticks or drawn in full color by computers, we now handle molecules as we do ordinary objects of the macroscopic world. We look at them, we weigh them, and in many other ways we size them up.

In chemistry one usually speaks of small molecules, large molecules, or even macromolecules. This implies a definition of molecular size. It goes without saying that molecules are very small objects in terms of the dimensions of the objects of ordinary life: just consider the fact that the cube of the unit of molecular dimensions, the Ångström unit, when expressed in cm^3 is 10^{-24}, or just the reciprocal of Avogadro's number. This means that as an order of magnitude, one mole of Ångström-sized objects fits into one cubic centimeter. A cubic centimeter contains an inordinately large number of molecules.

The simplest indicators of molecular size may be just the number of atoms in the molecule, N_M, or the total number of electrons, Z_M, a molecular equivalent of the atomic number, or the molecular mass, M_M, the sum of the masses of all protons and neutrons in the atoms of which the molecule is made. These simple indicators are useful for the rationalization of simple facts. For example, every chemist is familiar with the observation that the melting and boiling temperatures within a series of members of the same family of organic compounds increase with increasing number of carbon atoms. Since melting and boiling require the thermal activation of molecular objects, it is reasonable that more thermal energy is needed to activate larger objects.

This book is mainly concerned with the phenomena that occur when molecules aggregate. Molecular packing is defined here as the process by which an ensemble of molecules condense from a state of high dispersion, in which the distances between molecular centers of mass are much larger than molecular diameters, into a state where the space between molecules is smaller than the space occupied by the molecules themselves. The condensed system thus formed can be (paraphrasing Linus Pauling) anything of sufficient stability to make it convenient for the chemist to consider it as an independent state: from a Langmuir–Blodgett or epitaxial film to a liquid, a crystal, and anything in between. For the description of molecular packing one needs some indicators that describe the amount of space occupied by one molecule – spatial size descriptors – and possibly also some indicators of the distribution of this occupied space – shape descriptors.

To describe quantitatively the spatial size of an ordinary macroscopic object, it is customary to give three dimensions: length, width, and height. This is possible in principle also for a molecule, because atom–atom bond distances may be known to a high degree of accuracy by spectroscopic or diffraction experiments, and atom–atom distances can be added up to give overall molecular dimensions. If the exact bond distances and angles are not known for the particular molecule under examination, one can use the average values that are known for the most common chemical connectivities [5,6]. However, molecular objects are rather complex, including convex and concave regions, and there is no unique choice of three main directions along which the molecular dimensions can be measured. One possibility is the use of the reference frame of the three principal inertial axes of the molecule, a unique, easily identifiable frame. For a polyatomic molecule with atoms of mass m_i whose nuclear positions are designated bycartesian coordinates x_i, y_i, z_i in a generic reference frame, the inertial matrix, **I**, has the following form:

$$A = \sum m_i x_i^2, \quad B = \sum m_i y_i^2, \quad C = \sum m_i z_i^2, \quad D = \sum m_i x_i y_i,$$
$$E = \sum m_i x_i z_i, \quad F = \sum m_i y_i z_i$$

$$\mathbf{I} = \begin{bmatrix} B+C & -D & -E \\ -D & A+C & -F \\ -E & -F & A+B \end{bmatrix} \quad (1.1)$$

The eigenvalues of **I** are the principal moments of inertia. These moments for medium-large polyatomic molecules are relatively large, of the order of 10^{-44} kg m^2; this is why the spacings between the corresponding quantum mechanical rotational energy levels are very small (see Section 3.1.4). The origin of the inertial reference frame is taken at the molecular center of mass; then, the eigenvectors of **I** form the matrix that allows the rotation from the original reference frame to the inertial reference frame.

Once the nuclear coordinates are expressed in the inertial reference frame, one can locate the extreme points occupied by a nucleus in the molecule along each of the three inertial axes, thus defining three limiting molecular dimensions (Table 1.1 and Fig. 1.1). The radii of the peripheral atoms (see below) may be added to these dimensions. The axis corresponding to the highest moment is perpendicular to the plane of maximum spread of nuclear positions – like, for example, the molecular

Table 1.1 Principal moments of inertia of organic molecules (units of amu Å^2) and distance (Å) between extreme molecular points (nuclear positions plus 1.1 times the atomic radii in Table 1.2) along the directions of inertial axes. I (kg m^2) = I (amu Å^2) 1.66×10^{-47}

Molecule	I_x	I_y	I_z	d_x	d_y	d_z
benzene	178.4	89.2	89.2	3.9	6.7	7.4
biphenyl	1083.6	909.9	173.8	3.9	6.7	11.7
naphthalene	566.6	405.5	161.1	3.9	7.4	9.1
adamantane	325.2	291.2	258.3	7.3	6.7	7.1
n-hexane	456.4	440.1	34.0	4.2	5.0	10.5
n-heptane	714.4	694.3	41.4	4.2	4.8	11.8
cubane (Fig. 1.4)	146.0	146.0	146.0	6.9	6.9	6.9
C$_8$ isomer (Fig. 1.4)	747.8	723.4	24.4	3.9	5.6	11.8
cyclohexane	202.5	115.5	115.5	5.0	7.1	7.1

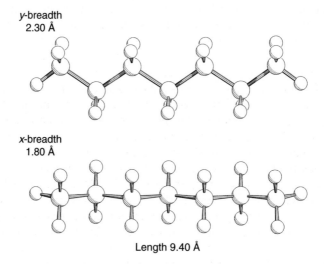

Fig. 1.1. The heptane molecule seen along the inertial axes, with molecular dimensions. The z-axis is the molecular length. Atomic radii may be added for the approximate estimation of the full molecular dimensions.

plane in naphthalene – and the extension of the molecule along this axis can thus be considered as a sort of molecular "thickness". The axis corresponding to the smallest moment runs along the direction of molecular elongation – like, for example, the carbon chain in an unbranched aliphatic hydrocarbon – so that the corresponding molecular dimension can be taken as the molecular "length". But if the molecular connectivity is too complicated, or if the molecule is highly branched and folded, these measures may be scarcely meaningful or even useless.

1.3.2 Atomic radii

Nuclear positions and atomic weights may be important, but molecules are made of an electron density distribution as well, and it is the distribution of outer electrons that determines the direction and the intensity of cohesive forces in condensed matter. A definition of molecular size, including the electron distribution, presupposes a definition of atomic size. An isolated atom can be considered as a very small positively charged nucleus at the center of a negatively charged electron cloud that, in principle, extends to infinity because the atomic electronic wavefunctions go to zero only at infinite distance from the nucleus. This definition is rigorous, but one wants a more practical picture, something like a sphere of electron density with the nucleus at its center. This requires the definition of an atomic radius. Such a radius is not a quantum mechanical observable, and therefore must be regarded as a model quantity that must be clearly defined in each context, while the given definition should afterwards be used only in that context.

The atomic radius can be defined as the distance from the nucleus at which the atomic electron density falls below a certain threshold. This operational definition implies that electron density elements below that threshold are irrelevant in the description of the phenomenon in which one is interested. This definition, based on the electronic properties of the atom, has some limitations and pitfalls: for example, a small change in threshold would cause a rather large variation in atomic radius; and it may be difficult to chose a unique threshold value that would give reasonable radii for very different atomic species, say hydrogen and platinum. On the other hand, the definition of an atomic radius in an experimental sense is impossible for an isolated atom. Even for the simplest chemical system, a noble gas, the concept of atomic radius must be correlated with some property or event that depends on more than one atom, like for example the collision frequency. In molecules and in condensed phases, where atoms and molecules come into close contact and aggregate, the concept of atomic radius depends on the concomitant recognition of the degree of bonding between neighboring atoms: carbon atoms in the benzene molecule have one radius with respect to other carbon atoms, another radius with respect to hydrogen, and another still with respect to intermolecular contact in a direction perpendicular to the aromatic ring in the benzene crystal. This is a major difficulty [7].

This book is mainly concerned with intermolecular effects, and therefore with atomic radii in connection with intermolecular contacts. These are sometimes called non-bonding radii, because they refer to contacts between atoms that are not joined by ordinary chemical bonds. For historical reasons they are also sometimes called van der Waals radii in honor of the famous pioneer of intermolecular interaction studies. A non-bonding atomic radius is rather vaguely defined as the radius of a sphere representing the usual space occupation by each atom in a molecule, so that the spheres of neighboring non-bonded atoms may not overlap. This refers both to intermolecular overlap in condensed phases, and to overlap between atoms in distant parts of the same molecule. A sensible procedure for the determination of these radii uses a careful analysis of the geometrical conditions of proximity between pairs of

Table 1.2 Atomic non-bonding radii (Å)

	Kitaigorodski [8]	Bondi [9]	Gavezzotti [10]	Rowland and Taylor [11]
H	1.17	1.2	1.17	1.10
C	1.80	1.70	1.75	1.77
N	1.57	1.55	1.55	1.64
O	1.36	1.50	1.40	1.58
F	–	1.47	1.30	1.46
Cl	1.78	1.75	1.77	1.76
Br	1.95	1.85	1.95	1.87
S	–	1.80	–	1.81
P	–	1.80	–	1.90
Ar	–	1.88	–	1.80
I	2.1	1.98	2.10	2.03

atoms in different molecules within structured media. In crystals, this would simply be an analysis of the distribution of intermolecular distances. In liquids, this could be an analysis of peak positions in radial distribution curves (see Section 9.4.1).

Table 1.2 presents atomic non-bonding radii obtained by surveys of atom-atom contacts in organic crystal structures [8–11]. What are these numbers? They are precisely what has just been said: one half of the statistically most frequent distance between pairs of atoms of each atomic species in crystals. What can they be used for? As Kitaigorodski proposed many years ago [8], the outer boundary of the molecule is the envelope surface of these overlapping spheres. In this way, a model of the molecule can be built that gives an immediate impression of space occupation, because neighbor molecules may come into close contact only to an extent that does not involve a large intermolecular overlap of the molecular envelopes. An additional bonus of this model is that molecular volume and the area of a molecular surface can be easily and accurately calculated (see below). These numbers are quantitative descriptors of the space size of a molecule.

The temptation of assigning an energetic significance to these surfaces is high, but seldom justified. When building atom–atom empirical parametric schemes to calculate the crystal packing energies of organic molecules (Section 4.9.1) it might be wise to keep the statistical information on distances at hand, and to model the empirical atom–atom interaction functions so that their minima be close to the most frequent values of the sum of the involved atomic radii. Statistical studies of interatomic contacts that reveal systematic deviations from the average atomic radii, towards closest contact, can sometimes be correlated with special bonding effects: typically, hydrogen bonding brings about a drastic reduction of O...H distances from the sum of average hydrogen and oxygen radii (see Section 8.5). But it may be very dangerous to apply this model out of the purely geometrical context in which it has been developed, or beyond the boundaries posed by its physical limits: there is no compelling information on interaction energies in a particular comparison between any given intermolecular distance and the sum of these atomic non-bonded radii. Changes in intermolecular distances between single nuclei can be as large as 0.2–0.3 Å without a significant energy expenditure, with small electron density changes diffuse over the entire molecules in contact. A particular interatomic distance that is less than the sum of these radii has in principle equal chances of being the result of an attractive interaction, of a repulsive interaction, or of no special interaction at all, the implied atoms being in that position as a result of stronger forces on other parts of the molecular system. As a general rule, atom–atom intermolecular contacts in organic condensed media are the result of diffuse interactions rather than of direct atom–atom bonding interactions.

1.3.3 Molecular volume and surface

In the study of the structure of condensed matter, the first step is the analysis of the ways in which molecular objects occupy the available space. One can define intensive quantities, the simplest and experimentally most accessible one being the mass per unit volume of the material, the density, D_X. Similar compounds with different chemical composition have different densities: for example, the average density of n-alkyl alcohols with 5–10 carbon atoms is 0.82 g cc^{-1}, the average density of the corresponding thiols is higher, 0.84 g cc^{-1}, notwithstanding the fact that the molar volume of the thiols is also higher. This is because on substitution of one oxygen atom by sulfur the percent increase in intrinsic molecular volume is much smaller than the percent increase in atomic weight. On the contrary, compounds of different

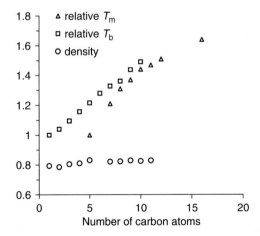

Fig. 1.2. Aliphatic alcohols, linear chain: relative melting and boiling temperatures (for T_m, $C_5 = 1$, for T_b, $C_1 = 1$) and density (g cc^{-1}) as a function of the number of carbon atoms in the chain.

molecular size but similar chemical composition have roughly the same density, as shown in Fig. 1.2.

Intrinsic molecular volume, or the volume of the envelope of atomic spheres, can easily be calculated. Let N be the number of atoms in a molecule, with nuclear positions x_i reckoned in some reference frame, say the inertial reference frame. Let R_i be the atomic intermolecular non-bonding radius of atom i, briefly called henceforth the atomic radius. Let r_{ij} be the distance between the nuclei of two atoms joined by a chemical bond. Whenever r_{ij} is smaller than the sum of atomic radii, the sphere of atom i cuts into the sphere of atom j a spherical cap of height h_{ij}. Molecular volume, V_M, can be calculated [8,10] by computing the total volume of the atomic spheres and subtracting the volumes of the intersecting caps:

$$h_{ij} = R_i^2 + r_{ij}^2 - R_j^2/(2r_{ij}); \quad V_i = 4/3\pi R_i^3 - \sum_{i,j,\text{bound}} 1/3\pi h_{ij}(3R_i - h_{ij}); \quad V_M = \sum_i V_i \tag{1.2}$$

This method runs into difficulties with highly branched or strained bond connectivities because several spherical caps may overlap and the cap volumes may be subtracted more than once. For example, the volume is underestimated by some 10% for cyclopropane ring systems. Alternatively, V_M can be calculated more accurately [10] by subdividing a parallelepiped box that contains the whole molecule into a very large number of small cubes, and counting the number of those elementary cubes that are inside at least one atomic sphere (N_{in}). If $V°$ is the volume of the cube (typically, 10^{-3} Å^3), one gets

$$V_M = N_{in} V° \tag{1.3}$$

The dimensions of the sides of the parallelepiped are conveniently taken as the difference between extreme atomic points along the inertial axes plus the largest atomic radius. An analytical calculation of molecular volume is also possible [12], although differences between the various methods of calculation usually do not exceed a few percent for non-strained molecules.

The density of a condensed phase is mainly determined by the weight of the nuclei, which take only a small fraction of space, and thus depends only marginally on space occupation. A molecular self-density, D_M, can be defined as the ratio of molecular mass to molecular volume [13]:

$$D_M = M_M/V_M \tag{1.4}$$

As seen in Fig. 1.3, crystal density correlates very strictly with molecular self-density, so that a high density in a crystalline material depends primarily on the heavy elements contained in the molecule, and only to a minor extent on efficient space occupation.

Although the molecular surface, as already discussed, is not a physical observable, it may be operationally defined as the surface that encloses all the nuclei and very nearly

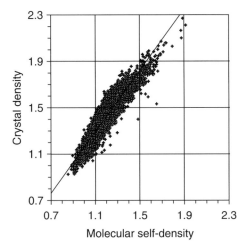

Fig. 1.3. The crystal density (g cc^{-1}) against molecular self-density, equation 1.4, for a large sample of crystalline compounds. The least-squares equation is $y = 1.33\,x - 0.17$.

all the electrons in the molecule. The area of the molecular surface is itself another obvious quantitative size descriptor, but many other properties of this surface, like its curvature, may be used in molecular modeling [14]. The surface area may include all the exposed surface of the atomic envelopes; otherwise, some of the smaller clefts in between atomic spheres may be neglected in the calculation of what is called a solvent-accessible surface. This last quantity can be estimated by letting a spherical probe of given radius roll on the molecular surface and evaluating the contact area. Empirical correlations can be established between this contact area and solvation energies.

There is a rich literature on the methods for calculating surface areas [15]. For instance, a simple method [16] involves sampling the surface of the atomic sphere of each atom i with a large number of points and counting those that are inside any other atomic sphere (N_I) and those that are not (N_O). The free surface of atom i, S_i, and the total molecular exposed surface, S_M, are then

$$S_i = 4\pi R_i^2 [N_O/(N_I + N_O)] \qquad (1.5)$$

$$S_M = \sum_i S_i \qquad (1.6)$$

Molecular volumes and surfaces can be subdivided into a number of contributions from standard chemical groups. These partial volumes and surfaces are constant for the same chemical group in different molecules, as long as the geometry of that group is not significantly altered by some particular bonding situation. Table 1.3 collects some of these group increments.

Table 1.3 Group increments to molecular volume and surface. All dangling bonds are to non-hydrogen atoms

Group	Volume (Å^3)	Surface (Å^2)
—CH$_3$	22	32
—CH$_2$—	16	20
>CH—	11	6
>C< quaternary	5	1
—H	1.5	5
HC≡C—	30	42
H$_2$C=CH—	34	47
—C$_6$C$_5$ phenyl	79	92
—C$_6$H$_4$—	74	79
—C≡N	25	36
—NO$_2$	31	43
—COOH	35	49
—CHO	27	38
—OH	15	25
>CO	22	27
>NH	10	14
—CONH$_2$	37	52
—O—	10	12
—S—	18	22
—SO$_2$	31	37
—F	9	20
—Cl	19	30
—Br	24	35
—I	32	42
—SH	23	34

As an example of the use of the data in Table 1.3, consider again the alcohol family. For a linear chain alcohol with n carbon atoms one gets:

$$V_M(\text{Å}^3) = 22 + (n-1)16 + 15 = 16n + 21$$
$$S_M(\text{Å}^2) = 20n + 37 \qquad (1.7)$$
$$M_M(\text{amu}) = 14n + 18$$

In these series at least, molecular volume, surface and weight are strictly correlated.

1.3.4 Size in terms of electron density

A somewhat more appropriate and physically acceptable definition of molecular volume and surface can be derived from the spatial extension of the electron density. The electron density $\rho(x, y, z)$ can nowadays be easily calculated on a three-dimensional grid with spacing d, so that (x, y, z) are the coordinates of the center of a small element of volume d^3. Using a rather curious neologism, this small element should be called an electron density "voxel", but in analogy with the two-dimensional case it will here be given the more familiar name of "pixel". The charge within the cubic volume of pixel i is $q_i = \rho_i(x, y, z)d^3$. Let N_{thr} be the number of pixels in which the electron density is larger than a preset threshold. Then the molecular volume and the total number of electrons within the molecular envelope can be calculated as:

$$V_{M,thr} = N_{thr} d^3 \qquad (1.8)$$

$$Q_{tot} = \sum_{i=1, N_{thr}} q_i \qquad (1.9)$$

If Q_{tot} is less than the sum of the atomic numbers of the atoms that compose the molecule, some electrons have been left out of the envelope.

Table 1.4 shows some typical results. It turns out that the molecular volumes calculated using Kitaigorodski's method (equation 1.2) and the atomic radii obtained from the analysis of close contacts between atoms in crystal coincide with the volumes calculated by the electron density method using a threshold of 0.02 eÅ^{-3}. On the one hand, this is a remarkable positive result of the rigid spheres model, because it means that these atomic radii have a consistent meaning in terms of electron density and that, broadly speaking, close contact between molecules in crystals has indeed to do with mutual avoidance of molecular envelopes. On the other hand, Table 1.4 shows that the number of electrons left out of the molecular envelopes is not negligible, going from 0.5 to 0.9 electrons for a small molecule, or, in terms of chemical bonding, a quarter to one half of a covalent chemical bond, a number that is only moderately sensitive to the extension of the basis set adopted in the calculation of the electron density. Moreover, the use of such a threshold screens out the peripheral electrons, which are the ones in most close proximity with neighbor molecules, and therefore are most likely to play a role in the weak intermolecular bonding that occurs between

Table 1.4 Molecular volumes in terms of electron densities. MP2/6-31G** wavefunction

Crystal	Threshold, (eÅ^{-3})	$V_{M,thr}$ (equation 1.8)	Electrons out	V_M (equation 1.2)
succinic anhydride	0.002	142.8	0.056	83.0
	0.010	99.5	0.26	83.0
	0.020	82.1	0.51	83.0
benzene	0.020	84.6	0.59	83.3
naphthalene	0.020	130.0	0.83	127.4
benzoic acid	0.020	112.9	0.69	113.6
1,4-dinitrobenzene	0.020	132.2	0.73	135.8
1,4-dichlorobenzene	0.020	117.0	0.79	112.3
n-hexane	0.020	114.8	0.82	106.1

Table 1.5 Overlap integrals between molecular electron densities and overlap volumes for the nearest neighbor molecules in the succinic anhydride crystal. MP2/6-31G** wavefunction

Distance between centers of mass (Å)	Overlap integral (10^{-3} e Å^{-3})	Overlap volume at 0.002 e Å^{-3} limit (Å^3)
4.51	2.37	8.4
5.43	1.57	5.2
6.08	0.98	4.8
6.15	0.75	4.4
6.26	0.80	3.5

organic molecules. Clearly, the rigid spheres model gives a crude account of intermolecular avoidance due to repulsion, but has nothing to say on the true mechanics of intermolecular bonding.

A sensitive indicator of the amount of overlap between neighbor molecules in a crystal is the actual overlap integral between electron densities:

$$S_{AB} = \int \rho_A(x,y,z)\, \rho_B(x,y,z)\, dv \tag{1.10}$$

The corresponding overlap volume is the sum of the elementary volumes of all space pixels in which the electron density of both molecules is above a given threshold. The case is illustrated in Table 1.5 for the succinic anhydride crystal. As expected, the overlap integral correlates with the distance between molecular centers of mass. Coherently with the results shown in Table 1.4, the overlap volumes are non-zero only if the electron density threshold is less than 0.02 e Å^{-3}. In Section 12.2 it will be shown that the overlap integral correlates with the repulsion energy between neighbor molecules, and that the outer electrons play a crucial role in defining the details of the directional intermolecular interaction in crystals.

1.4 Must a molecule have a shape?

The shape of a given object is a mental picture to most people, but few would be able to give a quick and satisfying definition of shape in words, even less in numbers. A blindfold test in which each member of a community is required to provide a definition of shape would no doubt result in as many different answers as the number of members of the community. Interestingly, in a standard English dictionary the definition of "shape" reads "outer form", and the definition of "form" reads "shape". Quantifying molecular shape is much more difficult than quantifying molecular volume, because the concept of shape is a purely non-digital, analogic one. Everybody would agree that the shape of the cubane molecule in Fig. 1.4 is very different from the shape of its linear chain unsaturated isomer, but providing a number or a mathematical formula

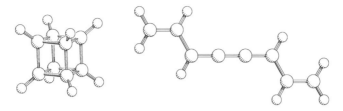

Fig. 1.4. Two C$_8$H$_8$ isomers with their molecular schemes: left, cubane, right, 1,7-octadien-4-yne.

that describes the difference is much more difficult. Many theories and methods have been proposed [17], especially in biological recognition studies where shape complementarity is, if not a positive recognizer, at least a powerful screener of impossible binding modes.

For the investigation of molecular aggregation, it is important to determine if and how molecular shape affects the ability of a given molecule to occupy space in an efficient manner. But what is meant by "efficient"? Condensed phases are, indeed, condensed, meaning that molecules come into close contact with one another, as implied by the previously given definition of molecular packing. The internal structure and compactness of a material is determined by a competition between the attractive influence of intermolecular potentials and the disrupting action of molecular kinetic energy. Intermolecular attractions fall off rather quickly with intermolecular distance: a crude model, not entirely devoid of scientific merit, builds upon this fact and just concludes that molecules will try to get as close as possible to one another – or, in evocative words, in condensed phases empty space is wasted space [18]. In this primitive picture the stability of a condensed molecular system increases with decreasing molar volume. In the rigid-sphere model, a descriptor of molecular shape should be an index that expresses the ability of a given molecule to fill space in such a way as to leave as little empty space as possible. For spherical objects the highest possible occupation factor is about 0.7, for parallelepipeds it is 1.0; however, no organic molecule resembles a parallelepiped (not even cubane) and there will always be bumps and clefts that will cause some empty space to be left in between approaching molecules. The problem of finding the most efficient arrangement for objects of arbitrary shape, as molecules are, has no known method of solution.

Some simple numerical indicators can be proposed. Consider again the numerical integration method that allowed the calculation of molecular volume in equation 1.3. If V_{box} is the total volume of the surrounding box and N_{tot} is the total number of sample points, a self-occupation coefficient can be calculated as:

$$C_{self,occ} = V_M/V_{box} = N_{in}/N_{tot} \qquad (1.11)$$

N_{in}/N_{tot} is the fraction of occupied space in the rectangular box and therefore $C_{self,occ}$ can be interpreted as an index of similarity of a certain object to a parallelepiped, and can be used as an index of the intrinsic ability of the molecule to occupy space

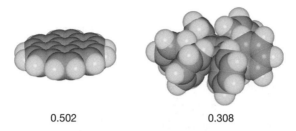

0.502   0.308

Fig. 1.5. The self-occupation coefficients, equation 1.11, for coronene (left) and hexaphenylbenzene (right).

efficiently. For example, the value for a flat, compact, almost brick-like molecule like coronene is much higher than that for a very twisted molecule like hexaphenylbenzene (Fig. 1.5). Some other indices can be proposed, using ratios of moments of inertia; for example, the three moments of inertia for a globular object are almost equal, while for an elongated (cylindrical) object one of the moments is much higher than the others (see Table 1.1). More indices can be proposed on the basis of atomic connectivity, of the amount of branching along aliphatic chains, on the number of rotatable single bonds, and the like [19].

While these definitions and descriptors may be useful in systematizing molecular conformations, no safe correlations have ever been found between any of them and solid state properties of organic compounds. The seemingly inevitable conclusion is that as far as the study of molecular aggregation in condensed phases is concerned, these descriptors are too qualitative to be of any real use. Condensed phase properties depend on subtle details of the arrangement of the electron density in the molecule and between molecules, so that the true physics of the interaction must be considered.

The shape of a molecule is not independent of experimental conditions. For molecules with torsional degrees of freedom, flexible molecules, shape becomes a temperature-dependent concept, because an actual sample of such a compound in the gaseous or liquid state or in solution will contain a variable mixture of all the thermally allowed conformers. The actual shape found in the crystal may depend on crystallization conditions, because torsional degrees of freedom sometimes readjust to comply with packing requirements. In a classical example, the angle between the molecular planes of the two benzene rings in biphenyl is 40° in the gas phase, but the molecule adopts a completely planar conformation in its crystal, paying the conformational energy price in order to facilitate crystal packing.

In terms of intermolecular cohesion, definitions of molecular shape can be introduced through some effects on experimental bulk properties. These operational definitions are particularly useful for comparisons among isomers rather than on an absolute basis. In isomers, the chemical composition is the same but the distribution in space of molecular components is different. A typical example is provided by disubstituted benzenes [20]: every practicing organic chemist knows that 1,4-isomers are almost invariably the highest melting ones. Figure 1.6 shows that differences in

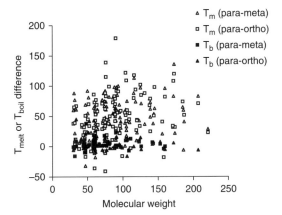

Fig. 1.6. Differences between melting temperatures (T_m) and between boiling temperatures (T_b) of disubstituted benzene isomers. Differences in melting temperatures are high and the 1,4-isomer is almost invariably the highest-melting one. Differences in boiling temperatures are small.

melting temperature between the 1,4 and the 1,3 isomers, or between the 1,4 and 1,2 isomers, are very high, while the corresponding differences in boiling temperatures are very small. This effect appears irrespective of the nature of the substituent, indeed the data in Fig. 1.6 are for pairs among a variety of about forty different substituents, including polar and apolar ones, hydrogen-bonding or not. For some reason that is not entirely clear, the molecular shape of the 1,4-isomer is more favorable for the construction of a crystalline phase than the shapes of the 1,2- or 1,3-isomers. This is a true solid-state effect because it does not appear in the liquids.

Another bulk property that can be related to molecular shape is molecular mobility in the solid state. In a medium where molecules are otherwise tightly locked together, large-amplitude rotations of some parts of the molecule, or of the entire molecule, are possible only if they occur without a substantial variation in the overall molecular shape. For example, rotation of the hydrogen atoms in methyl groups around the C–CH$_3$ axis is widespread in organic crystals, where methyl hydrogens are often found to be disordered. If the entire molecule is involved in the reorientation, the shape must be such that the rigid crystalline environment is not seriously perturbed by the rotation. Flat circular molecules perform large-amplitude rotational motions in their molecular plane and even complete reorientations in the crystal; Table 1.6 shows some potential energy barriers to these molecular rotations, as determined by temperature-dependent NMR spectroscopy. Globular molecules form rotationally disordered or plastic crystal phases, where only the periodicity in the position of the centers of mass is retained. The result is often an increase in molar volume of the crystal without a complete loss of crystal ordering. Table 1.7 collects some typical volume increments on melting for organic compounds: cyclohexane, a globular molecule, exhibits a very small volume change on melting. Other shape effects can be gleaned from the data in

Table 1.6 Experimental potential energy barriers to rigid-body reorientations in some organic crystals [21,22]

Compound	Barrier (kJ mol^{-1})	Compound	Barrier (kJ mol^{-1})
benzene	14	cyclohexane	23
dichlorotetramethylbenzene	32	adamantane	22
hexamethylbenzene	27	plastic furan	8
trichlorotrimethylbenzene	43	plastic neopentane	4
2-fluoronaphthalene	46		

Table 1.7 Volume changes on melting for organic compounds [22]

Compound	% volume change	Compound	% volume change
benzene	13.3	phenanthrene	8.7
nitrobenzene	10.9	chlorobenzene	7.1
cyclohexane	5.2	methylbenzene	11.3
biphenyl	12.4	1,2-dimethylbenzene	8.0
anthracene	16.8	1,4-dimethylbenzene	22.1
1,4-dihydroxybenzene	13.5	average, n-alkanes > C_6	11.3
alkyl acids > C_3	8.3–15.0		
average alkyl alcohols, odd C number	8.5	even C number	11.7

Table 1.7, for example, in the difference between the volume change of anthracene and phenanthrene. The more regular shape of anthracene is clearly more favorable for tight molecular packing in the crystal than the bent shape of its isomer phenanthrene, whose crystal also has a lower packing coefficient and a lower heat of sublimation. These observations and conclusions are however qualitative and sporadic, and can hardly be used in systematic theories of molecular packing.

1.5 Historical portraits: a chemistry course in the early 1960's

I was born in October 1944, on the same day that Marshal Rommel took his life, at the gentle urging of Mr Adolf Hitler. While I was trying to survive on the scarce food supplies then available, Mr Mussolini was wreaking havoc with the torn remains of his – our – brutally abused homeland. And yet, just six years later I rather comfortably entered primary school, perhaps out of sheer luck because both our family apartment house and the school building had miraculously survived the massive bombings of Milano and stood unscathed among piles of rubble all around. My father had been quickly acquitted on charges of having been an active member of the Fascist Party (like nearly everybody else), on the grounds that he had done nothing else in his life but work hard, and had just been caught (like nearly everybody else) in Mussolini's net. Although I still wore previously well worn second-hand clothes, morale was at its highest. When I entered high school in 1958, northern Italy was on the verge of becoming (guess what?) a rich country. By the time I finished, in 1963, my father had passed away, but from what he had left as a pension our family of three could afford expensive clothes and a car.

It was time to decide what to do with my life. My grandfather used to run a small wine shop in the outskirts of Milano, and had died of unknown causes, together with his wife, at an early age, probably out of malnutrition. My father had nearly escaped the same fate as a child; god only knows what were the health conditions in the italian countryside around 1910. None of my four grandparents had gone beyond the very first classes of elementary school, and mother had just been a lovely housewife after working a few years as a typist and shorthand clerk in an insurance company. No family member had ever shown an inclination for trade or money making. But there was a hint of cultural ambition in the family, something unheard of in 1935: an aunt (my father's sister) had graduated at the University of Milano and had become a school teacher. Our house was full of books. Somehow it went without saying that my brother and I, in spite of humble origins but taking full advantage of a relative economic well being, would go to University. The problem was, I had to decide what subject to take. I wanted something connected with the natural sciences (*naturalis philosophia*), away from competition with my brother who was inclined towards the humanities. Physics seemed too elitist, mathematics too evanescent. I wanted something long and tough, as I felt I should put my brain to tackling something less inane than the Latin and Greek nonsense I had been exposed to for the past eight years. In those days in Milano all these conditions meant either the School of Engineering (the *Politecnico*) or studying Industrial Chemistry. I chose the second with one thought at the back of my mind – an uncle was making money with a small chemical company, and one never knows.

In 1964, a large part of the Faculty of Natural Sciences of the University of Milano was housed in an old three-story building with a grim facade surmounted by an inscription in capital Roman letters that said "University" – the inscription was oddly shifted to the right, as the adjective "Royal" had been hastily scratched off a few years earlier (as had Italian royal fame, which had already gone down the drain when the royal family scrambled out of Rome in 1943, leaving country and army without any help or instructions). The building housed the whole of chemistry, mathematics, biology, geology. The industrial chemistry class comprised about a hundred pupils, who literally lived the whole day, eight to five, Monday morning to noon Saturday, in two classrooms and two laboratories. Physicists (always smarter than their chemist cousins) already had their brand new, streamlined building, which had just been erected by ploughing up an old field where I had played soccer only a few years earlier.

The approach to general chemistry teaching in these days was dismaying. I met with a lot of nineteenth century flavors, lists of old fashioned recipes and almost Rosacrucian procedures, not to mention that safety was a word no one seemed to have ever heard of as we stayed for hours in a narrow room pumping H_2S into our beakers and into our lungs as well. I now realize that I barely escaped the parachor (if not the phlogist) and qualitative analysis by tasting. I was taught Avogadro's law through the labyrinth of weight laws, before I was told what a molecule really was. Chemical bonding was a matter of placing small dots or small arrows in the right place between symbols for atoms. But there was some light in the darkness: an organic chemistry book by Morrison and Boyd, from which I picked up a correct approach to chemistry, that is, looking at molecules in three dimensions, and studying the way bonds are made and broken with due consideration of their directions in space. In partial justification of my teachers of those times, one should recall that there were no projection facilities available, no slides, no nothing, and ball and stick models were clumsy and expensive. So everything was in two dimensions on the blackboard, including those uncanny half circles with an oxygen atom in the middle that had to be drawn to schematize ring closure in sugars.

Chemical analysis was, for two full years, a matter of mixing two clear liquids and getting a precipitate or at least something coloured, and of laboring for hours with harmful chemicals to do something that even in those days could have been done in ten minutes on a machine. Again, there was a light in the darkness, in the form of a book by Brand and Eglinton where we were taught how to identify a chemical product by joint use of IR, UV, and NMR. In organic chemistry practice, Vogel's book was fun, pretty much in the same intellectual perspective as "*Il Talismano della Felicità*" ("The Talisman of Happiness", the best known Italian cookbook), but at least there I found the thrill of making matter transmute from one form to another and the satisfaction of observing the dry, neat crystals of a newly synthesized substance–incidentally, my 1965 organic chemistry laboratory was my first encounter with molecular aggregation.

Thermodynamics was given in two separate courses, one with the classical approach, which mainly consisted of mysterious cycles involving mysterious gases, and the second with statistical mechanics, given as a list of extremely complicated formulas. No one bothered to tell us that the two things led to the same target, and I totally missed the beauty of the agreement between Clausius' and Boltzmann's intellectual achievements. Besides, I took statistical thermodynamics before learning molecular spectroscopy, to what instructional purport, one can easily imagine.

On the other hand, in our first years we were exposed to a large amount of mathematics, including calculus in one and many variables, analytical geometry, and analytical and theoretical mechanics. This was music to my ears as I realized that the exertions of an intellect working within the boundaries of rigidly fixed rules were not a diminution, but an amplification of human capabilities. The idealistic (and provincial) notion that mathematical rules would only put a rein on one's imagination, one of the tenets of twentieth century Italian philosophy, was an outright lie, amplified through the years only because human arts speak in everyday words which anybody thinks they can understand, while mathematics requires an intellectual discipline which many lazy minds do not care to acquire. What I found even more fascinating was that the same mathematics I learned in mathematics courses proper was used also in physics courses, and came to results in numbers, which could be compared with numbers harvested from measurements of objective phenomena. This struck me as an exceptional discovery. I became omnivorous and began devouring books on almost anything from matrix algebra to astronomy to particle physics. I mentally cursed the high school teachers who had made us spend so much time on mediocre medieval poems and had not told us about Galileo.

References and Notes to Chapter 1

[1] Neugebauer, O. *The Exact Sciences in Antiquity*, Dover, New York, 1969.
[2] Levine, I. N. *Quantum Chemistry*, 5th edn, Prentice Hall, 2000, p. 474.
[3] Bader, R. F. W.; Fang, D.-C. Properties of atoms in molecules: caged atoms and the Ehrenfest force, *J. Chem. Theor. Comp.* 2005, **1**, 403–414.
[4] Pauling, L. *The Nature of the Chemical Bond*, Cornell University Press, 2nd edn, 1939, Chapter 1.
[5] Orpen, A. G.; Brammer, L.; Allen, F. H.; Kennard, O.; Watson, D. G.; Taylor, R. in: H.-B. Burgi; Dunitz, J. D. (Eds), *Structure Correlation,* VCH, Weinheim, 1994, Vol. 2, Appendix A.
[6] If a molecular model is not available, it can be built using known geometrical parameters, for example by joining together molecular fragments in standardized geometries: see e.g. Sadowski, J.; Gasteiger, J., From atoms and bonds to three-dimensional atomic coordinates: automatic model builders, *Chem. Rev.* 1993, **93**, 2567–2581.
[7] The atoms in molecules (AIM) theory (see Section 12.1.1) defines atoms in an unequivocal manner by the analysis of topological properties of the molecular electron density. AIM atomic basins are delimited by complex surfaces and are far from spherical.
[8] Kitaigorodski, A. I. *Organicheskaya Kristallokhimiya,* Translated from Russian in *Organic Chemical Crystallography*, Consultants Bureau, New York, 1961, p. 7.
[9] Bondi, A. van der Waals volumes and radii, *J. Phys. Chem.* 1964, **68**, 441–451. Bondi, A. *Physical Properties of Molecular Crystals, Liquids and Glasses*, John Wiley & Sons, New York, 1968, p. 2.
[10] Gavezzotti, A. The calculation of molecular volumes and the use of volume analysis in the investigation of Structured media and of solid-state organic reactivity, *J. Am. Chem. Soc.* 1983, **105**, 5220–5225.
[11] Rowland, R. S.; Taylor, R. Intermolecular nonbonded contact distances in organic crystal structures: comparison with distances expected from van der Waals radii, *J. Phys. Chem.* 1996, **100**, 7384–7391.
[12] Connolly, M. Computation of molecular volume, *J. Am. Chem. Soc.* 1985, **107**, 1118–1124.
[13] Dunitz, J. D.; Filippini, G.; Gavezzotti, A. Molecular shape and crystal packing: a study of $C_{12}H_{12}$ isomers, real and imaginary, *Helv. Chim. Acta* 2000, **83**, 2317–2335.
[14] Mezey, P. Molecular Surfaces, in *Reviews in Computational Chemistry* 1990, pp. 265–294, edited by K. Lipkowitz and D. Boyd. VCH, Weinheim.

[15] For some representative examples of the proposed methods of calculating molecular surface areas see: Connolly, M. L., Analytical molecular surface calculation, *J. Appl. Crystallogr.* 1983, **16**, 548–558 (solvent accessible surface, solvent modeled by a sphere); Stouch, T. R.; Jurs. P. C. A simple method for the representation, quantification, and comparison of the volumes and shapes of chemical compounds, *J. Chem. Inf. Comput. Sci.* 1986, **26**, 4–12 (point-by-point integration on a grid mesh); Meyer, A. Y. Molecular mechanics and molecular shape. V. On the computation of the bare surface area of molecules, *J. Comput. Chem.* 1988, **9**, 18–24 (point-by-point method); Silla, E.; Tunon, I.; Pascual-Ahuir, J. L. GEPOL: An improved description of molecular surfaces. II. Computing the molecular area and volume, *J. Comp. Chem.* 1991, **12**, 1077–1088 (tessellation methods: spherical surfaces are divided into triangular tesserae); Le Grand, S. M.; Merz, J. K. M. Rapid approximation to molecular surface area via the use of Boolean logic and look-up tables, *J. Comp. Chem.* 1993, **14**, 349–352 (matching of point-by-point representation of surface for correlation with hydration energies in fast molecular dynamics calculations); Perrot, G. B.; Cheng, K. D.; Gibson, J.; Vila, K. A.; Palmer, A.; Nayeem, B.; Maigret; Scheraga, H. A. MSEED: A program for the rapid analytical determination of accessible surface areas and their derivatives, *J. Comp. Chem.* 1992, **13**, 1–11 (accessible surface area for modeling the interactions with solvents in energy calculations); Petitjean, M. On the analytical calculation of van der Waals surfaces and volumes: some numerical aspects, *J. Comp. Chem.* 1994, **15**, 507–523 (complete analytical calculation of molecular surface and volume).

[16] Gavezzotti, A. Molecular free surface: a novel method of calculation and its uses in conformational studies and in organic crystal chemistry, *J. Am. Chem. Soc.* 1985, **107**, 962.

[17] For a general review, see Arteca, G. A. Molecular Shape Descriptors, in *Reviews in Computational Chemistry*, 1996, **9**, pp. 191–253, edited by K. B. Lipkowitz and D. B. Boyd, VCH, Weinheim. For other samples of the rich literature: Arteca, G. A.; Mezey, P. Shape characterization of some molecular model surfaces, *J. Comp. Chem.* 1988, **9**, 554–563 (topological method); Shoichet, B. K.; Bodian, D. L.; Kuntz, I. D. Molecular docking using shape descriptors, *J. Comp. Chem.* 1992, **13**, 380–397; Zachmann, C.-D.; Heiden, W.; Schlenkrich, M.; Brickmann, J. Topological analysis of complex molecular surfaces, *J. Comp. Chem.* 1992, **13**, 76–84 (calculation of the local and global curvature of molecular surfaces); Laskowski, R. A. SURFNET: A program for visualizing molecular surfaces, cavities and intermolecular interactions, *J. Mol. Graph.* 1995, **13**, 323–330; Malhotra, A.; Tan, R. K.-Z.; Harvey, S. C. Utilization of shape data in molecular mechanics using a potential based on spherical harmonic surfaces, *J. Comp. Chem.* 1994, **15**, 190–199 (development of a potential function that allows the use of topographical information in molecular modeling); Good, A. C.; Ewing, T. J.; Gschwend, D. A.; Kuntz, I. D. New molecular shape descriptors: application in database screening, *J. Comp.-Aided Mol. Des.* 1995, **9**, 1–12.

[18] Dunitz, J. D.; Gavezzotti, A. Attractions and repulsions in molecular crystals: what can be learned from the crystal structures of aromatic hydrocarbons?, *Acc. Chem. Res.* 1999, **32**, 677–684.
[19] Hall, L. H.; Kier, L. B., The molecular connectivity chi indexes and kappa shape indexes in structure-property modeling, in *Reviews in Computational Chemistry,* 1991, pp. 367–422, edited by K. B. Lipkowitz and D. B. Boyd. VCH, Weinheim.
[20] Gavezzotti, A. Molecular symmetry, melting temperatures and melting enthalpies of substituted benzenes and naphthalenes, *J. Chem. Soc., Perkin Trans.* **2**, 1995, 1399–1404.
[21] Gavezzotti, A.; Simonetta, M. Crystal chemistry in organic solids, *Chem. Revs.* 1982, **82**, 1–13.
[22] Ubbelohde, A. R. *The Molten State of Matter*, John Wiley & Sons, Chichester 1978, pp. 12–13, 148.

2

Molecular vibrations and molecular force fields

These simplifications ... too often so drastically alter the complexion of the problem as to render the answers either suspect or unreasonable. The basis of the present work is accordingly an effort to break through this barrier ... by employing machine calculation, thus allowing a far greater magnitude of mathematical effort in a reasonable time with untiring accuracy, and a consequent capability of a more intimate probing into these problems than is possible with hand calculation (the IBM 709 computer, used in the present work, is capable of 8000 additions, 4000 multiplications or 500 more complex functions per second).

Hendrickson, J. B. *J. Am. Chem. Soc.* 1961, **83**, 4537.

Quantitative interpretation of organic chemical phenomena requires the availability of accurate thermochemical information. Experimental data are sparse ... A method is needed whereby such energies can be estimated inexpensively, easily, and with high expectation of accuracy, even when the molecules of interest have not yet been prepared. At the present time, molecular mechanics calculations represent the best approach to a solution of this problem.

Schleyer, P.v.R., *et al. J. Am. Chem. Soc.* 1973, **95**, 8005.

2.1 Vibrational modes and force constants

The data and the methods discussed in Chapter 1 allow the preparation of beautiful detailed molecular models in many styles (wireframe, ball-and-stick, envelope) that originate from the knowledge of the position of atomic nuclei at equilibrium. All these models are deceiving in one crucial aspect: they depict a static molecule, without any trace of molecular vibrations. In fact, all atoms oscillate around equilibrium positions under the restraining action of the bonding potentials. These relentless motions, on a timescale of 10^{-13} s, are a large component of the internal kinetic and potential energy of a chemical system, and are described as displacements of the nuclei of all atoms in the molecule, without regard of the electron clouds, whose "motions" are on a much faster timescale. The molecular modeler who neglects molecular vibrations is like an entomologist who looks at his beautiful collection of embalmed butterflies and never realizes that these insects can fly.

In a classical mechanics approach, the description of motion requires a dynamic trace of the displacement of an object in time, a trajectory that collects the variation in time of some coordinates q. But which coordinates? For molecules, a first choice could be cartesian coordinates for each nucleus in a laboratory reference frame; the collection of all such coordinates is symbolized by the square brackets:

$$[q_C] = [x_i, y_i, z_i] \tag{2.1}$$

In general, molecular motion is a combination of translation, rotation, and vibration. If one is interested only in vibration, the translational and rotational components of motion must be deconvoluted out, and that requires a change in coordinates. Translation could be screened out by using a reference system with its origin at the center of mass, but that would still leave a mixture of rotation and vibration. What is needed in order to have pure vibrations is a set of coordinates that vary in time in such a way that there is no net angular velocity around the three inertial axes.

For an N-atom molecule there are $3N - 6$ internal degrees of freedom, so there must be $3N - 6$ independent vibrations and one needs as many independent coordinates for their description. Chemical intuition works in terms of chemical bonds, so that the most intuitive way of representing molecular vibrations is in terms of the variations in bond lengths, bond angles, and torsion angles. These are internal coordinates, $[q_I]$. Cartesian coordinates are independent by definition, but finding a set of truly independent internal coordinates for a large polyatomic molecule is all but a trivial matter [1].

Consider water as a first example: three obvious internal coordinates are the two O—H distances and the HOH angle. Trajectories in terms of these three coordinates would automatically screen out translation and rotation. But this is not yet enough. Elementary physics tells that two springs with equal force constant when joined together must couple their motions somehow, so that describing the vibration of water by separate R_{OH1} and R_{OH2} stretch is not proper. More generally, all "springs" within the molecule are coupled, and what is sought is a set of vibrational coordinates that explicitly include contributions from all these couplings. This has to be done by some sort of linear combination. A first step in this direction is the use of symmetry-adapted coordinates, q_S, while the final goal (as will be made clear in what follows) are the normal coordinates, q_N. In terms of internal coordinates:

$$q_{N,k} = \sum_j a_{kj} q_{I,j} \tag{2.2a}$$

$$q_{S,k} = \sum_j b_{kj} q_{I,j} \tag{2.2b}$$

The next problem is finding the coefficients of the linear combination, the a_{kj}s or b_{kj}s. Of course there are rigorous mathematical procedures, but it is often useful to have an intuitive guess at what the normal coordinates should look like, and here is where symmetry helps. If the molecule is symmetric so must be its vibrational potential. For example, in the water molecule the force needed for stretching each of the two bonds must be the same, and the bond stretching vibrations must couple in a symmetry-adapted fashion into a normal mode. This restricts the vibrational normal modes to simultaneous displacements along the bond directions by the same span (Fig. 2.1), although atoms may move in the same direction (down–down or up–up in

32   MOLECULAR VIBRATIONS AND MOLECULAR FORCE FIELDS

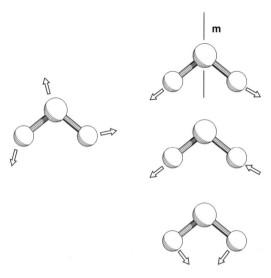

Fig. 2.1. A triatomic molecule, say water. Left: the overall molecular motion involves displacement of all atoms in a combination of translation, rotation, and vibration. On the right, the arrows sketch the displacements of the nuclei along normal coordinates, symmetry-adapted in terms of the molecular symmetry, represented by a mirror plane whose trace is denoted by **m**. In all normal coordinates, also the oxygen atom is slightly displaced so as to ensure that there is not net translational or rotational motion.

the symmetric mode, described by a coordinate $R_1 + R_2$) or in opposite directions (down–up or up–down in the antisymmetric mode, described by coordinate $R_1 - R_2$).

Thus, a rule of thumb says that valid normal coordinates must be either entirely symmetric or entirely antisymmetric over the most common symmetry operations: if m and n label any two internal coordinates related by symmetry, then either $a_{km} = a_{kn}$ or $a_{km} = -a_{kn}$. This rule of thumb holds also in the construction of the coefficients of atomic orbitals in molecular orbitals (see Section 3.5). The bending mode of water straddles the mirror plane and is already symmetry-adapted if the displacements of the two hydrogen atoms are equal and in opposite directions (if the motions were in the same direction, that would be a molecular rotation, not a vibration).

In the same way as molecular orbitals, normal coordinates must be normalized (excuse the pun), meaning that the coefficients of the linear combination in equation 2.2a or 2.2b must be such that the total number of dimensions of vibrational space is always $3N - 6$. Normal coordinates are also orthogonal, meaning that they fulfill the essential request that led to their definition: they do not overlap, and displacement of nuclei along one of the normal coordinates has no component along any other normal coordinate. Nuclei move along normal vibrational modes going through the equilibrium positions at the same time, that is, they all oscillate in phase, although with different amplitudes. This automatically incorporates symmetry requirements, so normal coordinates are by definition symmetry-adapted. Normal coordinates have many advantages, the price to be paid being a loss of immediate structural transparency in the large and delocalized combination of internal coordinates.

The vibrational potential energy, V, can be written as a function of any of the coordinate sets, $V = V(q_C)$ or $V = V(q_I)$ or $V = V(q_N)$, and coordinate sets transform one into another using linear combinations that involve geometrical relationships. The potential energy can be written as a Taylor series: the first derivatives are null at equilibrium, and if the vibrations are as a first approximation considered harmonic, for any set of coordinates $[q]$ the second derivatives are the force constants, $f(i,j)$, and higher-order derivatives are neglected:

$$2V = 2V° + 2\sum_i [\partial V(q)/\partial q_i]° q_i + \sum_i \sum_j [\partial^2 V(q)/\partial q_i \partial q_j]° q_i q_j \quad (2.3)$$

$$f(i,j) = [\partial^2 V(q)/\partial q_i \partial q_j]° \quad (2.4)$$

where the "°" superscript indicates that the derivatives must be calculated at vibrational equilibrium (zero forces). $V°$ can be taken equal to zero, since the choice of the energy zero is arbitrary.

In a normal mode vibration in general all nuclei are displaced from equilibrium, so that no molecular vibrational mode corresponds to only one bond stretching or bond bending. However, in some particular cases a normal mode may be almost entirely localized on a particular group; for example, in molecules containing a carbonyl group a frequency of 1700 cm^{-1} corresponds to a normal mode involving almost exclusively C=O bond stretching. The coupling between springs is high when intrinsic frequencies are similar, and small when intrinsic frequencies differ. The C=O bond "spring" is so stiff that its coupling with the other, much weaker springs in the molecule is very small. So one usually says that the 1700 cm^{-1} frequency is C=O bond stretching, but that is only an approximation because all atoms, to a smaller or larger extent, are involved in the vibration. For the same reason, coupling between bond stretches is larger than coupling between bond stretch and bond bending.

Force constants are a quantitative measure of the stiffness of the corresponding modes, and the couplings between modes are quantitatively expressed by off-diagonal force constants. Force constants over normal coordinates correspond to complex nuclear motions, which, as already mentioned, cannot be related to elementary bonding deformations. But since the coefficients of the combinations can be calculated, one can back-transform into force constants corresponding to displacements along internal coordinates, like pure bond stretching and pure angle bending (see the Appendix to this chapter). It takes some mathematics, but computers can do that in no time at all. Of course, force constants change as the coordinate system changes, but proper vibrational frequencies are always the same.

Molecular vibrations are excited by infrared radiation [2]. A typical flowchart of a vibrational spectroscopy experiment on a polyatomic molecule could be as follows.

1(a). Use cartesian coordinates for the atomic displacements; or
1(b). choose proper independent internal coordinates.

2(a). Find a geometrical transformation from cartesian to internal coordinates; or
2(b). prepare approximate normal coordinates in terms of internal coordinates with tentative coefficients for equation 2.2, obeying symmetry prescriptions.
3. Record the whole IR and Raman spectrum of the molecule using a spectrometer.
4. Assign bands in the spectra, that is, find which absorbed frequency corresponds to which normal mode (not always an easy task).
5. From the observed frequencies, calculate the corresponding force constants.
6. Use the transformations defined at point (2) above to obtain force constants over internal coordinates.

Force constants and vibrational frequencies must be known at the same time, so that steps 4–6 require some trial-and-error cycles before the whole set of assignments, frequencies, and force constants becomes self-consistent. In this kind of spectroscopic work much attention is paid to the accurate reproduction of experimental vibrational frequencies, and the derived force constants are very sensitive to fine detail in the chemical constitution and bonding conditions. Nevertheless, if one assumes that chemical bonding is very nearly the same for the same chemical group in different molecules, IR spectroscopy experiments on simple and representative molecules or on collections of similar molecules may provide average force constants for the stretching of the most common chemical bonds and for the bending of the most common bond angles (see some examples in Table 2.1). These data lend themselves to chemical interpretation; for example, it is understandable that the stiffness of the C—C bonds increases on going from the single bond in alkanes to the stronger bond in sp^3—sp^2 systems, to the one-and-a-half bond in benzene, to the full double bond in olefins, to the triple bond in alkynes.

The above treatment assumes harmonic vibrational potentials. A harmonic potential is parabolic and its second derivative, or the curvature of the potential energy function, or the force constant, is the same for compression or elongation of the bond distance or for small and large displacements from equilibrium. These assumptions may be realistic as long as the vibrational amplitude is small, but certainly are not valid on

Table 2.1 Some typical force constants over internal molecular coordinates. Units of N m^{-1} for stretching and 10^{-20} N m for bending [3][a]

C—C stretching, alkanes	430–450	C—C—C bending, alkanes	108
Csp^3—Csp^2	500	H—C—C bending, methyl	65
C—C stretching, benzene	625	H—C—H bending, methyl	54
C—C double bond stretching, olefins	910		
C—C triple bond stretching, alkyne	1300		
C—H stretching, alkanes	450–470		
C=O stretching	800	C—C—O bending, carbonyl	118
C—N stretching	650	C—C—N bending	91

[a] K (N m^{-1}) = 0.0166 K (kJ mol^{-1} Å^{-2}); K (N m) = 0.0166 10^{-20} K (kJ mol^{-1}).

the far right side of the minimum when the bond becomes longer and longer. As more and more vibrational energy is poured into the vibrating system, the bond length will increase to a point when the bond will eventually break, this being a prerequisite for chemical reactivity. In fact a system vibrating under a fully harmonic potential will oscillate forever but will never be able to escape from its harmonic potential well: evolution in chemistry, from simple thermal expansion to phase changes to chemical reactivity, depends on anharmonic behavior.

The force constants are conveniently represented in a matrix **F**, each of whose rows and columns corresponds to one of the coordinates of the chosen set. The **F** matrix is symmetrical; for water (see Fig. 2.1) the numbers are [3]:

$$\mathbf{F} = \begin{array}{c} \\ R_1 \\ R_2 \\ \theta \end{array} \begin{array}{ccc} R_1 & R_2 & \theta \\ 843 & -10 & 24 \\ -10 & 843 & 24 \\ 24 & 24 & 71 \end{array}$$

Given the definition of the force constants, the total vibrational potential energy is

$$2V(\text{vib}) = 843 \, \delta R_1^2 + 843 \, \delta R_2^2 + 71 \, \delta \theta^2 \\ + 2(24 \, \delta R_1 \delta \theta + 24 \, \delta R_2 \delta \theta - 10 \, \delta R_1 \delta R_2) \quad (2.5)$$

The first three terms involve the diagonal elements of **F**. The cross-terms represent the vibrational interaction between internal coordinates; had F_{12} been zero, there would have been no stretch–stretch coupling and the frequency of the normal modes for symmetric and antisymmetric stretching would have been the same. A broad conceptual similarity exists between coupling of modes by off-diagonal terms in molecular vibrations, and coupling of orbitals by overlap integrals in molecular orbital calculations. Off-diagonal **F** matrix elements are always much smaller than diagonal ones: many spectroscopic calculations take as a first approximation the diagonal force field, in which all off-diagonal elements of the **F**- matrix are zero.

The above treatment of bond stretch is a purely classical one, while molecular vibrations are quantized (Section 3.1.6). To show a link between the quantum mechanical concept of excited state and the classical concept of bond stretch, the expected variation in the length of a bond in a given vibrational state can be estimated by equating the quantum mechanical vibrational energy (equation 3.24) to the classical potential energy, $1/2k\delta x^2 = h\sqrt{(k/\mu)}(v + 1/2)$, and solving for the maximum displacement δx. For example, for $v = 0$ in CO one gets something like 0.05 Å. Clearly, δx increases with increasing quantum number.

2.2 Molecular mechanics

In the early 1950s, spectroscopic and diffraction data on molecular structure were beginning to accumulate, but there was no safe and sound method for predicting molecular structures by computational simulation. Quantum chemistry was at a rudimentary stage, because of the lack of suitable computers, so that the idea of using classical potentials seemed appealing. Such an approach required: (1) the definition of some "standard" bond lengths and angles, as obtained for example by averaging available structural data on similar chemical groups in different molecules, or by taking some wisely chosen chemical systems as reference states; (2) the assumption that the final structure of any complex molecular system would result from the action of a "strain" from these standards; and (3) the use of spectroscopic vibrational force constants to calculate the strain energies associated with the deformation of the actual molecule from the reference structure. Finding the equilibrium structure of any molecule then simply required minimizing the strain energy, the computational effort required to evaluate the simple harmonic potentials involved in the simulation being quite within range of the computing powers of the day. Theoretical organic chemists finally found themselves able to deal with the real world, instead of talking all the time about H_2 and carbon monoxide.

There are several conceptual arguments against this "molecular mechanics" construction, and there were many difficulties in its actual implementation. First, molecular vibrations are quantized, and are not near an equipartition regime at ordinary temperatures, so vibrational energies cannot be considered as a continuum. Second, organic molecules are multiform, while experimental force constants are available only for a handful of simple systems. Third, it must be assumed that equilibrium structural parameters and force constants are transferable among chemical systems, because if new parameters are to be found for each new molecule to be treated, the method loses much if not all of its practical appeal. Then, of course, the whole machinery rests upon a judicious choice of these transferable parameters.

The strain energy in terms of parameters related to vibrational spectroscopy contains terms for bond stretching, bond angle bending, and dihedral vibration (torsion),

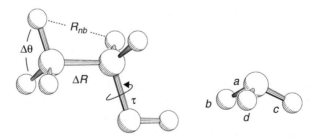

Fig. 2.2. The structural elements of vibrational force fields: on the left, stretch δR, bend $\delta \theta$, and torsion τ. Additional terms include non-bonded interactions, R_{NB}, and supplementary positional restraints like improper dihedrals (right), which do not have a physical interpretation. In the example given, the improper dihedral is the dihedral angle between planes *acd* and *bcd* and serves the purpose of forcing planarity of the entire fragment.

as shown in Fig. 2.2. A typical analytical expression is

$$E_{\text{strain}}(\text{vib}) = \sum_i 1/2\, k_i(g_i - g_i^\circ)^2 + \sum_{i,j} 1/2\, k_{ij}(g_i - g_i^\circ)(g_j - g_j^\circ) + \sum_k f(\tau_k) \quad (2.6a)$$

$$f(\tau) = \sum_n [V_n^\circ + V_n[1 + \cos(n\tau)]] \quad (2.6b)$$

where the ks are the force constants, the gs are the actual values of some internal coordinates, related to various kinds of molecular deformations, and g° is the reference, or "strainless" value of each of these deformation coordinates. There is some freedom in the choice of the number and type of these terms, as there is some freedom in the choice and type of internal coordinates. Ideally, one should pick a complete set of truly independent internal coordinates, but the definition of such a set becomes rather difficult for large symmetry-less molecules. Redundancy in the number of coordinates brings about redundant energy terms, but this is of minor concern since these energies are empirical functions anyway.

In early applications of the molecular mechanics method, an evident difficulty was that when the simple expression 2.6a was used, force constants were not transferable even among closely related systems. It was soon realized that the influence of some interaction between atoms not bound together but bound to the same atom (the geminal or 1–3 interactions) or, more generally, also between all other non-bonded atoms, had to be somehow considered. These supplementary non-bonded terms were to represent the change in chemical environment that could not be accounted for when using completely transferable force constants. The Urey–Bradley force field adds these interactions to a valence force field that includes only diagonal terms [4]. For application to molecular mechanics calculations, these non-bonded interactions were simulated by some function of the interatomic distance:

$$E(\text{non bonded}) = \sum_{i,j} f(R_{ij}) \quad (2.7a)$$

$$f(R_{ij}) = AR_{ij}^{-n} + BR_{ij}^{-m} + C\exp(-DR_{ij}) \quad (2.7b)$$

where R_{ij} is the distance between two atomic nuclei, $A, B, C, D \ldots, n, m, \ldots$ are empirical disposable parameters, typical values being $n = 6$ for the attractive and $m = 12$ for the repulsive branch.

These additional terms became known as "steric interaction" terms. It was initially thought [5] that the necessary parameters might be extracted from second virial coefficients or noble gas interaction functions, but it soon became evident that the best attitude was to consider them as fully adjustable parameters. Moreover, it was not even clear if all non-bonded interactions should be considered, rather than just a few contributions from neighboring hydrogen atoms. [6]

The first applications of the molecular mechanics method were to hydrocarbons, for the good reason that spectroscopic force constants were more readily available. When heteroatom-containing molecules began to be considered, with presumable permanent electrostatic polarizations, a further elaboration of the formulation led to the inclusion of coulomb-type interactions:

$$E_{\text{coul}} = \sum_{i,j} 1/(4\pi\epsilon) q_i q_j (R_{ij})^{-1} \quad (2.8)$$

where ϵ is the permittivity of the medium, which in an orthodox treatment should be equal to the vacuum permittivity $\epsilon°$, but has sometimes been used as an independent parameter; q_i are point charges positioned at atomic nuclei or at other strategic molecular locations, and R_{ij} is the distance between the interacting charges. While the coulombic term improves the flexibility and hence the performance of the method, it introduces further problems. The charges are not real observables, so that they are to be considered as more parameters to be determined somehow, with the additional difficulty that they are by definition not transferable and must be determined anew for each and every new molecule, mostly by some empirical scheme [7] or by a quantum chemical calculation (see Section 4.2 for alternatives). In addition, everyone can appreciate how far from reality is a model that places partial charges, sometimes negative, at the positions of atomic nuclei.

The expression for the total molecular energy then becomes:

$$E_{\text{tot}} = E_{\text{strain}}(\text{vib}) + E(\text{non bonded}) + E_{\text{coul}} \quad (2.9)$$

The earliest applications of the molecular mechanics method involved the calculation of the strain energy of some fixed conformation. The automatic minimization of the energy was introduced as soon as computer resources permitted [8], with great care being taken to observe symmetry requirements and to optimize coordinate transformations, as appropriate in times when the calculation of the derivatives of the potential for a molecule like cyclododecane took 22 minutes on the fastest computer then available (this would today take some micro-, if not nanoseconds). The theory and efficiency of minimization procedures are rather technical subjects, which will not be treated here [9], but the true physical problem is made difficult for large molecules by the fact that the number of minima in the potential energy hypersurface increases with molecular complexity, and so does the risk of landing in metastable minimum energy conformations.

When a minimum is reached, the resulting equilibrium structure can be compared with experimental spectroscopic or electron diffraction and X-ray diffraction results. Isomerization energies can be calculated as differences between the strain energies of the isomers. From the second derivatives of the potential at the minimum, which must all be positive if the state is a true minimum, one can recalculate the molecular normal vibration frequencies as predicted by the molecular mechanics run, and from them, by a simple statistical mechanics calculation, the vibrational contribution to the heat capacity, which can also be compared with experiment (Section 7.3). The heat of formation can be estimated by introducing fixed contributions from each type of bond, to be added to the molecular mechanics strain energy:

$$\Delta H_\mathrm{f} = \sum_{\mathrm{k.bonds}} E_\mathrm{k} + E_\mathrm{tot} \tag{2.10}$$

The E_ks are further parameters to be empirically determined. The estimated ΔH_f values can be compared with experimentally available heats of formation from heat of combustion experiments. All these comparisons between calculated and experimental molecular properties can in principle be exploited in the calibration of force field parameters.

Spectroscopic analyses were the starting point for force field calibration[10], but it soon became obvious that further elaboration was required. Several schools of thought and method began to emerge. On one side, the Bartell–Kuchitsu school concentrated on preserving as much as possible the physical significance of the potential parameters [11] and the adherence to spectroscopic principles, for the accurate calculation of structural parameters to be compared with the results of equally accurate structural determinations carried out in the same group. The Lifson school proposed a consistent force field, in which parameters, although of almost completely empirical derivation, would be chosen so as to reproduce a complete set of molecular properties, including vibrational frequencies [12]. Other approaches [13] seemed to care more for improvement in the accuracy of the results at the expense of an increased number of parameters, whose adherence to first principles became less and less obvious. Further elaborations came from groups led by Scheraga [14] and Boyd [15]. By 1973, an extensive and critical review [16] could conclude that, at least for hydrocarbons, bond lengths could be calculated with an accuracy of 0.01 Å, bond angles with an accuracy of 1–2° while the standard deviation between observed and calculated heats of formation was of the order of 4 kJ mol^{-1}. The conclusion was that "... the molecular mechanics method, in principle, must be considered to be competitive with experimental determination of the structures and enthalpies of molecules."

2.3 Evolution of molecular force fields

When the molecular mechanics method is applied in its full functional extension, it has moved a long way from the initial assumption of a minimization of vibrational strain energy, and there are a large number of parameters to be adjusted. Once one starts adjusting parameters, the method becomes parametric, and there is no point in defending first-principle derivations or in preserving a strict correspondence between molecular mechanics force constants and spectroscopic force constants. The most sensible attitude is to readjust all parameters all together, and this is what has been done, since the pioneer years, in more recent times for the preparation of the most popular and widely applied force fields. One is not restricted to functional forms that have some resemblance to spectroscopic potentials, either, but new computational machines can be invented for specific problems. For example, the so-called "improper dihedrals" have been devised as a simple method for constraining a given molecular fragment to be planar, a typical instance being aromatic trigonal centers. Calling a the central atom and b, c and d the three atoms bound to it, the improper dihedral "energy" is defined as (Fig. 2.2):

$$E_{\text{imp.dih.}} = 1/2\, K(\tau - \tau^\circ)^2 \tag{2.11}$$

where K is a sort of force constant and τ is the dihedral angle between planes a-b-c and b-c-d. The strain energy of glorious memory has now become a sort of penalty function that the system has to pay for escaping from some constraint, imposed on the basis of structural or thermochemical evidence. Its only relationship to true molecular energies is that it is expressed in kJ mol^{-1} (just for compatibility, because it is summed into the total molecular energy). The molecular mechanics method is today a fully parametric computational machine to generate new chemical knowledge from previously well digested knowledge. Much as the neologism is repulsive (as if there were methods based on ignorance) one must call the molecular mechanics method a knowledge-based method, where the principles of theoretical chemistry have very little to say. As a consequence, the physical meaning of each of the parameters and of each of the functional forms dissolves; for example, there is little hope that the coulombic terms do indeed describe the coulombic interactions. The whole molecular energy is fitted by the whole set of empirical parameters.

Also the target of molecular mechanics has greatly changed. In the 1970s, the chemical synthesis of strained hydrocarbon molecules and the prediction of their bonding modes and stability was a great challenge, even though the compounds had no immediate practical use or application. The prediction of the structure and heat of formation of such systems is no longer a frontier achievement, and the long battles over the physical interpretation of bond length variations of a few hundredths of an Ångström are things of the past. Besides, the prediction of equilibrium structures of small molecules is nowadays quite feasible by highly refined quantum mechanical methods, with which molecular mechanics cannot compete. Much more relevant to contemporary chemistry is, for example, the prediction of the structure and stability of some key drug molecules, or of the molecular complexes formed by these drugs and their receptor, protein or DNA sites. Since biologically active organic molecules invariably include complex and multiform chemical functions, this task requires reasonable potential models for widely variable chemical environments. For such systems, even a rough estimation of the minimum energy conformation(s) over a few rotatable single bonds, or an approximate idea of the overall molecular shape, may be more important than the accurate prediction of some bond lengths. At this point, much can be conceded, in a trade of rigor for performance. The other frontier topic is the prediction of the folding modes of peptides and proteins, which may well involve the minimization of the total energy of a 2,000-atom system, and there too one sees little room for minute detail.

As a consequence of this evolution in scope and targets, rougher but more widely applicable force fields have been advocated [17]; the philosophy is to use general force constants and geometry parameters based on some generalized effective atomic properties, or on hybridization, rather than specialized parameters that depend on the exact nature of the atomic species and on the chemical environment. Non-bonded interaction parameters also depend on some easily generalized (although less and less easily explained in terms of first principles) quantities like atomic radii or "atomic" van der Waals energies [18]. And since the results are more immediately applicable to practical problems, competition has shifted from academic dispute to at least a good part of copyright protection [19].

If the target is practical performance rather than the advancement of academic knowledge, anything that works is acceptable: molecular mechanics methods have been supplemented by analyses and exploitations of other molecular properties and quantifiers, including global descriptors of molecular shape, polarity, etc. These techniques have been called Quantitative Structure-Activity Relationships (QSAR), quantitative structure-property relationships (QSPR) and drug discovery or computer-aided drug design (CADD) [20]. These approaches and the ensuing research lines are in many ways offspring of the molecular mechanics approach to structure and energy predictions. These topics are however outside the scope of the present book.

In essence, the mechanics of a modern force field is not too different from the mechanics of older ones. The basic equation is still very similar to equations 2.6–2.9, with few additions and extensions, like improper dihedrals. Of course, the manifold increase in computing power between 1965 and 1985 allowed an incredible number of sophisticated manipulations, besides enlarging the scope of the method from 20-atom to 2,000-atom molecules. Extensive optimization led to a massive increase in the number of parameters. In a typical example, the CHARMM force field [21], developed in the 1980s by the Karplus groups, accounted for 29 atoms or atom groups, and already provided standard parameterization for 59 types of bond stretching, about 150 bond bending types, 38 torsion types, and 42 improper torsion types. Of course, the force constants and equilibrium values were not all different, the same parameter serving sometimes for an entire group of strain situations. Non-bonded interaction parameters were also generalized; for example, for interaction between atomic species k and m the CHARMM force field proposed the Slater–Kirkwood equation:

$$f(R_{ij}) = A^{km}(R_{ij})^{-12} - B^{km}(R_{ij})^{-6} \tag{2.12a}$$

$$B_{km} = 3/2[1/(4\pi\epsilon^\circ)]^{1/2}(ehm_e^{-1/2}\alpha_k\alpha_m)/[(\alpha_k/N_k)^{1/2} + (\alpha_m/N_m)^{1/2}] \tag{2.12b}$$

$$A_{km} = 1/2 B_{km}(R_k + R_m)^6 \tag{2.12c}$$

in which ϵ° is the vacuum permittivity, e and m_e are the electron charge and mass, h is Planck's constant, and for atomic species k, α_k is the atomic polarizability, R_k is the atomic radius, and N_k is the effective number of outer shell electrons. With this kind of approach the parameterization of n atomic species requires $3n$ parameters instead of $n(n+1)/2$ parameters. In the Universal Force Field UFF [18b] each atom in the periodic table is characterized by six parameters: a bonding radius, a bonding angle, a non-bonded distance, non-bonded energy and scale parameter, and an effective charge, and all force field parameters are derived from combinations of these numbers (there are number for einsteinium and fermium, should one really need to model those atoms in a molecular mechanics run).

A special problem, of crucial importance in applications of molecular mechanics to biological problems, is the treatment of the hydrogen bond, an interaction which is halfway between a true chemical bond and a non-bonded interaction. This complication was tackled in an entirely practical attitude, using empirical formulas; for a given A...H–D interaction A being the acceptor atom and D the donor, the CHARMM force field used the following:

$$E_{HB} = [A^{AD}(R_{AD})^{-n} - B^{AD}(R_{AD})^{-p}](\cos^2 \theta_{AHD}) \qquad (2.13)$$

where n and p are positive integers, usually in the range 10–12, A and B are given parameters for each pair of donor-acceptor types, and the entire contribution is switched off as soon as the hydrogen-bonding angle θ_{AHD} falls below 90°. Other force fields use similar approaches, with the added difficulty that contributions over hydrogen bonding functions may or may not superimpose to ordinary or damped non-bonded or coulombic contributions over the same involved atoms.

Another innovation introduced by modern thinking in molecular mechanics is the united atom. The molecules treated may be very large, or even if the molecular species considered is small, the number of molecules in a typical dynamic run (Chapter 9) may be very large, so that the total number of atoms in the system becomes enormously large. Since any force field requires the calculation of all distances between all interaction sites, some atom groups are represented by one site only, typical examples being the methyl or methylene groups, so that, for example, n-hexane is treated as a six-site molecule. This approximation saves computing times, but has adverse effects on the calculation of some fine detail properties and is quite untenable in calculations involving molecular crystals, where hydrogen positions play a crucial role. All-atom formulations tend to replace older united-atom formulations, as for example in a later variation of the same CHARMM force field [22] optimized also for the treatment of nuclei acids. Experimental data on molecular geometries, vibrational spectra, base-pairing energies and heats of sublimation of crystals were extensively used to calibrate the force field parameters. The same path has been followed in the development first of an united-atom [23] and then an all-atom [24] version of Kollman's AMBER force field. The GROMOS96 force field developed in the van Gunsteren groups [25] offers both alternatives.

All these force fields are embedded into computer packages and appropriate parameters are automatically called for when the user specifies the desired molecule. Many of the computer packages so far mentioned are available to academic users for a nominal fee, others appear in commercial packages that may sell for very high prices.

Of course the force fields can be to some extent manipulated by the user, so it is not difficult to imagine that many variants of these force fields have been developed and used for special purposes. One of the biggest problems in such applications is invariably finding the appropriate point charge parameters for use in equation 2.8. In recent times, accurate quantum mechanical calculations have become feasible, and results on recognition energies in prototypical dimers and oligomers have also been successfully used in the calibration of force field parameters [26], while atom centered charges have been derived by fitting the calculated electrostatic potential (the so-called ESP charges) [24].

In a series of papers dealing with different types of organic families, Jorgensen and his groups have developed the optimized potentials for liquid simulation (OPLS), a force field construction accurately calibrated to reproduce the structure and energetics of organic liquids in Monte Carlo simulations; these are reviewed in Section 9.5.2. Another attempt at deriving an empirical force field for molecules in isolation and in condensed phases has been presented by Sun [27]. A compilation of parameters for force field calculations is available [28].

2.4 Appendix: an example of coordinate transformation

Consider a bent triatomic molecule such as water, Fig. 2.1. Each atom is described by a set of three cartesian coordinates, which form an array of nine numbers x_i, $i = 1$ to 9, and each displacement during a vibration is called δx_i. The atomic mass corresponding to each coordinate is called m_i. Defining mass-weighted coordinates q_i and writing the total kinetic energy T for the motions during the displacements yields:

$$q_i = \sqrt{m_i} \; \delta x_i \qquad (2.14)$$

$$2T = \sum_i (\partial q_i / \partial t)^2 \qquad (2.15)$$

Using equations 2.3 and 2.4:

$$2V = \sum_{i,j} f_{ij} q_i q_j \qquad (2.16)$$

The sum of kinetic and potential energy for a classical dynamic system is called the lagrangian, $L = T + V$, and Newton's equations of motion, one for each coordinate q_k, are written as

$$d/dt[\partial L/\partial \dot{q}_k] + \partial L/\partial q_k = 0 \qquad (2.17)$$

$$-\partial^2 q_k / \partial t^2 = \sum_j f_{kj} q_j \qquad (2.18)$$

Equation 2.17 is a multidimensional expression of Newton's law $F = ma$, and equation 2.18 is the equivalent of Hooke's potential $ma = -kx$. As usual for oscillatory motions, the trajectories are sinusoidal functions:

$$q_k = A_k \cos(2\pi \nu_k t + \phi); \quad \partial^2 q_k/\partial t^2 = -A_k \omega_k \cos(2\pi \nu_k t + \phi) \quad (2.19)$$

where the As are amplitudes, ϕ is the phase of the oscillation, ν_k is the frequency, and $\sqrt{\omega_k} = 2\pi \nu_k$. Substituting equation 2.19 into 2.18:

$$A_1(f_{11} - \lambda) + A_2 f_{12} + \cdots + A_9 f_{19} = 0$$
$$\cdots \quad (2.20)$$
$$A_1 f_{91} + A_2 f_{92} + \cdots + A_9(f_{99} - \lambda) = 0$$

This is a system of nine linear equations in nine unknowns, the A_is. It has non-trivial solutions (solutions for which not all A_is are zero) only if the following condition is satisfied:

$$\text{Det}(\mathbf{F} - \lambda \mathbf{I}) = 0 \quad (2.21)$$

where the force constants are arranged in a 9×9 matrix, $\mathbf{F}$, and $\mathbf{I}$ is a 9×9 identity matrix. This is a standard secular equation, and diagonalization of the $\mathbf{F}$ matrix yields nine frequency values. The vibrations in a triatomic molecule are just three, not nine, so six roots of the ninth-degree secular equation will be zero (pure translations or pure rotations).

For the i-th eigenvalue λ_i, back substitution into the system 2.20 gives a set of nine coefficients, A_{ik}. These are the coefficients of the linear combination of the starting mass-weighted cartesian coordinates, giving the normal coordinates Q:

$$Q_i = \sum_k A_{ik} q_k \quad (2.22)$$

The secular equation 2.21 shows the relationship between vibrational frequencies and force constants: given the observed frequencies and some algebra, the value of the force constants can be derived; and vice versa, if force constants are known, the vibrational frequencies can be calculated.

These force constants are however over cartesian displacements, and not over stretching or bending in internal coordinates. Consider now another linear transformation that combines cartesian displacement coordinates q_i into internal coordinates S_k:

$$S_k = \sum_j B_{kj} \delta x_j \quad (2.23)$$

$$G_{kn} = (1/m_i) \sum_i B_{ki} B_{ni} \quad (2.24)$$

$$\mathbf{F}' = \mathbf{G}^\text{T} \mathbf{F} \mathbf{G} \quad (2.25)$$

The $\mathbf{B}$ matrix is a rectangular, 9×3 matrix; the $\mathbf{G}$ matrix is a 3×3 matrix. Equation 2.23 brings the nine-coordinate cartesian basis set into a three-coordinate internal basis set.

Finding the elements of matrices **B** and **G** is a purely geometrical problem. The force constant matrix **F'** is now a 3 × 3 matrix over the stretching and bending internal coordinates. A little algebra shows that the corresponding secular equation is now

$$\text{Det}(\mathbf{GF'} - \lambda \mathbf{I}) = 0 \qquad (2.26)$$

This equation has three solutions for the λ values corresponding to the same three non-zero vibrational frequencies obtained from equation 2.21, but is much more practical to work with.

After going through the same steps as in equations 2.20–2.22, the solution of the secular equation 2.26 also yields the transformation from internal to normal coordinates, with a new set of coefficients:

$$Q_i = \sum_k C_{ik} S_k \qquad (2.27)$$

For very large molecules finding the elements of the **G** matrix becomes hopelessly complicated. Alternatively, one prepares symmetry coordinates and then proceeds directly to the solution of the secular equation and to finding normal coordinates by trial and error.

2.5 Historical portraits: Got a force constant?

At the end of 1968, my graduation thesis, reporting the painstaking task of determining one crystal structure (see Chapter 5), was ready for discussion. I was an average student with an average curriculum, but my grades rose somewhat because the evaluation committee was mildly happy about my computer programming jobs. The top feature was a routine I wrote partly in Fortran and partly in symbolic language for the IBM 1620 (see Chapter 10) to map a Fourier synthesis on the planes of the aromatic rings rather than on the conventional crystallographic planes. Within a few month's it proved quite obsolete, but it served the purpose on graduation day.

Computer programming was then a fascinating opportunity to use a brand new and revolutionary piece of scientific equipment. Nowadays, computer programming is considered a sort of side opportunity in scientific work, and some people think its intellectual relevance is akin to solving crossword puzzles. To some extent, that is a respectable opinion, although the same could be said of some of the long, patient and scarcely scientific practical tasks that must be carried out in an organic chemistry synthesis lab. Nevertheless, an undisputable merit of learning to computer program is that one learns the importance of being absolutely exact in all formulations, and one learns it the hard way, witness the long hours spent in fighting bugs that might depend on just one forgotten index, or mis-typed character, or missing flag or initialization. That exercise has left an indelible mark on all my successive scientific endeavors: I am not happy until I am satisfied that every single detail has been perfectly taken care of.

In those days, people in our group were busy with the problem of why biphenyl changes its conformation from twisted in the gas phase to flat in the solid state. Understanding such a process required understanding the molecular mechanics of the relative orientation of aromatic rings. After graduation there was no real chance and hence no immediate promise for my permanence in the research group, as it was customary for fresh graduates to go out into the world of industry, which in those times was particularly hungry for chemists. Nevertheless, I had already made up my mind that I would spend the rest of my life in chemical research in academia, and even if there was no positive opportunity at that moment, nobody objected to my hanging around either, under the pretext of finishing up the thesis work. Of course that was a pretext, because no research job is ever finished. So I started a project on developing my own computer program for molecular mechanics, just as many other groups were then doing because that was the hot development in theoretical chemistry. The big problem was that the calculation of potentials required cartesian coordinates for each atomic position, while molecular structure was better visualized in terms of internal coordinates, and it was not easy to program the conversion for a completely general molecular structure. One simple way of dealing with such a problem was of course to write a separate subroutine for each molecule with the appropriate conversion formulas; but in those days writing a program that was good only for one molecule was thought to be a shame, because compiling a Fortran code was a time- and money-consuming task. So I spent a lot of time writing a complicated general routine that recognized bits and pieces of algebraic and trigonometric functions in alphabetic form, probably spending much more time and money in debugging that monster than would have been spent recompiling the program for every molecule. Fortunately, by 1970 the computing center already had a small machine which, although very slow in actual calculations, did have a Fortran compiler and was available to the general public on a self-service basis. So I spent days and days with my pack of punched cards, fishing out the single wrong ones and re-punching them with the correct instruction as the debugging process proceeded.

The transition from determination of molecular structure by X-ray diffraction to molecular mechanics for conformational analysis, as the calculation of strain energies was called in those times, is a very natural one, both being aspects of the transition from two-dimensional formulas on paper to three-dimensional models in space. But such an evolution of chemical thought was not yet obvious. A few months later, I presented a short communication at the annual meeting of the Italian Crystallographic Association, with a description of our computer program and the application to polyphenyl compounds. The proceedings of the meeting allowed a few minutes' time for each author to defend his paper. While I was answering questions from the audience, which included members of the group of Corradini, Edoardo Giglio and Alfonso Maria Liquori in Rome, at the time involved in a similar enterprise, I remember a thunderous voice, coming from the back of the hall, shouting: "Ladies and gentlemen, I would like to remind you that this is a Congress of *Crystallography!*" The offended gentleman, probably a mineralogist, was maybe right after all, but he certainly did not appreciate the contribution of X-ray crystallography to stereochemistry.

Whenever a new project in molecular mechanics was started, the main problem was to find the appropriate parameters, because in spite of the efforts on the part of many authors to offer a complete coverage of chemical environments, one's molecule always had that particular chemical bond or that particular bond angle that was not included in the list of standard ones for which force constants were supplied. So those were days of a quest for force constants, leafing through back issues of spectroscopic journals or just by analogy between more or less similar chemical groups. We were also forced to patch together bits and pieces of different force fields, in spite of strong advice against that. Fortunately, the essential results were hardly sensitive to small changes in one or a few parameters, and I remember our astonishment when we got quite reasonable results on one occasion when we had by mistake changed to positive the sign of the attractive term in the non-bonding functions! The reason for this is that a roughly reasonable molecular structure is obtained by putting the proper restraints at the proper places, even though the relative intensities of the restraints are only guesswork. And, of course, that the intramolecular attractive non-bonded term is hardly important in determining molecular structure.

Circulation of computer codes had also begun to some extent. One time we received a printout of some 2,000 Fortran instruction, and, since there was no hope of getting a hardware copy of any sort, we set to the task of punching them one after another on the card puncher. An even bigger job was checking it out for misprints afterwards.

In the winter of 1973 I spent a four month period in Orsay at the Centre Europeen de Calcul Atomique et Moleculaire (CECAM), then under the firm managing hand of Carl Moser. I traveled there in my small car, carrying a heavy load of punched cards that I managed to slip through the iron meshes of French customs by pretending I did not speak any French and letting the officer understand that the cards were of no value because they were full of holes. The Center was jointly subsidized by the research Councils of several European countries, but scientists from all over the world were welcome there. The place was in the outskirts of Paris, and the forty minutes or so daily commuting journey from downtown was a good opportunity to read and work, in spite of the crowd and the din of the trains – strange how the nearness of a familiar person in a silent room can be more disturbing than the closeness of two hundred shouting strangers among the shrieking of strained iron. The Centre offered each guest a (modest) stipend, office space, and free computing time. The atmosphere was good, especially for young people who were able to be in touch with more experienced scientists. Carl Moser's dogs, two pugs named Uxel and Ulfila, were not to everybody's liking, but a few grunts and sniffs were about the only real inconvenience of their frequentation of the Center. The food at the *cantine* was great.

In Orsay I met Larry Bartell, who was also spending a sabbatical there. He was the father of structural force fields and of gas-phase electron diffraction. Gas-phase ED was for some time considered as the mainstream to molecular structure determination,

48   MOLECULAR VIBRATIONS AND MOLECULAR FORCE FIELDS

and in fact it was until X-ray crystallography became a routine application of a simpler technique and cheaper apparatus, cheap at least in terms of practical considerations because X-ray diffraction machines can be bought, while ED machines have to be built more or less from scratch. Besides, electron diffraction experiments can be carried out only on relatively volatile, and hence small, molecules. Gas-phase ED still remains the only experimental technique that can give detail of the structure and to some extent of the conformational dynamics of small molecules free from interaction with their neighbors.

Every day, at mid-afternoon, Larry and I sat for an hour or so over tea, and he would produce a pad, which he decorated in his minute handwriting with a well sharpened pencil, as he explained to me the bits and pieces of molecular mechanics and, at the same time, of general molecular physical chemistry. I have kept those sheets for years. I still have the habit of doing the same thing whenever I teach something to somebody, so he or she can keep a written record of what has been said. Larry also gave me some computer codes, which I managed to transfer to a magnetic tape, the big advancement in data storage at the time: giving something that weighed a kilogram and a half to store a few thousand Fortran instructions. One day I mustered enough courage to ask him what was the very first origin of the non-bonded energy parameters in his force field for hydrocarbons: he straightened his right arm, closed his fist and made a move downwards as if pulling something out of the sky. And that was precisely what he meant. I also learned from him the proper English expression for that: "educated guessing". I knew, however, that he and his group had spent years building ED machines, performing very careful experiments to determine accurate equilibrium molecular geometries that his force field accurately reproduced. Spending my time with him taught me the difference between scientific work and shooting in the dark; one has experimental data, builds a mathematical model, and then guesses in an educated way the numerical parameters on which the mathematical model relies. That was education indeed.

References and Notes to Chapter 2

[1] For the transformations between cartesian and internal coordinates see Dunitz, J. D.; Buergi, H.-B. Molecular structure and coordinate systems, in *Structure Correlation,* edited by H.-B. Buergi and J. D. Dunitz, VCH, Weinheim, 1994, Vol. 1, Chapter 1.

[2] In chemical spectroscopy, IR frequencies are usually expressed in reciprocal centimeters, cm^{-1}. This is a very convenient unit, because it does away with large powers of ten. The true frequency in Hz is obtained by multiplying by the speed of light, $29,979,245,800 \, cm \, s^{-1}$. For example the stretching frequency of the CO molecule is $2143 \, cm^{-1}$ or 6.424×10^{13} Hz.

[3] Usually, the units of the F-matrix elements are as follows: stretch and stretch–stretch, $N \, m^{-1}$; stretch–bend, 10^{-10} N; bend and bend–bend, $10^{-20} N \, m$. Linear displacements are measured in meters, angular displacements in radians (pure

numbers), so that energies always come out in Nm=J. A useful unit is the attojoule, aJ, or 10^{-18} J.

[4] Urey, H. C.; Bradley, C. A. The vibrations of pentatonic (*sic*) tetrahedral molecules, *Phys. Revs.* 1931, **38**, 1969–1978. One reads there: "It seems that there may be repulsive forces between the corner atoms ... The introduction of terms in the potential energy proportional to $1/r_j^n$ ... makes it possible to secure very good agreement between calculated and observed frequencies...The calculated frequencies are not very sensitive to the value of n which may be anywhere from 5 to 9."

[5] Westheimer, F. H.; Mayer, G. E. The theory of the racemization of optically active derivatives of diphenyl, *J. Chem. Phys.* 1946, **14**, 733–738.

[6] Hendrickson, J. B. Molecular geometry. I. Machine computation of the common rings, *J. Am. Chem. Soc.* 1961, **83**, 4537–4547.

[7] For a simple scheme see Gasteiger, J.; Marsili, M. Iterative partial equalization of orbital electronegativity – a rapid access to atomic charges, *Tetrahedron* 1980, **36**, 3219–3228.

[8] Wiberg, K. B. A scheme for strain energy minimization. Application to cycloalkanes, *J. Am. Chem. Soc.* 1965, **87**, 1070–1078.

[9] The steepest descent method is very convenient for a fast approach to the minimum region from a very approximate trial structure. This method requires the evaluation of the first derivatives of the potential. When the system is close to a minimum, the convergence of the steepest descent cycles becomes very slow, and the Raphson–Newton method is preferred. This method requires first and second derivatives of the potential. Alternatively, the Symplex method can be used: it does not require the calculation of derivatives, but requires a very high number of evaluations of the response function. This is not a real problem when the response function is as simple as equation 2.9. See e.g. Press, W. H.; Teukolski, S. A.; Vetterling, W. T.; Flannery, B. P. *Numerical Recipes, The Art of Scientific Computing,* 1992, Cambridge University Press, Cambridge.

[10] Schachtschneider, J. H.; Snyder, R. G. Vibrational analysis of the n-paraffins-II. Normal co-ordinate calculations, *Spectrochim. Acta* 1963, **19**, 117–168.

[11] Bartell, L. S.; Kuchitsu, K. Derivation of physically significant nonbonded interaction constants in hydrides and modified Urey–Bradley analysis, *J. Chem. Phys.* 1962, **37**, 691–696; Jacob, E. J.; Thompson, H. B.; Bartell, L. S. Influence of non-bonded interactions on molecular geometry and energy: calculations for hydrocarbon based on Urey–Bradley field, *J. Chem. Phys.* 1967, **47**, 3736–3753; Fitzwater, S.; Bartell, L. S. Representation of molecular force fields. 2. A modified Urey–Bradley field and an examination of Allinger's gauche hydrogen hypothesis, *J. Am. Chem. Soc.* 1976, **98**, 5107–5115.

[12] Lifson, S.; Warshel, A. Consistent force field for calculations of conformations, vibrational spectra and enthalpies of cycloalkane and n-alkane molecules, *J. Chem. Phys.* 1968, **49**, 5116–5129; Ermer, O.; Lifson, S. Consistent force field calculations. III. Vibrations, conformations and heats of hydrogenation of nonconjugated olefins, *J. Am. Chem. Soc.* 1973, **95**, 4121–4132.

[13] Allinger, N. L.; Tribble, M. T.; Miller, M. A.; Wertz, D. H. Conformational analysis. LXIX. An improved force field for the calculation of the structures and energies of hydrocarbons, *J. Am. Chem. Soc.* 1971, **93**, 1637–1648.

[14] Scott, R. A.; Scheraga, H. A. Conformational analysis of macromolecules. II. The rotational isomeric states of the normal hydrocarbons, *J. Chem. Phys.* 1966, **44**, 3054–3069.

[15] Boyd, R. H. Method for the calculation of the conformation and minimum potential-energy and thermodynamic functions of molecules from empirical valence-force potentials – application to the cyclophanes, *J. Chem. Phys.* 1968, **49**, 2574–2583.

[16] Engler, E. M.; Andose, J. D.; Schleyer, P. v.R. Critical evaluation of molecular mechanics, *J. Am. Chem. Soc.* 1973, **95**, 8005–8025.

[17] Davies, E. K.; Murrall, N. W. How accurate does a force field need to be? *Computers Chem.* 1989, **13**, 149–156.

[18] (a) Mayo, S. L.; Olafson, B. D.; Goddard, W. A. III, DREIDING: a generic force field for molecular simulations, *J. Phys. Chem.* 1990, **94**, 8897–8909; (b) Rappe, A. K.; Casewit, C. J.; Colwell, K. S.; Goddard, W. A. III; Skiff, W. M. UFF, a full periodic table force field for molecular mechanics and molecular dynamics simulations, *J. Am. Chem. Soc.* 1992, **114**, 10024–10035, and the two successive papers with more detailed description of the application of UFF; (c) Clark, M.; Cramer, R. D. III; van Opdenbosch, N. Validation of the general purpose TRIPOS 5.2 force field, *J. Comp. Chem.* 1989, **10**, 982–1012.

[19] Some purists may think that this is a sort of decadence of the ideal of theoretical chemistry, not without reason. On the other hand, a chemical modeler may nowadays find him- or herself in the place of somebody who, with a quick and cheap calculation, can steer the way to the development of a useful compound for a pharmaceutical company, if not by predicting the best compound, at least by screening out some percentage of very unlikely candidates. That can be worth a very large amount of money in terms of spared resources and manpower. If publishing his or her results and letting the whole scientific community freely benefit from them is the proper thing to do in an ideal world, supplying money for free to a pharmaceutical company is not the proper thing to do in the real one. Where is the divide between scientific deontology and just being a fool? No one can tell, but the fact is that molecular mechanics force fields are nowadays mostly encapsulated in computer packages that sell for a very high price, and whose parameters and procedures are concealed to prevent improper re-use or duplication. As a consequence, these packages are often, if not always, black-box ones, in the sense that the user has all that is needed for any molecule, at the price of not always knowing exactly what the program is doing or from where the actual parameters come.

[20] A paper by Hopfinger: Hopfinger, A. J. A QSAR investigation of dihydrofolate reductase inhibition by Baker triazines based upon molecular shape analysis, *J. Am. Chem. Soc.* 1980, **102**, 7196–7206, claims to be "the clearest instance to date

where the consideration of molecular shape/conformation leads to an improved quantitative description of drug potencies". The correlations are established on the basis of shape descriptors and the experimental concentration needed for 50% in vitro inhibition of activity. Some other entry points to the literature on these subjects are: Kuntz, I. D.: Meng, E. C.; Shoichet, B. K. Structure-based molecular design, *Acc. Chem. Res.* 1994, **27**, 117–123; Vedani, A.; Dobler, M.; Zbinden, P. Quasi-atomistic receptor-surface models: a bridge between 3D QSAR and receptor modeling, *J. Am. Chem. Soc.* 1998, **120**, 4471–4477; Griffith, R.; Luu, T. T. T.; Garner, J.; Keller, P. A. Combining structure-based drug design and pharmacophores, *J. Mol. Graph.* 2005, **23**, 439–446. Recently, a compilation of some 1700 molecular descriptors has appeared: Todeschini, R.; Consonni, V. *Handbook of Molecular Descriptors*, 2000, Wiley-VCH, Weinheim.

[21] Brooks, B. R.; Bruccoleri, R. E.; Olafson, B. D.; States, D. J.; Swaminathan, S.; Karplus, M. CHARMM: a program for macromolecular energy, minimization, and dynamics calculations, *J. Comp. Chem.* 1983, **4**, 187–217.

[22] MacKerell, A. D.; Bashford, D.; Bellott, M.; Dunbrack, R. L.,Jr.; Evanseck, J. D.; Field, M. J.; Fischer, S.; Gao, J.; Guo, H.; Ha, S.; Joseph-McCarthy, D.; Kuchnir, L.; Kuczera, K.; Lau, F. T. K.; Mattos, C.; Michnik, S.; Ngo, T.; Nguyen, D. T.; Prodhom, B.; Reiher, W. E. III; Roux, B.; Schlenkrich, B.; Smith, J. C.; Stote, R.; Straub, J.; Watanabe, M.; Wiorkiewicz-Kuczera, J.; Yin, D.; Karplus, M. All-atom empirical potential for molecular modeling and dynamics studies of proteins, *J. Phys. Chem. B* 1998, **102**, 3586–3616.

[23] Weiner, S. J.; Kollman, P. A.; Case, D. A.; Chandra Singh, U.; Ghio, C.; Alagona, G.; Profeta, S.; Weiner, P. A new force field for molecular mechanical simulation of nuclei acids and proteins, *J. Am. Chem. Soc.* 1984, **106**, 765–784.

[24] Weiner, S. J.; Kollman, P. A.; Nguyen, T. A.; Case, D. A. An all atom force field for simulations of proteins and nuclei acids, *J. Comput. Chem.* 1986, **7**, 230–252; Cornell, W. D.; Cieplak, P.; Bayly, C. I.; Gould, I. R.; Merz, K. M.; Ferguson, D. M.; Spellmeyer, D. C.; Fox, T.; Caldwell, J. W.; Kollman, P. A. A second generation force field for the simulation of proteins, nucleic acids and organic molecules, *J. Am. Chem. Soc.* 1995, **117**, 5179–5197.

[25] van Gunsteren, W. F.; Billeter, S. R.; Eising, A. A.; Hunenberger, P. H.; Kruger, P.; Mark, A. E.; Scott, W. R. P.; Tironi,I. G. *Biomolecular Simulation: The GROMOS96 Manual and User Guide*, 1996; BIOMOS b.v., Zurich, Groningen and Hochschulverlag AG an der ETH Zurich; Scott, W. R. P.; Hunenberger, P. H.; Tironi,I. G.; Mark, A. E.; Billeter, S. R.; Torda, A. E.; Huber, T.; Kruger, P.; van Gunsteren, W. F. The GROMOS biomolecular simulation program package, *J. Phys. Chem.* 1999, **103**, 3596–3607.

[26] Hwang, M. J.; Stockfisch, T. P.; Hagler, A. T. Derivation of Class II force fields. 2. Derivation and characterization of a class II force field, CFF93, for the alkyl functional group and alkane molecules, *J. Am. Chem. Soc.* 1994, **116**, 2515–2525; Halgren, T. A. Merck molecular force field. I. Basis, form, scope, parameterization, and performance of MMFF94, *J. Comp. Chem.* 1996, **17**, 490–519, and subsequent papers in the same issue.

[27] Sun, H. COMPASS: an ab initio force field optimized for condensed-phase applications – Overview with details on alkane and benzene compounds, *J. Phys. Chem.* 1998, **B102**, 7338–7364.

[28] Jalaie, M.; Lipkowitz, K. B. Published force field parameters for molecular mechanics, molecular dynamics and Monte Carlo simulations, in *Reviews in Computational Chemistry* edited by K. Lipkowitz and D. Boyd. Wiley-VCH, John Wiley & Sons, New York, 2000, 14, pp. 441–486.

3

Quantum chemistry

"God does not play dice."

> Attributed to Albert Einstein, speaking of quantum mechanics; if true, it shows that even geniuses may sometimes be wrong.

3.1 Some fundamentals of quantum mechanics

In the early decades of the last century, a large and convincing body of experimental evidence pointed to the fact that atomic particles share some properties of massive bodies and some properties of waves. The founders of the new mechanics of the microscopic world, in search of the appropriate equations for the description of what seemed then a very weird universe, turned to the equations of vibrating strings, material waves whose state and energy change in leaps with the number of nodes. There is a certain resemblance between the quantum mechanical Schrödinger equation and the classical dynamic equation of vibrating strings.

The new theory that helped play the new music was then called wave mechanics or quantum mechanics, the first name being obvious, the second coming from Planck's nomenclature. Its founders were many, and all men, according to the 99.9% chauvinistic view of science then fashionable – the remaining 0.1% being on account of Maria Slodowska. Along with Niels Bohr, Louis de Broglie, Werner Heisenberg, and others, the main contributor was perhaps Erwin Schrödinger, who also put the quantum mechanical master equations in the appropriate form for chemical applications. One outstanding physicist who reportedly did not like the new thing was Albert Einstein, much to his dismay, because the success of the new discipline was rapid and triumphant. Quantum mechanics was not only a new theory in physics, but indeed an entirely new *weltanschauung*.

A complete study of quantum mechanics (QM) would require a knowledge of higher mathematics and a course in theoretical physics, but the derivation of some basic equations for the calculation of molecular energies requires no more than a little algebra and calculus [1].

3.1.1 Dynamic variables, wavefunctions, operators

In quantum mechanics dynamic variables are the same as in classical mechanics: time (t), positional coordinates, $\mathbf{x} = (x, y, z)$, linear momentum $\mathbf{p} = m\mathbf{v}$, and angular momentum $\mathbf{M} = \mathbf{R} \wedge \mathbf{p}$. Energy plays a leading role in QM just as it does in classical mechanics.

In classical mechanics, dynamic differential equations are solved to obtain trajectories. In QM the behavior of bodies is described by a state function or wavefunction, $\Psi(x, y, z, t)$. The quantity

$$P(d\tau) = [\Psi\Psi^*]dx\,dy\,dz\,dt \tag{3.1}$$

is interpreted as the probability that a body be found in the infinitesimal volume of time-space defined by $dx\,dy\,dz\,dt = d\tau$. Ψ^* is the complex conjugate of Ψ. While a Newtonian trajectory states that a body will be found at a position $\mathbf{x}$ at a given time t, no equation in QM will say more than what equation 3.1 has to say about the position of a body; newtonian mechanics is deterministic, QM is probabilistic.

The wavefunction must be single-valued and continuous, and its derivatives must also be continuous. Another important condition is:

$$\int \Psi\Psi^* d\tau = \text{a finite value} \tag{3.2}$$

The integral represents a summation of infinitesimal probabilities over the entire space, and therefore cannot be infinite – actually, there are ways of ensuring that its value be equal to 1, as an integral probability should (the normalization of the wavefunction).

A pragmatic definition of an operator is: a procedure that transforms a function into another function. Thus, "3" is a number but "3 times" is an operator that transforms the function $3x + 1$ into the function $9x + 3$. The operator "derivative with respect to x", $\partial/\partial x$, transforms the function $\cos(x)$ into the new function $-\sin(x)$. When two operators act upon a given function, the result depends in general on the order of application; if not, it is said that the two operators commute.

To prepare quantum mechanical equations, dynamic variables must be associated with operators according to the rules given in Table 3.1. The classical hamiltonian, or the sum of kinetic and potential energies of a body, transforms into the hamiltonian operator.

3.1.2 The Schrödinger equation and stationary states

The master equation for chemical applications of QM is the time-dependent Schrödinger equation:

$$\hat{H}\Psi = -h/i(\partial\Psi/\partial t) \tag{3.3}$$

where $\hat{H}$ is the hamiltonian operator, h is Planck's constant, and i is the imaginary unit (i = $\sqrt{-1}$). When the potential energy is independent of time, the problem

SOME FUNDAMENTALS OF QUANTUM MECHANICS

Table 3.1 Rules for the association of dynamic variables to operators. The square of a variable corresponds to double application of the operator; "times" means that the variable transforms into a multiplier. h is Planck's constant, italic $h = h/2\pi$; $i = \sqrt{-1}$, $i^2 = -1$. m_P is the mass of the particle

Dynamic variable	Operator
Position x, y, z	x "times", y "times", z "times"
Linear momentum p_x, p_y, p_z	$-h/i\, \partial/\partial x, -h/i\, \partial/\partial y, -h/i\, \partial/\partial z$
Angular momentum $M_z = xp_y - yp_x$	$-h/i\,[x\,\partial/\partial y - y\,\partial/\partial x]$
Kinetic energy $T = 1/(2m_P)p^2$	$-[h]^2[1/(2m_P)][\partial^2/\partial x^2 + \partial^2/\partial y^2 + \partial^2/\partial z^2] \equiv -[h]^2[1/(2m_P)]\nabla^2$
Potential energy V	V "times"

can be reduced to a considerably simpler form: the wavefunction factorizes into a time-dependent part, f(t), and a time-independent, position-dependent part, $\psi(\mathbf{x})$:

$$\hat{H}\psi(\mathbf{x}) = E\psi(\mathbf{x}) \tag{3.4}$$

$$\Psi(\mathbf{x}, t) = f(t)\psi(\mathbf{x}) \tag{3.5}$$

A system that obeys equation 3.4 is said to be in a stationary state. Chemistry is mostly concerned with stationary states of atomic and molecular systems.

The language of QM has only a limited correspondence with the language of common objects or even with the specialized language of classical dynamics. In the QM equations for a hydrogen atom, for example, there are no explicit forces or motions, no trajectory of the electron around the nucleus, and indeed there are no material bodies (although there still is mass). The only quantum mechanical reality is a wavefunction that gives the probability of finding the electron at a position $\mathbf{x}$ around the nucleus.

Equation 3.4 is a particular case of a more general rule: the only possible (that is, observable) values for a dynamic variable a whose associated operator is $\hat{A}$ are the solutions of the equation

$$\hat{A}\psi = a\psi \tag{3.6}$$

The numbers a are called eigenvalues of operator $\hat{A}$, and ψ is also called the eigenfunction associated with the eigenvalue. Thus, equation 3.4 states that the only observable values for the energy of a system in a stationary state are the eigenvalues of the hamiltonian operator.

From equation 3.4, multiplying on the left by $\psi^*(\mathbf{x})$, and integrating over all space, one gets:

$$<E> = \int \psi^*(\mathbf{x})\hat{H}\psi(\mathbf{x})d\tau / \int \psi^*(\mathbf{x})\psi(\mathbf{x})d\tau \quad (3.7)$$

If the wavefunction is normalized, the denominator is unity and the expectation value of the energy is written as:

$$<E> = \int \psi^*(\mathbf{x})\hat{H}\psi(\mathbf{x})d\tau \quad (3.8)$$

$<E>$ can be interpreted as the most likely value that would be measured for the energy, in a probabilistic sense, as an average over a very large number of observations of the system under consideration.

3.1.3 Linear momentum

The simplest system consists of just one particle of mass m_P, with momentum p, in a one-dimensional space along coordinate x. The kinetic energy is $E_{\mathrm{kin}} = (p)^2/(2m_P)$. The hamiltonian T+V is transformed into its quantum mechanical operator form ($V = 0$). Using the prescriptions of Table 3.1, the Schrödinger equation for such a system is:

$$-[h^2/(2m_P)][d^2\psi(x)/dx^2] = E\psi(x) \quad (3.9)$$

This differential equation is promptly solved as follows:

$$\psi(x) = A\exp(ikx); \quad d^2\psi(x)/dx^2 = -k^2\psi(x)$$
$$[h^2/(2m_P E)]k^2\psi(x) = \psi(x) \quad (3.10)$$
$$k = (2m_P E)^{1/2}/h = p/h$$

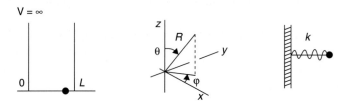

Fig. 3.1. Three key situations for quantum chemistry. Left: a particle, black dot, in a linear space along x between two walls of infinite potential. Center: angular momentum, a particle at the end of vector **R** rotates around the z axis. Polar coordinates are R, θ, ϕ. Right: a particle, black dot, oscillating against a rigid wall under the restraint of a spring whose stiffness is described by constant k.

An interesting development occurs if the particle is constrained in the range of x-space enclosed between two points $x = 0$ and $x = L$ (Fig. 3.1), which represent walls of infinite potential. The potential is zero anywhere except at the walls. The particle cannot go through the walls because of the infinitely high potential barrier, a condition that mathematically translates into $\psi(x) = 0$ for $x = 0$ or $x = L$. In this case, a real wavefunction $\psi(x) = A \sin(kx)$ will do: it is zero at $x = 0$. Writing the boundary condition for $x = L$ and rearranging:

$$\psi(L) = A \sin(kL) = 0$$

$$kL = n\pi \quad (n = 0, 1, 2 \ldots; \text{the quantum number}) \quad (3.11)$$

$$n\pi/L = [(2m_P E)^{1/2}]/h$$

$$E(n) = n^2 h^2/(8m_P L^2)$$

The energy of the particle is thus quantized. Quantization arises as a mathematical consequence of the boundary conditions. In the same way, a string vibrates in overtones only if it is held fixed at its two ends.

The waverfunction can be normalized as follows:

$$\int \psi\psi \, dx = 1 = A^2(L/2) \quad (3.12)$$

$$\psi_n(x) = (2/L)^{1/2} \sin(n\pi x/L)$$

The particle cannot occupy x-positions where the wavefunction is zero. Such places are called the nodes of the wavefunction. The particle exists at the right and at the left of the node, without being able to exist at the node, but it cannot be said that it "goes through" the node. The quantum number n cannot be zero, otherwise the wavefunction would be zero for any x, and this is obviously not acceptable because the particle must be somewhere. So the energy of the particle cannot be zero, and the energy corresponding to the $n = 1$ level is called the zero-point energy. It is an aspect of the uncertainty principle: if the particle were at rest and its energy were exactly zero, both quantities would be known with infinite accuracy, while the uncertainty principle states that the product of the uncertainty of energy and the uncertainty of position must be a finite quantity.

The three-dimensional equivalent of equation 3.11, for a cubic box of side L, is

$$E(n, m, l) = (n^2 + m^2 + l^2)h^2/(8m_P L^2) \quad (3.13)$$

where n, m, l are three integer quantum numbers.

3.1.4 Angular momentum

The quantum mechanical treatment of a particle whose classical trajectory is not linear considers a particle possessing non-zero angular momentum. Let rotation be in the $x - y$ plane at a distance R from the rotation axis, z (Fig. 3.1). The moment of inertia is $I = m_P R^2$, the angular momentum is $\mathbf{M} = \mathbf{p} \wedge \mathbf{R}$, where the linear momentum is $p = m_P v$, v being the tangential velocity, and the kinetic energy is $E = M_z^2/(2I)$. In polar coordinates (R, θ, φ) these conditions transform into R = constant, $\theta = 90°$, so that the only degree of freedom is along ϕ. Equation 3.6 can be used to find the eigenvalues and the wavefunctions, $\psi(\phi)$ of the angular momentum operator defined in Table 3.1:

$$\hat{M}_z \psi(\phi) = -h/i[x\partial/\partial y - y\partial/\partial x]\psi(\phi) = M_z \psi(\phi) \tag{3.14}$$

The angular momentum operator has a much simpler form in polar coordinates (Fig. 3.1):

$$(h/i)d[\psi(\phi)]/d\phi = M_z \psi(\phi) \tag{3.15}$$

Since the particle is spatially constrained on a circular trajectory, $\psi(\phi)$ must include quantum conditions and the energy must be quantized. The wavefunction can be written as $\psi(\phi) = \exp(im\phi)$ so one easily gets:

$$d\psi/d\phi = im \, \psi(\phi) \tag{3.16}$$

$$hm = M_z \tag{3.17}$$

where m is a disposable constant. The wavefunction must be single-valued and it must take the same value after an integer number of periods. In other words, the wavefunction, and hence the probability of finding the particle at a given location, must be the same at ϕ and at $\phi + 2\pi$:

$$\exp(im\,\phi) = \exp[im(\phi + 2\pi)]$$
$$\exp(im2\pi) = 1 \quad m = 0, 1, 2, 3, \ldots \tag{3.18}$$

Boundary conditions on the wavefunction force m to be an integer (the rotational quantum number). The rotational kinetic energy is quantized as follows:

$$E = M_z^2/(2I) = (h^2 m^2)/(2I) \quad m = 0, \pm 1, \pm 2, \pm 3, \ldots \tag{3.19}$$

For an arbitrary rotational motion, the full vector expression of the angular momentum along any direction in space must be considered, with its three components:

$$M^2 = M_x^2 + M_y^2 + M_z^2 \tag{3.20}$$

The system can be visualized as a particle of mass m_P moving on the surface of a sphere. The conceptual development is the same as before, but the algebra now becomes very complicated. Nevertheless, the final result has a really simple form:

$$M = h[l(l+1)]^{1/2} \quad l = 0, 1, 2 \ldots \tag{3.21a}$$

$$M_z = hm \quad -l \le m \le +l \tag{3.21b}$$

$$E = M^2/(2I) = (1/2I)h^2[l(l+1)] \tag{3.21c}$$

A new quantum number, l, appears; the energy is quantized, and there is a new relationship between the quantum number for total angular momentum l, and the quantum number for rotation in the x-y plane, m. For any given value of the total angular momentum, its projection on the z-axis can only have a number of discrete values, so equation 3.21b is a vectorial, or directional, quantization. For example, for $l = 1$, the total momentum is $h\sqrt{2}$, and the z-component can only be $-h, 0,$ or $+h$. Calling α the angle between the angular momentum and its z-component, the only allowed values are $\cos(\alpha) = -1/\sqrt{2}, 0,$ or $+1/\sqrt{2}$.

The above results for the quantization of angular momentum have been derived for a single particle, but the same results hold for two masses (say, a diatomic molecule) rotating around their center of mass, when I is the moment of inertia of the system. This opens the way to the quantum mechanical treatment of rotating molecules.

3.1.5 Spherical harmonics

In a one-dimensional vibrating string, a node is a point where the vibration amplitude is zero. In a flexible two-dimensional membrane, like the skin in a drum, a node is a line where the displacement is zero. The three-dimensional picture of a node requires a little more imagination, and could be viewed as the compression or expansion of a rubber sphere, where the loci of zero displacement, the nodes, are surfaces. The nodes have one dimension less than the oscillating system to which they refer (points on a line, lines on a surface, surfaces in a volume). A general principle built into the mathematics and the physics of the problem, a rule that has no exceptions, says that the energy of the vibrating system increases with the number of nodes. In quantum mechanics, the energy eigenvalue associated with any eigenfunction is higher when the number of nodes in the wavefunction is higher.

The generalized wavefunctions of the angular momentum operator are called spherical harmonics, and their symbol is Y_{lm}. They depend on two quantum numbers, and are eigenfunctions of the angular momentum and of its projection on the z axis. They obviously are three-dimensional functions, and their nodes are surfaces. In analogy with equation 3.21:

$$M^2 Y_{lm} = h^2 l(l+1) Y_{lm}, \quad l = 0, 1, 2 \ldots, \quad -l \leq m \leq +l \quad (3.22)$$

$$M_z Y_{lm} = hm Y_{lm}$$

In spite of the apparent complexity, the functional form of the spherical harmonics (Table 3.2) is quite simple. The (0,0) harmonic is a constant. The (1,0), (1,1), and (1,−1) harmonics have a nodal plane each, the $z = 0, x = 0,$ and $y = 0$ plane (the xy, yz, and xz plane) respectively (Fig. 3.2). A study of the harmonics for $l = 2$ reveals that they have two nodal planes each. The higher harmonics for $l = 3$ are a bit more complicated, but they have three nodal planes each.

Table 3.2 The spherical harmonics (unnormalized); $r^2 = x^2 + y^2 + z^2$

l	m	Y_{lm}, polar	Y_{lm}, cartesian
0	0	1	1
1	0	$\cos\theta$	z/r
1	1	$\sin\theta\cos\varphi$	x/r
1	−1	$\sin\theta\sin\varphi$	y/r
2	0	$3\cos^2(\theta) - 1$	$(3z^2 - r^2)/r^2$
2	1	$\sin\theta\cos\theta\sin\varphi$	yz/r^2
2	−1	$\sin\theta\cos\theta\cos\varphi$	xz/r^2
2	2	$\sin^2\theta\sin 2\varphi$	xy/r^2
2	−2	$\sin^2\theta\cos 2\varphi$	$(x^2 - y^2)/r^2$

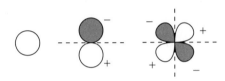

Fig. 3.2. Spherical harmonics with increasing quantum number. Left to right: no nodal planes, one nodal plane, two nodal planes.

3.1.6 *The harmonic oscillator*

The potential energy of a body moving under the action of a harmonic restraining force (Fig. 3.1) is $V = 1/2 kx^2$, where k is the force constant. The quantum mechanical equation for the stationary states of a harmonic oscillator of mass m_P is:

$$-(h^2/2m_P)d^2[\psi(x)]/dx^2 + 1/2\, kx^2\psi(x) = E\psi(x) \qquad (3.23)$$

The algebra for the solution of this equation is long and complex, but the final result for the quantized energies turns out again in a simple and elegant form:

$$E = h(k/m_P)^{1/2}(v + 1/2) = h\nu(v + 1/2), \quad v = 0, 1, 2, \ldots \qquad (3.24)$$

The classical vibrational frequency is denoted by ν and the vibrational quantum number by v.

A classical oscillator in its lowest energy state is at rest at $x = 0$. A quantum mechanical oscillator can never be constrained at $x = 0$ because the wavefunctions are non-zero over the whole range. Besides, the energy is not zero even when $v = 0$; the vibrational zero-point energy is $E° = 1/2 h\nu$. As for linear momentum, this zero-point energy is ultimately connected with the uncertainty principle.

3.2 The hydrogen atom and atomic orbitals

The reference frame for the quantum mechanical treatment of the hydrogen atom has a motionless nucleus at the origin, while one electron of mass m_e and charge e spans the surrounding space with linear and angular momentum. The potential energy between the $+e$ charge of the nucleus and the $-e$ charge of the electron at a distance r has the coulombic form $V = -e^2/r$ (the term $1/(4\pi\varepsilon_0)$ will be considered implicit). The total hamiltonian is transformed as usual into a hamiltonian operator, whose kinetic energy part has all its three components in three-dimensional space. The final QM equation for the stationary states of the hydrogen atom is:

$$-\hbar^2/(2m_e)\nabla^2\psi - (e^2/r)\psi = E\psi \qquad (3.25)$$

The Es are the allowed electronic energies, and the corresponding wavefunctions give the probability of finding the electron at any point in space.

The solution of equation 3.25 requires a few pages of algebra and will not be given in detail here. The conceptual framework of the procedure is as follows.

1. Equation 3.25 is transformed from cartesian to polar coordinates, so that electronic states are described by wavefunctions $\psi(r, \theta, \phi)$.
2. Since each term in the hamiltonian depends either on the radial coordinate r or on the angular coordinates θ and ϕ, the total wavefunction can be factorized into a product of a radial part, $R(r)$, that depends only on the nucleus–electron distance, and an angular part, $S(\theta, \phi)$, that depends only on angular coordinates:

$$\psi(r, \theta, \phi) = R(r)S(\theta, \phi) \qquad (3.26)$$

3. Acceptable radial wavefunctions must as usual be well behaved (single-valued, continuous, etc.), and, since the electron is radially constrained by the electrostatic potential, quantum conditions apply with a radial quantum number, n. A few of these radial wavefunctions are shown in Table 3.3.
4. The angular wavefunctions are the spherical harmonics of Table 3.2.
5. Radial and angular quantum conditions coexist, and this brings about a coupling of radial and angular quantum numbers:

$$n = 1, 2, 3, \ldots; \quad 0 \le l \le n-1; \quad -l \le m \le +l \qquad (3.27)$$

The global wavefunctions, equation 3.26, are called orbitals. The etimology of this term goes back to "orbit", or something connected with revolution of one body around another, but it is by now clear that the picture of an electron "orbiting" around the

Table 3.3 The unnormalized radial functions in atomic hydrogen-like orbitals (equation 3.26). The distance r is measured in units of the Bohr radius, 0.529 Å

n	l	$R(r)$	Radial nodes at:
1	0	$\exp(-r)$	none
2	0	$(1 - r/2)\exp(-r/2)$	$r = 2$
2	1	$r\exp(-r/2)$	$r = 0$
3	0	$(1 - 2/3r + 2/27r^2)\exp(-r/3)$	$r = 9/2(1 \pm 1/\sqrt{3})$
3	1	$r(1 - r/6)\exp(-r/3)$	$r = 0; r = 6$
3	2	$r^2\exp(-r/3)$	$r = 0$

nucleus is not entirely consistent with the quantum mechanical picture. A probabilistic interpretation of the wavefunction, useful in chemistry, implies instead that the square of the orbital gives the probability of finding the electron at any point (x, y, z) or (r, θ, ϕ). The electron is thus represented by a probability "cloud" around the nucleus. All $R(r)$'s decrease with increasing r, and orbitals go to zero for $r \to \infty$; any electron density is in principle diffuse over the entire universe. Their overall nodes and shape are similar to those of the spherical harmonics, Fig. 3.2, but modulated by the radial dependence. Some of the radial functions are zero for some values of r: these are called radial nodes. The total number of nodes is the sum of the number of radial and angular nodes. Finally, for historical reasons, orbitals with $l = 0, 1, 2 \ldots$ are called $s, p, d \ldots$ orbitals.

If the zero of the electronic energy scale is taken as the state in which the electron is at an infinite distance from the nucleus, all eigenvalues for the stationary states of the hydrogen atom correspond to stabilizing energies:

$$E = (-1/h^2)m_e\, e^4 (1/n^2) \tag{3.28a}$$

$$\Delta E(n_2 - n_1) = 1/h^2\, m_e\, e^4 [(1/n_1^2) - (1/n_2^2)] \tag{3.28b}$$

The energy depends only on n, so for an electron in a hydrogen-like orbital, the energy is the same for s, p or d levels with the same principal quantum number, and for each n there are n^2 states with the same energy. In quantum chemical jargon, states of the same energy are called degenerate. The energy differences calculated by equation 3.28b match perfectly the experimental electronic excitation energies, and this is one of the triumphs of quantum mechanics, since no classical theory had ever been able to reproduce the experimental spectroscopic results.

The classical picture cannot explain why the electron does not fall on the nucleus under the influence of the attractive electron–nucleus electrostatic force. QM explains this very naturally, because as the electron comes closer to the nucleus, its potential energy becomes more stabilizing, but the electron is more and more restricted in space and its kinetic energy increases (recall equation 3.13 with decreasing L). Thus the probability of finding the electron has a maximum at a certain distance from the nucleus, which results from the balance between potential and kinetic energies.

3.3 Spin

Of all un-classical properties of matter, spin is the least classical of all. In fact, if the QM treatment of the linear and angular momentum of an electron, and of the electron–nucleus electrostatic potential, to some extent descend from classical analogs, there is no classical equivalent of spin.

No one can propose an actual, or even a mental, picture of a spinning electron. Quantum mechanics deals with objects of the microscopic world, objects that one cannot see or feel with human senses. The justification for the laws and formulas of quantum mechanics comes from agreement of the predicted quantities with the results of subtle experiments: the necessity of spin emerged when it was observed that some energy splittings in the fine structure of atomic spectra could not be reproduced by equations which included only kinetic energy and electrostatic potential. Paul Adrien Maurice Dirac and Wolfgang Pauli led the way into this development and worked out the mathematical equations. They were awarded Nobel prizes in 1933 and 1945, respectively.

Not only the electron, but also the proton, or hydrogen nucleus, and other atomic nuclei, have their own spin. Spin is a quantum property, and therefore must be described by a wavefunction and by quantum numbers, and the energies of spin states must be quantized. The usual procedure of writing down a classical hamiltonian and then transforming into the QM hamiltonian operator cannot be followed since there is no classical spin energy. The theory proceeds from a number of postulates that will now be discussed.

- Postulate (a): the spin angular momentum, SAM, is described by a QM operator $\hat{S}$, similar to the angular momentum operator:

$$\hat{S}^2 = \hat{S}_x^2 + \hat{S}_y^2 + \hat{S}_z^2 \tag{3.29}$$

- Postulate (b): each particle has its own spin quantum number, s; for an electron or a proton, the only possible value for s is 1/2.
- Postulate (c): there are only two spin quantum states and two eigenfunctions, called α and β, such that:

$$\hat{S}^2 \alpha = s(s+1)h^2 \alpha; \quad \hat{S}^2 \beta = s(s+1)h^2 \beta \tag{3.30}$$

$$\text{SAM} = [s(s+1)]^{1/2} h$$

- Postulate (d): the operator for the z-component of SAM, $\hat{S}_z$, has the same eigenfunctions as $\hat{S}^2$, with a new quantum number, called m_s:

$$-s \leq m_s \leq +s$$
$$\hat{S}_z \alpha = 1/2 h\alpha \quad \hat{S}_z \beta = -1/2 h\beta \tag{3.31}$$

The first of these equations is general, the second holds only for electrons, protons, and all particles with $s = 1/2$.

The above equations suggest a consideration of spin as if it were a vector in the same manner as angular momentum. One can therefore calculate the angle between the spin "vector" and its z "component", by using the standard geometrical relationship $S_z = S \cos \theta$:

$$1/2 = [1/2(1/2 + 1)]^{1/2} \cos \theta$$
$$\theta = \arccos(1/\sqrt{3}) = 54.7° \text{ or } 125.3° \tag{3.32}$$

An electron may exist in two states, state α with spin quantum number $1/2$ and $\theta = 54.7°$, or state β with spin quantum number $-1/2$ and $\theta = 125.3°$. For a pictorial touch, the state with $s = 1/2$ is sometimes called "spin up" and the state with $s = -1/2$ is called "spin down". In keeping with the probabilistic interpretation of QM, spin wavefunctions must be two-valued functions, being equal to one whenever the spin conditions are fulfilled, and equal to zero whenever they are not.

The above equations seem so far quite arbitrary. A further postulate provides the necessary connection with experimental evidence.

- Postulate (e): electrons behave like magnets with a magnetic dipole moment, μ, whose modulus is given by:

$$|\mu| = 2(\text{SAM})e/(2c\, m_e) = 2[s(s+1)]^{1/2} he/(2c\, m_e) = 2[s(s+1)]^{1/2} \beta_M \tag{3.33}$$

where c is the speed of light, m_e is the electron mass, e is the electron charge, and β_M is a measurable quantity called the Bohr magneton. The two spin states correspond to two possible orientations of the magnetic dipole with respect to any arbitrary z-direction. Using equation 3.32:

$$\mu_z = |\mu| \cos \theta = \pm \mu (1/\sqrt{3}) = \pm \beta_M \tag{3.34}$$

In the absence of external perturbations the two spin states are degenerate, but when an electron is placed in an external magnetic field the two spin states are separated by an energy difference that depends on the product of the external field and the component of the magnetic dipole, and can be measured experimentally. Nuclear spin is at the root of nuclear magnetic resonance, perhaps the most versatile analytical technique in chemistry.

The total wavefunction for an electron in the hydrogen atom, or in any other electronic system, must result from a combination of atomic orbitals and spin wavefunctions. The appropriate combination is a product because the probabilistic interpretation requires that a total wavefunction must be zero whenever any of its parts is zero. The product function is called a spin-orbital. For example, the single electron in the 1s orbital of hydrogen may be described by two such spin-orbitals:

$$\chi_\alpha = \psi(1s)\alpha; \quad \chi_\beta = \psi(1s)\beta \qquad (3.35)$$

3.4 Many-electron systems

Not considering the rather improbable hydrogen anion, helium is the smallest system with more than one electron around the nucleus. The helium wavefunction must depend on the coordinates of both electrons, formally labeled 1 and 2. The kinetic energy operators are as usual, and there are now three coulombic terms, representing the interaction of each of the two electrons with the nucleus, now with charge +2, and the electrostatic interaction between the two electrons, at a distance r_{12}. The Schrödinger equation is

$$-h^2/2m_e \nabla_1^2 \psi - (2e^2/r_1)\psi - (h^2/2m_e)\nabla_2^2 \psi - (2e^2/r_2)\psi + (e^2/r_{12})\psi = E\psi \qquad (3.36)$$

This differential equation cannot be solved analytically. The insurmountable difficulty is that the e^2/r_{12} term depends on the coordinates of both electrons, and there is no way of breaking down the total wavefunction into a radial and an angular part.

A very productive approximation, one that has been successfully used even for complex molecules, is to build complex wavefunctions from combinations of simple spin-orbitals. Consider again the 1s orbital for the hydrogen atom, now with two electrons in it, and let $(1s)_1$ and $(1s)_2$ be the orbital written in the coordinates of electrons 1 and 2, respectively. There are four possible spin-orbitals:

$$(1s)_1\alpha_1 \qquad (1s)_1\beta_1 \qquad (1s)_2\alpha_2 \qquad (1s)_2\beta_2 \qquad (3.37)$$

The total wavefunction to be constructed from these basis functions might then be made of sums or products of spin-orbitals. But there is a new prescription, coming again from mathematical speculation based on well established experimental facts, having to do with the multiplicity of atomic spectral lines. This prescription is known as the Pauli principle: acceptable total wavefunctions must be anti-symmetric, that is, they must change sign on permutation of the coordinates of any pair of electrons in the system, that is, when electron 1 formally becomes electron 2 and vice versa. This seems a very odd request indeed, and has no correspondence with classical terms. The Pauli principle must be accepted as a powerful mathematical sieve that restricts our choice of wavefunctions to the ones that correctly describe the behaviour of atomic systems, including the number of lines observed in their adsorption or emission spectra.

An easy way of ensuring that a combination of the four basis functions 3.37 is antisymmetric is to arrange them in a matrix and take the total wavefunction as the determinant of that matrix. N being a normalization factor, one obtains:

$$\psi = N \, \text{Det} \begin{bmatrix} (1s)_1 \alpha_1 & (1s)_2 \alpha_2 \\ (1s)_1 \beta_1 & (1s)_2 \beta_2 \end{bmatrix} = N(1s)_1(1s)_2[\alpha_1 \beta_2 - \alpha_2 \beta_1] \qquad (3.38)$$

This wavefunction is zero if the two electrons in the 1s orbital have the same spin, because it is either $\alpha_1 = \alpha_2 = 1$ (in which case $\beta_2 = \beta_1 = 0$) or $\beta_1 = \beta_2 = 1$ (in which case $\alpha_1 = \alpha_2 = 0$). Only when one electron is α and one is β does the spin wavefunction survive. This rule is fundamental for understanding how atoms and molecules are built, and can be generalized as follows: no two electrons with the same four quantum numbers n, l, m, and m_s can be found in any system. If they stay in the same orbital, their spins must be opposite (different m_s): paired spins are called a singlet state. If they stay in different orbitals, their spins can be any, since there is anyway a difference in the n, l, or m quantum numbers. In particular, if the spins are the same (unpaired spins) the state is called a triplet state. Systems in which all electron spins are paired are called closed shells, systems with unpaired spins are called open shells.

In classical terms, when two bodies or entities cannot coexist one says that a repulsive force acts between them. The incompatibility of electrons carrying the same quantum numbers is sometimes said to be due to "Pauli forces", or, for reasons that should be clear from the above discussion, "exchange forces". They are not mechanical or electrostatic forces, however. The energy of the system is not directly affected by some mysterious "Pauli potential", but changes through the change in electrostatic energy brought about by the different electron arrangement enforced by spin prescriptions. In a metaphoric sense, Pauli "forces" act on electrons like traffic lights act on cars – to prevent them from colliding.

Further advances in the theory of the electronic structure of atoms reveal that when there are many electrons, the energies of states with the same principal quantum number but different angular momentum are different, so that, for example, in the carbon atom the energetic level of a 2s orbital is different from that of a 2p orbital. The electronic structures of atoms may be built by assigning two electrons with paired spins to each atomic orbital, starting from the lowest-energy ones. This *aufbau* principle naturally produces periodically similar electronic structures. Mendeleev was indeed realizing the periodic repeat of partially filled outer shells in atomic systems.

3.5 Molecular orbitals: The Fock and Roothaan equations

In principle, a description of the electronic structure of many-electron atoms and of polyatomic molecules requires a solution of a Schrödinger equation for stationary states quite similar to equation 3.36 [2]. Even for a simple molecule like, say, methane, however, such an equation would be enormously more complicated, because the hamiltonian operator would include kinetic energy terms for all electrons, plus coulombic terms for the electrostatic interaction of all electrons with all nuclei and of all electrons with all other electrons. The QM hamiltonian operator for the electrons in a molecule reads:

$$\hat{H} = \sum_k \hat{H}(k) + e^2 \sum_{k>j} 1/r_{kj} \qquad (3.39)$$

where $\hat{H}(k)$ is the part for the kinetic energy of electron k plus its coulombic interactions with all the nuclei, and the second summation corresponds to the electron–electron coulomb potential. Since a full analytical solution of the Schrödinger equation is impossible even for a helium atom, the case with a hamiltonian like 3.39 is plainly hopeless.

A number of approximations are needed. Consider as a first step the construction of an appropriate wavefunction for a many-electron atom. Appropriate basis functions, or spin-orbitals, must be chosen in equation 3.26. The spherical harmonics are unchanged, since they carry the imprint of angular momentum. The problem is the radial part, $R(r)$: the hydrogen-like radial functions of Table 3.3 would be a good choice, but their analytical form is too complex for easy calculation of the necessary integrals. A practical attitude suggests a very general and flexible analytical form, in terms of a linear combination of basis functions chosen according to computational convenience. It turns out that the ideal basis function, the workhorse of quantum chemistry, is the gaussian basis function:

$$g(r) = N\alpha^{(2n+1)/4} r^{(n-1)} \exp(-\alpha r^2) \qquad (3.40)$$

in which r is the distance of the electron from the nucleus, α is called the exponent of the gaussian, n is 1, 2 or 3 for s, p or d functions respectively, and N is the normalization constant. The radial part of the i-th atomic orbital is then written as:

$$R_i(r) = \sum_j a_{ij} g_j \qquad (3.41)$$

The single gaussians 3.40 decay to zero as r tends to infinity because the inverse exponential decreases faster than the $(n-1)$-th power increases, but the whole function has no radial nodes, and the necessary nodal properties must be built into the overall radial functions by an appropriate choice of negative and positive coefficients a_{ij}. The complete expression for an atomic orbital χ_i is:

$$\chi_i = N'R_i(r)Y_{lm}(\theta, \phi) \tag{3.42}$$

where the Ys are the spherical harmonics in Table 3.2, and N' is an overall normalization constant. The real problem is the determination of the best values for the coefficients a_{ij}.

For a single atom the nucleus is fixed at the origin of coordinates. For molecules, a big simplification results from the fact that electrons rearrange so much faster than nuclei that the positional coordinates of the nuclei can be kept fixed in the calculation of electronic energies, and the molecular wavefunction Φ depends only on the coordinates of electrons. This is called the Born–Oppenheimer assumption (there are no potential energy terms for nuclei in the hamiltonian 3.39). The total electronic energy is the expectation value of the hamiltonian operator, equation 3.8:

$$\int \Phi^* \hat{H} \Phi d\tau = <E> \tag{3.43}$$

In an obvious conceptual development, molecular orbitals should be obtained as some form of combination of atomic orbitals [3]. A convenient approximation is to take a product of many functions, each of which contains the coordinates of one electron only, i.e. a product of one-electron orbitals:

$$\Phi(1, 2, 3, \ldots N) = \varphi_1(\text{electron } 1)\varphi_2(\text{electron } 2)\varphi_3(\text{electron } 3)\ldots \varphi_N(\text{electron } N) \tag{3.44}$$

This is a very severe approximation, because this wavefunction cannot describe the effects of electron correlation, or the simultaneous displacement, or other simultaneous change in properties, of many electrons at a time.

To ensure the required antisymmetry, so that the Pauli exclusion principle is obeyed, use is made again of the spin functions α and β and of a determinant form like equation 3.38. Let $\varphi(i, j)$ be the i-th orbital in the coordinates of the j-th electron. The antisymmetrized function is a sum of products of one-electron spin-orbitals and is called a Slater determinant:

$$\Phi(1, 2, 3, \ldots N) =$$

$$= (N!)^{-1/2} \text{Det} \begin{bmatrix} \varphi(1,1)\alpha & \varphi(1,1)\beta & \varphi(2,1)\alpha & \varphi(2,1)\beta \ldots \\ \varphi(1,2)\alpha & \varphi(1,2)\beta & \varphi(2,2)\alpha & \varphi(2,2)\beta \ldots \\ \ldots \ldots \ldots \ldots \ldots \ldots \\ \varphi(1,N)\alpha & \varphi(1,N)\beta & \varphi(2,N)\alpha & \varphi(2,N)\beta \ldots \end{bmatrix} \tag{3.45}$$

MOLECULAR ORBITALS

For a single atom, the φs are the atomic orbitals as 3.42. Molecules are made of atoms, and each molecular orbital (MO) is taken as a Linear Combination of Atomic Orbitals in the so-called LCAO-MO method:

$$\varphi_i = \sum_j c_{ij}\, \chi_j \qquad (3.46)$$

This equation is formally similar to equation 3.41, and the problem is still how to find the most appropriate coefficients for the linear combinations, those that will produce a wavefunction that correctly reproduces total energies, ionization potentials, and spacings between electronic levels. The problem for isolated atoms is only formally different from the problem for molecules, because in both cases the calculation is eventually carried out by choosing a number of primitive gaussian radial functions, by mixing them in the appropriate Slater determinant, and by finding the best coefficients for the combination into the final atomic or molecular orbitals (the orbitals of an isolated atom are the "molecular orbitals" for a one-atom molecule) [4]. But do not confuse basis orbitals and complete orbitals: for example, the 3s atomic orbital of a chlorine atom may be a combination of 1s, 2s, and 3s gaussians.

The mixing of AOs into MOs is restricted only by the nodal properties of the orbitals and by symmetry: the 3s orbital of a chlorine atom may not contain contributions from any of the p gaussians, because the s—p overlap between AO's centered on the same atom is zero. This can be easily checked by mentally overlapping the two spherical harmonics in Fig. 3.2, where the $(++)$ overlap is equal and of opposite sign to the $(+-)$ overlap. In the same way, the p_z AOs of the ethylene carbon atoms do not overlap with any of the s-type orbitals in the rest of the molecule, and mix as a separate subset of AO's into the π-MOs [5]. These restrictions ultimately stem from the angular momentum of electrons.

3.5.1 *The variational principle*

To solve the problem of finding the best coefficients for the combinations in equations 3.41 and 3.46 one must accept here another non-demonstrated truth, called the variational principle. It states that whenever the true wavefunction is approximated by some incomplete function that depends on a number of parameters, the expectation value of the energy, equation 3.43, is higher than the expectation value that competes to the exact wavefunction. Briefly, whenever the wavefunction is wrong, the energy goes up.

The procedure now requires a calculation of the expectation value of the energy, equation 3.43, given the hamiltonian of equation 3.39, the wavefunction in the form of a Slater determinant, 3.45, and the molecular orbitals in the LCAO approximation, equation 3.46. The solution of this problem required some genius and a lot more patience. It is in fact mostly a matter of algebra and integration, although of very tedious algebra and integration. The final equations are called the Fock equations, and their expression in LCAO terms goes under the collective name of Roothaan's equations (see Box 3.1). From these equations it appears that the c_{ij} coefficients are needed in order to calculate the energy. But these coefficients are just what one is looking for when calculating the energy. This obstacle is overcome by making use of a self-consistent field (SCF) approach, a recursive trial-and-error mechanism. The whole procedure for the calculation of the best energy and the best wavefunction for a given molecule goes through the following steps.

1. Prepare a molecular model with the positions of all nuclei, kept fixed because of the Born–Oppenheimer approximation.
2. Choose a set of atomic orbitals χ_i, for each atom in the molecule, from repertoires available in the literature. These orbitals are given as a list of exponents in equation 3.40 and coefficients in equation 3.41.
3. Calculate all the necessary overlap, H-type, coulomb and exchange integrals (Box 3.1); in a way, classical energies transform into quantum mechanical integrals over the corresponding parts of the hamiltonian operator (although there is no classical equivalent of the exchange energy).
4. Guess a tentative set of coefficients c_{ij}, for example by diagonalizing the overlap matrix alone, and prepare tentative molecular orbitals. Symmetry plays a role here, because any molecular orbital must be either entirely symmetric or entirely antisymmetric with respect to symmetry elements in the molecule (remember what was said about normal coordinates in Section 2.1).
5. Compute the elements of the P matrix.
6. Solve the secular equation and obtain a new set of coefficients (eigenvectors) and molecular orbital energies ε_i (eigenvalues), and a new value of the total electronic energy.
7. repeat steps 5 and 6 until the energies and coefficients converge to the best values and the total energy reaches a minimum (the SCF procedure).

Step 2 is easy because much effort has been devoted to the development of standard basis sets. The symbols summarize the number and type of primitive gaussian type orbitals (GTO) included in the basis set: 3G means that each atomic orbitals is expanded in three GTOs; split basis sets are composed of some GTO's for the core part of the AO and some other GTOs for the outer part, e.g. 4-31G, 1 GTO for the core, 3 GTO's for the outer part. To add flexibility, the so-called polarization functions may be added, like d-type orbitals on p AOs; these are indicated by asterisks, e.g. 4-31G*. Step 3 is very time consuming because the number of two-electron integrals increases approximately with the fourth power of the number of atomic orbitals employed, but modern computers can handle this task for very large molecules in a quite reasonable time. For an order of magnitude, a benzene calculation with 72, 126, 168, or 222 primitive GTO's requires the calculation of 3.4, 32, 100, and 306 million integrals, respectively.

MOLECULAR ORBITALS

Box 3.1 The essential terms in LCAO-MO equations

For N atomic orbitals, $N/2$ electrons, a closed-shell molecule:

$$\text{Molecular orbitals:} \quad \varphi_i = \sum_j c_{ij}\, \chi_j \tag{a}$$

$$\text{One-electron integrals: overlap:} \quad S_{pq} = \int \chi_p \chi_q\, d\tau \tag{b}$$

One-electron integrals: kinetic energy and electron–nucleus coulomb:

$$H_{pq} = \int \chi_p [-\hbar^2/(2m_e)\nabla^2 - \sum_{\text{atoms}} (Z\, e/r)] \chi_q\, d\tau \tag{c}$$

Two-electron integrals; electrons labeled 1 and 2:

$$\text{coulomb:} \quad <pq|rs> = \iint \chi_{p(1)} \chi_{q(1)} [1/r_{12}] \chi_{r(2)} \chi_{s(2)}\, d\tau_1\, d\tau_2 \tag{d}$$

exchange: $<pr|qs>$

$$\text{The Fock matrix, } \mathbf{F}: F_{pq} = H_{pq} + \sum_r \sum_s P_{rs}(<pq|rs> - 1/2 <pr|qs>) \tag{e}$$

$$\text{density matrix} \quad P_{rs} = 2 \sum_{(i,\text{occ})} c_{ri}\, c_{si} \tag{f}$$

(occ=doubly occupied orbitals only)

The variational principle:

$$\text{solve the secular equation} \quad \text{Det}(\mathbf{F} - \varepsilon_i \mathbf{S})\mathbf{c}_i = 0 \tag{g}$$

Diagonalization gives N sets of ε_i, eigenvalues (MO energies) and $\mathbf{c}_i$, eigenvectors with the LCAO coefficients for each molecular orbital.

The electronic energy:

$$E_{\text{EL}} = \sum_r \sum_s P_{rs} H_{rs} + 1/2 \sum_{p,q,r,s} P_{pq}\, P_{rs}(<pq|rs> - 1/2 <pr|qs>) \tag{h}$$

Nucleus A to nucleus B classical coulombic energy: $E_{\text{N,AB}} = e^2\, Z_A Z_B (r_{AB})^{-1}$

$$\text{The total energy:} \quad E(\text{tot}) = E_{\text{EL}} + \sum_{A,B} E_{\text{N,AB}} \tag{i}$$

$$\text{The electron density:} \quad \rho(x,y,z) = \sum_r \sum_s P_{rs} \chi_r(x,y,z) \chi_s(x,y,z) \tag{j}$$

3.5.2 Why do a molecular orbital calculation?

The following is a brief summary of the main applications of the results of a MO calculation.

The total molecular wavefunction

The wavefunction allows a calculation of the molecular electron density $\rho(x, y, z)$. In a way, this is the answer to the question "where are the electrons?". The electrostatic potential generated by the molecule can be calculated from the density, and this is useful in predicting what a molecule will do when it comes into close contact with another molecule; the electrostatic potential is at the origin of control of mutual recognition in molecular complexes and condensed phases, as well as of some directional aspects of reactivity (Section 4.2).

Is there a trace of atomic electron densities in a molecular electron density? Of course the electron density in the molecular space close to an atomic nucleus resembles the density of the isolated atom, while the deformations are larger where the bonding effects are larger, and the problem arises of assigning each electron density element to each of the atoms in the molecule. Richard Bader has developed a rigorous method [6] for the definition of atomic basins in molecular electron densities. The subdivision is based on the topological properties of the electron density $\rho(\mathbf{r})$, its gradient vector $\nabla \rho(\mathbf{r})$, and the matrix of second derivatives, its Laplacian $\nabla^2 \rho(\mathbf{r})$, and allows a precise (at least, as precise as the wavefunction is) calculation of atomic properties such as atomic volumes and atomic charges. In this respect, Bader's atoms in molecules (AIM) method is an extremely refined substitute for the rough Mulliken population analysis (see below). In addition, the method allows a determination of some lines, called bond paths, and some critical points, called bond critical points, through the electron density.

Critical points are points where $\nabla \rho(\mathbf{r}) = 0$. There are four types of critical point, depending on the rank (number of non-zero eigenvalues of the hessian matrix) and the signature, or the algebraic sum of the signs of the eigenvalues of the hessian matrix: type 1: all curvatures negative $(3, -3)$, corresponding to a local maximum in the density distribution in all three directions; in a molecule, these occur only at the nuclear positions; type 2, two curvatures negative, one positive $(3, -1)$, corresponding to saddle points in the density distribution (maximum in two directions, minimum in the third); these occur between each pair of nuclei linked by a chemical bond; type 3, two curvatures positive, one negative $(3, +1)$; these occur inside a ring of bonded atoms; type 4, all curvatures positive $(3, +3)$, corresponding to a local minimum in the density distribution in all three directions, as found inside molecular cages. $\rho(\mathbf{r})$ can be partitioned uniquely into sub-systems bounded by specific surfaces, the zero-flux surfaces in the gradient field $\nabla \rho(\mathbf{r})$. At any point in $\rho(\mathbf{r})$ the associated gradient vector $\nabla \rho(\mathbf{r})$ points along the direction of greatest increase in $\rho(\mathbf{r})$ and so defines a path that terminates at a $(3, -3)$ or $(3, -1)$ critical point, i.e., at a local maximum or saddle point in the density distribution. The zero-flux surface perpendicular to this path at its saddle point effectively defines the boundaries. These atomic basins are the AIM definition of the atoms. An atomic property is obtained by integrating the corresponding property density over the atomic basin, and any physical property of the molecule, e.g. its energy, is the sum of the properties of the component atoms.

Two nuclei are linked by a chemical bond if the corresponding $(3, -3)$ critical points are connected by a bond path, that is a line along which $\rho(\mathbf{r})$ is a maximum with respect to any neighboring line. The strength of the bonded interaction can then be characterized by the values of the electron density ρ_b and its Laplacian $\nabla^2 \rho_b$ at the bond critical point. A negative value of $\nabla^2 \rho_b$ indicates a charge concentration in the internuclear region of covalent bonding, while a positive value corresponds to charge depletion. If one accepts the basic principles of AIM, these are uniquely defined and objective methods for identifying a chemical bond. For very weak intermolecular bonds, however, the results of AIM analysis and bond paths may be less certain or very sensitive to the quality of the wavefunction, as might have been expected.

Molecular orbital energies

For N AO's and N electrons, N MOs are obtained. Two electrons of opposite spin are assigned to each molecular orbital, so that in a closed-shell system only $N/2$ MO's are occupied. In a zero-order assumption, when an electron leaves its orbital no rearrangement occurs, so that the energy of the HOMO is related to the molecular first ionization potential (Koopman's theorem) and the energy difference between the highest occupied molecular orbital (HOMO) and the lowest unoccupied molecular orbital (LUMO) is equal to the energy required for an electronic excitation of the molecule. The latter energy differences fall in the UV-visible region of the electromagnetic spectrum, and this is the basis for the comparison between MO results and results of electron spectroscopy.

The eigenvectors

Each eigenvector is a collection of coefficients for the LCAO that form the corresponding MO. Using this information, a visualization of the shape and directional properties of molecular orbitals becomes possible. Reactivity is related to electronic redistribution and, in particular, to electron exchanges between HOMO and LUMO of reacting partners; from the topology and symmetry properties of these frontier orbitals it is sometimes possible to obtain a qualitative prediction on the stereochemical course of chemical reactions [7].

Population analysis

A simple, old but still popular method of apportioning the electron density among atoms is Mulliken population analysis [8]. Consider any two AO's on two atoms i and j joined by a chemical bond, and a LCAO-MO of the form: $\phi = c_i \chi_i + c_j \chi_j$. Let the total population of that MO be N electrons (mostly, $N = 2$), to be somehow apportioned among the two atoms. Recalling that both the MO and the AO's are normalized, consider the following expression:

$$N \int \phi^2 d\tau = N \int [(c_i \chi_i)^2 + (c_j \chi_j)^2 + 2c_i c_j \chi_j \chi_i] \, d\tau \tag{3.47a}$$

$$N = Nc_i^2 + Nc_j^2 + 2Nc_i c_j S_{ij} \tag{3.47b}$$

The first two terms of the last equation are assumed to be the electron population on atom i and atom j, respectively, while the third term is called the overlap population. When there are many atoms, many electrons, and many occupied MO's the appropriate sums over all coefficients of the LCAO must be performed. Not surprisingly, it turns out that the higher the overlap population, the stronger the bond between atoms i and j. As a further approximation, the total overlap population is assigned half to atom i and half to atom j. In this way, a total electron charge population on atoms can be calculated. If the atomic number is Z, then Z less this population is interpreted as a net atomic charge. The method includes some really gross approximations and is extremely sensitive to the details of the MO calculation, and yet Mulliken point charges often give surprisingly good results in modeling the molecular electrostatic properties.

The bonding energy

The difference between the total molecular energy and the sum of total energies of constituting atoms, all from equation (i), Box 3.1, is interpreted as the total chemical bonding energy, or the negative of the atomization energy [9].

Conformational analysis and force constants

The total molecular energy can be minimized with respect to nuclear positions, to find the most stable conformation for the molecule (do not confuse this minimization, where nuclear positions are changed, with the minimization after the variational principle, where the best wavefunction is obtained for a given conformation). If a bond length or a bond angle or a torsion angle is varied in steps, an energy profile for the stretching, bending or torsional vibration can be obtained, and the curvature will yield the corresponding force constant [10]. When the wavefunction and the energies are of good quality, this procedure is a reliable alternative to the spectroscopic procedures described in Chapter 2.

3.6 Approximate quantum chemical methods: NDO and EHT

In 1960 the calculation of the integrals involved in a MO calculation was still a prohibitive task. Even a single overlap integral required a substantial amount of time, and one can imagine how difficult it may have been to obtain all the two-electron integrals required for a molecular orbital calculation, even for as small a molecule as methane and with a severely restricted basis set. The first years of application of the MO theory in the form shown in Box 3.1 were years of search for shortcuts through the deep and thick forest of the necessary integrals. The other rate-determining step was the diagonalization of the Fock matrix. In an extreme example, a study of the guanine-cytosine dimer by Enrico Clementi at the IBM San Jose research center required some 70 billion integrals for a total computing time of 8 days [11]. However, already in 1970 electronic computers could handle something like 100,000 integrals per second! The following is a brief chronological overview:

1926	Schrödinger equation
1931	Hückel method
1951	Roothaan equations
1953	Semiempirical π-orbitals methods (Pariser–Pople–Parr PPP)
1963	Extended Hückel
1965	CNDO
1970	Complete ab initio calculations for organic molecules (Pople)

A first approximation was to use valence orbitals only, in the assumption that core atomic and molecular orbitals would contribute very little to the properties and the reactivity of organic molecules. This assumption is, in general, a valid one. A further very popular approximation was the neglect of differential overlap (NDO), developed by the Dewar school [12]. The integrals of the type $<pq|rs>$ are small if $p \neq q$ and $r \neq s$, and one assumes also that the corresponding overlap integrals are small. The NDO conditions were then written as:

$$\int \chi_p \chi_q d\tau = \delta_{pq}$$
$$<pq|rs> = \delta_{pq}\delta_{rs} <pp|rr> \qquad (3.48)$$
$$<pp|rr> = \gamma_{AB}$$

where orbital p is on atom A and orbital r is on atom B; γ_{AB} is an empirical parameter that depends on atomic species and not on particular orbitals. Other restrictions are imposed on the core integrals:

$$H_{pq} = \beta^\circ_{AB} S_{pq} \quad \text{orbital } p \text{ on atom A, orbital } q \text{ on atom B} \qquad (3.49)$$
$$<p|V_B|q> = V_{AB}$$

β°_{AB} and V_{AB} are other empirical parameters. The first formulation was called complete NDO, or CNDO, while successive modifications led to partial or intermediate neglect of differential overlap, as computer resources became more and more generous.

A much more drastic approximation involves the complete neglect of electronic repulsion, in the so-called Extended Hückel Theory (EHT) approach developed in the 1960s by Roald Hoffmann [13]. All two-electron integrals are neglected, and the EHT recipe can be written as follows:

$$F_{pq} = H_{pq} = 1/2K(H_{pp} + H_{qq})S_{pq} \qquad (3.50)$$

where the H_{ii} matrix elements are approximated by the valence orbital ionization potential (VOIP) of the corresponding orbital, available from any atomic calculation. K is a universal empirical constant. EHT is simple enough for application to rather large molecules, say benzene derivatives, since it only requires the calculation of overlap integrals and works without a SCF procedure, with just one matrix diagonalization.

This method is called Extended Hückel because it is in fact an extension, including all the σ valence orbitals, of a much earlier method that considers only π-electron systems in which each atom is represented by one electron in one p_z orbital. Matrix elements are obtained as $H_{pp} = \alpha$ and $H_{pq} = \beta$, where α and β are empirical parameters for each atomic species. In the simple Hückel formulation butadiene is a four-orbital system and the secular equation can be solved by hand.

The Hückel and Extended Hückel total electronic energies are taken as just sums of MO energies:

$$E_{tot} = \sum_{(i,occ)} 2\varepsilon_i \quad \text{(occ = occupied orbitals only)} \qquad (3.51)$$

This is a terrible approximation, so that EHT total energies are worth very little in an absolute sense. The method, however, gives a reasonably good wavefunction, and it was used, in Roald Hoffmann's skilled hands, as a help to develop the symmetry arguments in molecular reactivity that led him to the Nobel prize in 1981 (together with Kenichi Fukui).

3.7 Evolution of quantum chemical calculations: Beyond Hartree–Fock

In the 1980s, a molecular orbital calculations was still an enterprise that required much careful and tedious planning: the choice of the basis set, the preparation of line input files with long lists of coefficients and exponents of the necessary gaussians, and so on. As computer resources expanded and developed, all these tasks began to be automatically incorporated in large computer packages that came with built-in basis functions and provided the user with a large number of options that could be invoked almost at the touch of a finger. Nowadays, a package like GAUSSIAN [14], for example, includes all the most fashionable basis sets, and provides analytical derivatives of the total energies so that automatic optimization of molecular geometry is possible. All the user has to do now is specify a certain number of keywords. If the expenditure of computer time may still be considerable, all these simplifications save a very large amount of the much more precious human time. The full optimization of the gas-phase structure of an organic molecule with 30–40 atoms can be carried out with ordinary computer equipment, while parallel computing and other sophisticated technologies allow the treatment of even larger systems. Parametric force field methods, at least for the purpose of finding the best molecular structure, are by now outdated. John Pople received the Nobel Prize in 1998 for his contributions to quantum chemistry and for his efforts in making quantum chemical methods available to the general public of chemists, instead of only to the restricted community of theoreticians.

In spite of all the evolution, there are still important problems for the calculation of intermolecular energies. Hartree–Fock (HF) methods use one-electron orbitals and therefore cannot account for those phenomena that depend on the simultaneous behavior of several electrons. Thus, HF energies may correctly represent the kinetic energies of electrons and the electrostatic effects between electrons and nuclei, but cannot take into account electron correlation. The results obtained at the limit of a complete (i.e. infinitely rich) basis set are called HF-limit energies and wavefunctions, the ideal best that can be obtained with one-electron orbitals. This intrinsic limitation forbids the treatment of dispersion energy, a crucial part of the intermolecular potential (see Chapter 4). Thus, for example, HF methods are intrinsically unsuitable for the calculation of the lattice energies of organic crystals.

3.7.1 *Configuration interaction and Møller–Plesset perturbation methods*

In the basic HF treatment, the total wavefunction is represented by one Slater determinant 3.45. If the total number of electrons is even, all electrons may be placed in molecular orbitals with paired spins, for a closed-shell configuration. If the number of electrons is odd, as in a radical, or if the total number of electrons is even but one wishes to unpair the spins of two electrons, a Slater determinant is written for an open-shell configuration (unrestricted Hartree–Fock, UHF) and the corresponding Roothaan equations are then solved. For a better approximation to the total energy one could simultaneously take into account all the possible configurations, the ground state with no excitation, plus single and double or even triple excitations. The total wavefunction may then be taken as a linear combination of all the resulting Slater determinants. The final equations become terribly complicated and the number of integrals to be evaluated increases enormously, but such calculations can be managed. This is in essence the philosophy of the so-called configuration interaction (CI) methods. It turns out that the inclusion of excited states is a way of representing electron correlation.

A powerful technique in quantum chemical manipulations is called perturbation theory. In many cases one has to deal with a hamiltonian operator for which the quantum chemical equations are too difficult or impossible to solve. A simpler hamiltonian may then be used to provide a zero-order solution, and then a perturbation operator is introduced, whose effect on the final results of the calculation can be obtained as a separate correction to the zero-order approximation. In Møller–Plesset (MP) perturbation theory, the Fock operator is the zero-order hamiltonian (equation (c) in Box 3.1) and a Slater determinant is the zero-order wavefunction. The zero-order energy is the sum of the energies of the occupied orbitals, and the sum of the zero-order and first-order energies is the HF energy, equation (h) in Box 3.1. At the n-th level (MPn) n-fold excited configurations are automatically taken into account. The total electronic energy is expanded in a sum of perturbation terms of n-order, and the sum of the terms for $n > 1$ is the correlation energy:

$$E = E^0 + E^1 + E^2 + E^3 + E^4 + \cdots \quad (3.52)$$

The energies that include successive terms in the expansion are called MP1, MP2, MP3, MP4 ... energies. The MP2 energy is always lower than the MP1 energy because the double excitation terms are stabilizing. The derivation of explicit expressions for the integrals and the energies is a matter of much algebra and patience [15].

3.7.2 Density functional theory (DFT) methods

The DFT method [16] rests upon a number of concepts and assumptions that will be briefly summarized in the following with a minimum amount of technical detail. Walter Kohn received the Nobel prize in 1998, jointly with John Pople.

1. The Hohenberg–Kohn theorem states that the energy of a system of electrons in an external potential $V(R)$ is a functional of the electron density (a function is a mathematical entity that takes a certain value for each specified value of a variable, a functional is a mathematical entity that takes a certain value for each specified value of another function). Thus, the ground state energy of a many-electron system is uniquely determined by the ground-state density.
2. In the Kohn–Sham assumption, for a system of interacting electrons moving in a potential $V(R)$ (i.e. a molecular electron cloud in the field of the nuclei) a local potential, $V_{KS}(R)$, can be introduced, such that a system of non-interacting electrons moving in the $V_{KS}(R)$ field will have the same density as the exact density of the interacting electron system.
3. The Kohn–Sham local potential contains a term for the electron–nuclei coulombic attraction, a term for the electron–electron coulombic repulsion, and an exchange term, which incorporates all the exchange and electron correlation effects:

$$V_{KS}(R) = V_{eZ}(R) + V_{ee}(R) + V_{exch}(R) \quad (3.53)$$

4. The Kohn–Sham hamiltonian operator (compare with the HF hamiltonian of equation 3.39) and the stationary state equation are then written:

$$\hat{H}_{KS} = -1/2\nabla^2 + V_{KS}(R) \quad (3.54)$$

$$\hat{H}_{KS}\, \varphi_i = \varepsilon_i\, \varphi_i \quad (3.55)$$

$$\rho(R) = \sum_i n_i\, \varphi_i\, \varphi_i^* \quad (3.56)$$

where n_i is the occupation number of the KS eigenstate with eigenfunction φ_i.

5. A determinantal wavefunction is prepared in terms of the Kohn–Sham orbitals and the expectation value of the energy is written as:

$$\int \Phi^* H \Phi \, d\tau = <E_{KS}> \tag{3.57}$$

$$E_{KS} = T_{KS} + \int \rho(R) \, V_{eZ}(R) dR + \int \rho(R) V_{ee}(R) \, dR + E_{exch,KS}$$

$$= T_{KS} + U_{eZ} + U_{ee} + E_{exch,KS} \tag{3.58}$$

where T_{KS} is the kinetic energy over the Kohn–Sham orbitals, the second term is the electron–nuclei attraction, the third term is the electron–electron repulsion, and $E_{exch,KS}$ is the exchange energy over the Kohn–Sham orbitals.

6. The exchange energy $E_{exch,KS}$ is replaced by an effective contribution $E_{ex,eff}$ that also includes correlation effects as best as possible. This will also be a functional of the density, $\varepsilon_{ex}(R)$. One obtains:

$$E = T_{KS} + U_{eZ} + U_{ee} + E_{ex,eff} = T_{KS} + U_{eZ} + U_{ee} + \int \rho(R)\varepsilon_{ex}(R) \, dR \tag{3.59}$$

The true expression for the total energy as the sum of the total kinetic energy plus the electron–nuclei attraction plus the total electron–electron interaction energy can now be compared with the KS expression for the total energy:

$$E = T + U_{eZ} + U_{ee,tot} \tag{3.60}$$

$$E_{ex,eff} = (T - T_{KS}) + (U_{ee,tot} - U_{ee}) \tag{3.61}$$

Thus, the effective correction energy has a term in the difference between the true kinetic energy and the kinetic energy of the non-interacting electrons, and a second term in the difference between the total electron interaction energy and the coulombic electron–electron repulsion. This difference incorporates the correlation energy.

The functional $\varepsilon_{ex}(R)$ is the key to the whole business: the exchange-correlation energy. Whether E does contain appropriate electron correlation contributions depends on whether $\varepsilon_{ex}(R)$ is indeed a realistic energy density. A first crude approximation, whose partial success was a surprise even to its inventors, is to use the exchange-correlation energy of an electron gas of uniform density, equal to the local density in the molecular system at point R (the local density approximation, LDA). Other more sophisticated approaches have been proposed, hence the various names appended to a DFT calculation (B, BP, B-LYP, and the like). The technical details for these formulations [15,16] are beyond the scope of this book.

7. In the LCAO implementation of DFT, the wave function is expressed as a linear combination of gaussians, and the expansion coefficients are determined by minimizing the total energy with respect to the density. The rest of the procedures

for the solution of the resulting secular equation (eigenvalue equation) are not too different from the procedure for the solution of the Hartree–Fock equations.

In a (rather crude) summary, it might be said that DFT does not rigorously solve the problem of correlation energy, but displaces the problem to a well defined location, the $\varepsilon_{ex}(\rho)$, where a parametric ambush can be set for its solution. The method, whose computational demand is similar or sometimes more modest than those of MO methods, has been very successful in applications to isolated molecules, where the introduction of electron correlation corrections has been seen to improve, for example, calculated optimized molecular geometries, but has not yet proved completely satisfactory for the calculation of intermolecular interaction energies in systems where coulombic contributions are not overwhelmingly dominant. Molecular crystals are a typical example.

3.7.3 Other beyond-Hartree–Fock methods

Several other methods for quantum chemical calculations that approach the correlation energy problem have been proposed, but their detailed examination would be out of place in this book [17]. They all require a substantial increase in computational effort with respect to HF and to MPn or DFT. Even these more refined methods stumble upon the fundamental physical obstacle represented by the fact that these intermolecular interaction energies are often very small (see Chapter 4). For intermolecular selectivity, sometimes the decision has to be made at the level of less than

Table 3.4 Some properties of the molecular wavefunction of ethylene at different levels of MO theory (HF, Hartree–Fock; MP2, second-order Møller–Plesset; B3LYP, DFT with a fashionable exchange functional; 3-21G, 42 gaussians; 6-31G**, 84 gaussians; D95**, 88 gaussians; cc-pVDZ, 94 gaussians)

Basis set	3-21G		6-31G**		D95**		cc-pVDZ	
HOMO energy (hartree)								
HF	−0.3740		−0.3689		−0.3726		−0.3724	
MP2	−0.3740		−0.3735		−0.3727		−0.3724	
B3LYP	−0.2687		−0.2656		−0.2714		−0.2722	
C atom charge (electrons)[a]								
HF	−0.43	−0.38	−0.26	−0.35	−0.26	−0.41	−0.10	−0.35
MP2	−0.37	−0.32	−0.22	−0.32	−0.27	−0.36	−0.09	−0.30
B3LYP	−0.37	−0.34	−0.20	−0.31	−0.34	−0.38	−0.04	−0.30
ρ at 1.54Å (10^{-4} e Å^{-3})								
HF	7.8		6.8		6.9		7.0	
MP2	7.3		6.4		6.4		6.4	
B3LYP	7.2		6.1		6.3		6.2	

[a] First entry: Mulliken population analysis; second entry: ESP electrostatic potential best fit, see Section 4.3.

3.7.4 Ethylene in perspective

Table 3.4 shows the results of molecular orbital and DFT calculations [14] for a simple molecule, ethylene, at different levels of theory and with different basis sets. The HOMO energy changes dramatically on going from MO to DFT. Atomic charges from Mulliken population analysis can be nearly anything, depending on the approximations made in the calculation of the wavefunction. Much more stable are the atomic charges derived from the best fit to the electrostatic potential, which, as expected, changes much less dramatically with changes in level of theory. This should be kept in mind when selecting atomic charge parameters for force field calculations (Section 4.9). The last set of entries in Table 3.4 show the change in the value of the electron density at a point 1.54 Å above the plane of the ethylene molecule, at the midpoint of the carbon–carbon double bond. The change, from 7.8 to 6.1×10^{-4} e Å^{-3}, may not seem much, but it is more than enough to change in a rather substantial way the picture of the π-interaction between two parallel stacked double bonds.

3.8 Dimerization energies and basis set superposition error

The calculation of lattice energies of molecular crystals by ab initio molecular orbital methods, or by any other quantum chemical method such as DFT, is still a formidable task because it requires either an evaluation of the interaction energies in a cluster of molecules representing the crystal structure, or a periodic orbital calculation (see Section 6.2). Both approaches imply a very high computational demand. A great amount of chemical information on intermolecular forces and energies can, however, be gathered by sample calculations on molecular dimers or small clusters. The intermolecular cohesion energy for a dimer formed by fragments A and B is given by

$$E_{\text{inter}} = E_{AB} - E_A - E_B \qquad (3.62)$$

The energies of the two separate fragments are subtracted from the total energy of the dimer. A key problem for the success of such a calculation is the choice of the basis set. A dimer calculation done by a method that does not go beyond the HF limit will only represent those energy contributions that do not stem from electron correlation, that is, mainly the purely electrostatic contributions. In fact, the difference between the energy calculated by a correlated wavefunction and those obtained by a HF wavefunction is often taken as a measure of the dispersion energy contribution.

The application of equation 3.62 suffers from another technical difficulty, called basis set superposition error (BSSE). For the calculation of the total energy of the dimer, E_{AB}, obviously the basis set is the sum of the basis sets of the separate fragments. In this way, each fragment in the dimer is allowed to use the external parts of the atomic orbitals of the other fragment, so that the quantum chemical identity of each fragment in the dimer is different from that of the isolated fragment. It is therefore customary to correct for this superposition effect by subtracting the energies of the separate fragments as obtained using the full basis set of the dimer: these energies are calculated by adding to each of the separate fragments some shadow atoms, carrying orbitals but no electrons, at the atomic positions of the other fragment in the dimer. This is the so-called counterpoise correction of Boys and Bernardi [18]. If these corrected fragment energies are called E'_A and E'_B, the counterpoise correction energy, E_{CP}, and the corrected interaction energy $E_{inter,CP}$ are given by:

$$E_{CP} = (E'_A - E_A) + (E'_B - E_B) \tag{3.63}$$

$$E_{inter,CP} = E_{inter} - E_{CP} = E_{AB} - E'_A - E'_B \tag{3.64}$$

This seemingly simple and intuitive approach is often questioned, and it has been argued that at least in some cases the counterpoise correction might in fact be an overcorrection [19]. The basis set superposition problem is entangled with the problem of the choice of the basis set; clearly, a very diffuse basis set better represents the actual electron distribution, but implies a higher BSSE. The problem is made more acute by the fact that very often the correction energy E_{CP} is of the same order of magnitude as the total intermolecular energy itself, so that a completely different picture of intermolecular bonding may results from the consideration of corrected or uncorrected dimerization energies. This is a major difficulty.

3.9 Historical portraits: early experiences in quantum chemistry

One day around 1975 Massimo Simonetta, then director of the Institute of Physical Chemistry at the University of Milano, summoned a few young would-be scientists to his room. I was among them. We all sat rather uneasily on our chairs in a half circle around his large desk. He said he wanted to start a project using the then available quantum chemical methods for the calculation of the energies and wavefunctions of large organic molecules, to be developed later into a project for the estimation of the interaction energies of smaller organic molecules with transition metal surfaces. He politely but firmly asked who, among the attendees, would be willing to take up such a task. The other people in the room had at least a rudimentary training in theoretical chemistry, appreciated how cavalier that purpose might be, and, equally politely if less firmly, declined. I was trained mainly as a crystallographer, and I had no idea of what would be ahead. Besides, I was the last in the line, and I knew very well that Simonetta would hardly take no for an answer. This is how I embarked on a many-year project that left me without firm absolute energy results, but with a hands-on experience in atomic orbitals, orbital symmetry, and molecular orbital calculations.

The first thing to do was to obtain a copy of Roald Hoffmann's Extended Hückel program. This was quite easy through a memorable institution called the Quantum Chemistry Program Exchange (QCPE)[20]. We wrote a letter, and after a few weeks we received a magnetic tape carefully wrapped in a cardboard parcel. The next thing to do was to adapt the program to our computers, possibly rewriting the code here and there to adjust core memory requirements and to speed things up a bit, if possible. In those days, source codes were stored on magnetic tapes that were kept in the computer room, and disk space was practically nil, so large parts of the integrals and of the necessary matrices were also written on magnetic tape as a temporary data storage device, to be retrieved in due course during the execution of the program. The calculation had to stop with a console instruction for the operator to mount a given tape on the tape drive, and a "rewind" instruction really meant a rewind operation, including the appropriate swishing sound and the ever looming danger of damage to the magnetic tape when it had not been mounted properly, with the consequent loss of weeks of hard work when no backup had been kept (backup was a clumsy and expensive business requiring the investment of a second magnetic tape).

The key to the whole business was of course the subroutine that calculated the overlap integrals. It consisted of a first part, which computed an integral between parallel orbitals at a given distance, and a second part, which did the necessary projections along the three directions in space for the calculation of the actual overlap integrals between atoms in any three-dimensional geometry. The task was rather easy for p orbitals, a bit more complicated for d orbitals. In any case I checked the whole thing through, learning a lot especially about orbital symmetry and orthogonality. We also experimented with block-diagonalizations, with little luck because the time it took the computer to transfer the data into the auxiliary blocked-out matrices overcompensated for the time saving due to the smaller size of the matrices.

In 1978 I spent a research term with Larry Bartell at the Department of Chemistry of the University of Michigan (I cannot call it a postdoc because I do not have a PhD, and hence could not be a postdoctoral student). The research project involved modifying the existing Extended Hückel programs for the use of hybrid orbitals. My experience with the overlap routine was essential because I knew how to do the necessary orthogonal transformations. One day, Roald Hoffmann gave a talk at the Michigan State University in Lansing, and we drove there from Ann Arbor to listen to the not yet Nobelized but already famous man. In the evening we drove back together with Roald to have dinner at Larry's home. As we drank beer on the patio, I was anxious to hear what Roald might have to say on my project, and I casually dropped the words "hybrid orbitals". Equally casually, Roald said, "I don't like hybrid orbitals". The conversation continued on another subject (I forget which).

A year later I returned to Ann Arbor to work on another project that involved the use of the POLYATOM [21] code with a pseudopotential [22], which was necessary because we wanted to work on iodine hexafluoride and the calculation was to include only valence orbitals. POLYATOM was already a well developed ab initio program including all the integral calculation and storage gear, plus the routines for doing the SCF procedure and population analysis. I used gaussian basis functions, which one had to punch in one after another, exponent, and normalized coefficient. When I got back to Milano I carried the POLYATOM package with me to work on the famous Simonetta project with metal surfaces, and, as usual, started fiddling with the Fortran source code to improve it. After a while, a substantial amount of results had accumulated in long and expensive production runs, which sometimes involved reserving whole nights of computer time, nights which I spent playing checkers with the operators in the computer room while the machine was running. One day, while checking things out, I discovered that some of the overlap integrals over d orbitals that should have been zero by symmetry were not zero by some tens of thousandths. That was one of the first times that the program had been used for d orbitals, and it occurred to me that there might be something not quite right in the part of the code that dealt with them. After long and tedious debugging sessions I discovered that when the program transformed from the independent d_{x2} and d_{y2} orbitals into the more common d_{x2-y2} combination, there was a misprint in one of the normalization constants, which I duly corrected. At that point, I had two choices: one, to forget about it, forget about the past, and carry on with the corrected code; two, to repeat all the calculations that had already been done. I chose the second option, and spent several weeks only to obtain very marginal changes in the results, as expected. Had it been worth my time and trouble? I still have doubts.

References and Notes to Chapter 3

[1] Textbooks of quantum mechanics and quantum chemistry are legion. A good and rigorous student's textbook is Kauzmann, W. *Quantum Chemistry, An Introduction*, 1957 Academic Press, New York, to which the reader is directed for more detail. A more recent book, which also includes extensive discussions of progress in quantum chemical calculations is Levine, I. N. *Quantum Chemistry*, 5th edition, 2000, Prentice-Hall, Upper Saddle River.

[2] A good introduction to the basic aspects of MO theory is in Hehre, W. J.; Radom, L.; Schleyer, P. v. R.; Pople, J. A. *Ab Initio Molecular Orbital Theory*, 1986, Wiley, New York.

[3] There is no fundamental reason why this should be so – indeed, there is nothing fundamental about orbitals; they are just convenient mathematical functions that help in one kind of approach to the approximate solution of the Schrödinger equation. Orbitals have become so popular in chemistry because they can be plotted numerically to help visualizing where electrons "are" in a molecule, or the strength of chemical bonds, but some schools in theoretical chemistry and nearly all of theoretical physics dispense with orbitals.

[4] The electronic structure of isolated atoms was determined by quantum chemical calculations by Clementi and Roetti: Clementi, E.; Roetti, C. *Atomic Data and Nuclear Data Tables* 1974, **14**, 177.

[5] Qualitative group or molecular orbitals can often be drawn on symmetry considerations alone, at least for simple molecules. See for example Jorgensen, W. L.; Salem, L. *The Organic Chemist's Book of Orbitals*, 1973, Academic Press, New York.

[6] Bader, R. F. W. *Atoms in Molecules: A Quantum Theory*, 1990, Oxford University Press; Bader, R. F. W.; Anderson, S. G.; Duke, A. J. *J. Am. Chem. Soc.* 1979, **101**, 1389–1395.
[7] Woodward, R. B.; Hoffmann, R. *The Conservation of Orbital Symmetry*, 1970, Academic Press, New York; Hoffmann, R.; Woodward, R. B. *Acc. Chem. Res.* 1968, **1**, 17–22.
[8] Mulliken, R. S. Electronic population analysis on LCAO-MO molecular wave functions. I. *J. Chem. Phys.* 1955, **23**, 1833–1840.
[9] Ditchfield, R.; Hehre, W. J.; Pople, J. A. Self consistent molecular-orbital methods. IX. An extended Gaussian-type basis for molecular-orbital studies of organic molecules, *J. Chem. Phys.* 1971, **54**, 724–728.
[10] See an early example in Stevens, R. M., Geometry optimization in the computation of barriers to internal rotation, *J. Chem. Phys.* 1970, **52**, 1397–1402.
[11] Clementi, E.; Mehl, J.; von Niessen, W. Study of the electronic structure of molecules. XII. Hydrogen bridges in the guanine–cytosine pair and in the dimeric form of formic acid, *J. Chem. Phys.* 1970, **54**, 508–520.
[12] Dewar, M. J. S.; Thiel, W. Ground states of molecules. 38. The MNDO method. Approximations and parameters, *J. Am. Chem. Soc.* 1977, **99**, 4899–4907. The paper immediately following this one has lots of detailed results.
[13] Hoffmann, R. An Extended Hückel theory. I. Hydrocarbons, *J. Chem. Phys.* 1963, **39**, 1397–1412.
[14] Frisch, M. J.; Trucks, G. W.; Schlegel, H. B.; Scuseria, G. E.; Robb, M. A.; Cheeseman, J. R.; Montgomery, J. A. Jr.; Vreven, T.; Kudin, K. N.; Burant, J. C.; Millam, J. M.; Iyengar, S. S.; Tomasi, J.; Barone, V.; Mennucci, B.; Cossi, M.; Scalmani, G.; Rega, N.; Petersson, G. A.; Nakatsuji, H.; Hada, M.; Ehara, M.; Toyota, K.; Fukuda, R; Hasegawa, J.; Ishida, M.; Nakajima, T.; Honda, Y.; Kitao, O.; Nakai, H.; Kiene, M.; Li, X.; Knox, J. E.; Hratchian, H. P.; Cross, J. B.; Adamo, C.; Jaramillo, J.; Gomperts, R.; Stratmann, R. E.; Yazyev, O.; Austin, A. J.; Cammi, R.; Pomelli, C.; Ochterski, J. W.; Ayala, P. Y.; Morokuma, K.; Voth, G. A.; Salvador, P.; Dannenberg, J. J.; Zakrzewski, V. G.; Dapprich, S.; Daniels, A. D.; Strain, M. C.; Farkas, O.; Malick, D. K.; Rabuck, A. D.; Raghavachari, K.; Foresman, J. B.; Ortiz, J. V.; Cui, Q.; Baboul, A. G.; Clifford, S.; Cioslowski, J.; Stefanov, B. B.; Liu, G.; Liashenko, A.; Piskorz, P.; Komaromi, I.; Martin, R. L.; Fox, D. J.; Keith, T.; Al-Laham, M. A.; Peng, C. Y.; Nanayakkara, A.; Challacombe, M.; Gill, P. M. W.; Johnson, B.; Chen, W.; Wong, M. W.; Gonzalez, C.; Pople, J. A. *Gaussian 03, Revision A.1*, Gaussian, Inc., Pittsburgh PA, 2003.
[15] For technical details see e.g. Clementi, E. *MOTECC, Modern Techniques in Computational Chemistry*, ESCOM Science Publishers, Leiden 1991, p. 486ff.
[16] Rather than going through the original references by Kohn and Sham, dense with mathematical detail, an interested chemist might wish to see e.g. Bickelhaupt, F. M.; Baerends, E. J. Kohn–Sham density functional theory: predicting and understanding chemistry, *Reviews in Computational Chemistry*, Vol. 15, edited by K. B. Lipkowitz and D. B. Boyd, 2000 Wiley-VCH, New York, pp.1–86;

Bartolotti, L. J.; Flurchick, K. An introduction to density functional theory, *Reviews in Computational Chemistry*, Vol. 7, edited by K. B. Lipkowitz and D. B. Boyd, 1995, VCH, New York pp. 187–216; Ziegler, T. Approximate density functional theory as a practical tool in molecular energetics and dynamics, *Chem. Revs.* 1991, **91**, 651–667; ref. [15], density functionals at p. 321ff.

[17] For applications of these methods to a classic system, the benzene dimer, see for example Sinnokrot, M. O.; Valeev E. F.; Sherrill, C. D. *J. Am. Chem. Soc.* 2002, **124**, 10887; Hobza, P.; Selzle, H. L.; Schlag, E. W. *J. Phys. Chem.* 1996, **100**, 18790.

[18] For the Boys–Bernardi procedure, see a discussion in ref. [15], p. 316*ff*.

[19] Basis set superposition error depends on the chemical constitution of the interacting moieties at short range, so that its entity may be different even for two different orientations of the same dimer. Usually, counterpoise-corrected energies are considered to be more reliable, but, for example, it has been suggested that the full counterpoise correction may in some cases overestimate the error and that the use of 50% correction may be advisable: see Tarakeshwar, P.; Choi H. S.; Kim, K.; S. Olefinic vs. aromatic π-H interaction: a theoretical investigation of the nature of interaction of first-row hydrides with ethane and benzene, *J. Am. Chem. Soc.* 2001, **123**, 3323–3331; Kim, K. S.; Tarakeshwar, P.; Lee, J. Y. Molecular clusters of π-systems: theoretical studies of structures, spectra, and origin of interaction energies, *Chem. Revs.* 2000, **100**, 4145–4186.

[20] QCPE, Quantum Chemistry Program Exchange, Department of Chemistry, Indiana University, Bloomington, Indiana (www.qcpe.indiana.edu). Founded in 1962 with modest help, this institution has been distributing, and promoting the enhancement of, theoretical chemistry programs for decades, for nothing more than nominal distribution fees. It currently holds (2005) about 770 computational chemistry systems, programs, and routines. Theoretical chemistry has benefited enormously from its actions.

[21] Newmann, D. B.; Basch, H.; Kornegay, R. L.; Snyder, L. C.; Moskowitz, J.; Hornback, C.; Liebman, P. POLYATOM, program No. 199, Quantum Chemistry Program Exchange, Department of Chemistry, Indiana University, Bloomington, Indiana.

[22] In a pseudopotential molecular orbital calculation, core electrons are represented by an effective potential, also expressed in gaussian form, and atomic valence orbitals only are explicitly included in the MO calculation, while core-valence interactions are still expressed as appropriate integrals over gaussian-type functions. This allows a large saving in computing resources, and is almost a must if heavy metal atoms are to be considered. See Bartell, L. S.; Rothman, M. J.; Gavezzotti, A. Pseudopotential SCF-MO studies of hypervalent compounds. IV. Structure, vibrational assignments and intramolecular forces in IF_7, *J. Chem. Phys.* 1982, **76**, 4136–4143.

4

The physical nature and the computer simulation of the intermolecular potential

Why, when we lift one end of a stick, does the other end come up too?
Rowlinson, J.S. Cohesion, a scientific history of intermolecular forces,
Cambridge University Press, Cambridge, 2002, p.1.

We need not know much quantum mechanics in order to discuss our simple model. We only need to know that in quantum mechanics the lowest state of a harmonic oscillator of the proper energy v has the energy $E° = 1/2hv$, the so-called zero-point energy.
London, F. The general theory of molecular forces,
Trans. Farady Soc. 1937, **33**, 8.

4.1 Experimental facts and conceptual framework

Everyday objects display an almost incredible variety of properties that may be felt with animal senses and that human intellect (the human race at present being thought to be the only species of animal on Earth capable of posing such questions) wishes to apprehend, categorize, understand, and, possibly, predict. We have gaseous bodies such as the Earth's atmosphere, liquid bodies such as a drop of water, solid bodies such as glasses or crystals, semi-solid bodies such as gels, or even such exotic states of condensation as liquid crystals. Each body may have a color, an odor, a density, an electric conductivity and a magnetic susceptivity, a hardness, compressibility, malleability or brittleness, a Young modulus and a shear modulus, a viscous or an elastic response to mechanical stress, a solubility in water or in organic solvents; some liquids spread easily, some flow only after vigorous external stress, others are thyxotropic; while liquid mercury stubbornly refuses to surrender its spheres. Some liquids evaporate quickly, while the saturated vapor pressure of some solid metals calls for no more than one atom in a cubic kilometer of vapor phase; complex bodies like animal and vegetal bodies are composed of delicate substructures such as cells or vesicles. All these properties are nothing but external aspects of internal electronic properties and of the ways in which molecules recognize their partners and aggregate. Frontier chemistry is nowadays the chemistry of molecular aggregation.

Consider a mole of pure liquid acetic acid. If heated above 392K at atmospheric pressure, the liquid will boil off, because its vapor pressure overcomes the restraining power of the mechanical "lid" constituted by the atmosphere and massive transfer into the vapor phase occurs. If cooled below 290K, the liquid will arrange into a crystalline structure (after a certain induction period due to the necessary kinetic adjustments). There is little arguing about the fact that condensation and crystallization result from some attractive power among the atoms and molecules that constitute the body. The fact that one does not fall through the floor and one cannot pass through a solid wall demonstrates that some repulsion also operates among atoms and molecules. Matter is made of atoms, that is protons and electrons, and all interactions in atoms and molecules are fundamentally electric in nature: in fact, the potential part of the quantum mechanical hamiltonian contains only electrostatic terms. But the term "electrostatic" will be used for the total, complex potential energy interactions within the electron system, while the term "coulombic" will be reserved for those terms that can be described by pure coulomb-law potential terms, those with an R^{-1} dependence on the distance between charge centers. In describing intermolecular phenomena, one often speaks only of potentials, but total electronic energies also include a term that results from the kinetic energy operator (Box 3.1). This is an intrinsic, temperature-independent electronic kinetic energy and has nothing to do with the molecular kinetic energy associated with nuclear vibrations and molecular motion.

Cohesion energies are measurable on macroscopic bodies, for example, through enthalpies of sublimation and of fusion. But a description of the essential physical phenomena in molecular recognition and aggregation can be better illustrated using simpler model systems, like atomic or molecular dimers. Consider first two separate systems, A and B, each one with its nuclei and electrons. A quantum mechanical calculation can be performed for each of them by writing the appropriate hamiltonian and solving the resulting Schrödinger equation. This calculation provides a complete description of the two separate molecules, including the atomization energy and the equilibrium structure. A calculation could then be performed on the dimer, the only difference being in the enhanced complexity of the hamiltonian, which now contains terms that describe the joint behavior of the electrons of the two molecular systems. In principle, a perfect solution of the bimolecular Schrödinger equation may find the best approach geometry of the dimer, and provides an undisputable value for the cohesive energy, defined as the total energy of the dimer less the sum of the separate energies of the monomers (Section 3.8).

There is no doubt that a perfect quantum mechanical calculation would take into account all the electronic effects inherent to molecular aggregation. Aside from the fact that such a perfect calculation is not feasible in practice, chemists want to understand the stabilization or destabilization of molecular aggregates in terms of chemical concepts like structure or charge distribution, and, even more ambitiously, they want to predict the recognition modes among molecules without recourse to laborious quantum chemical calculations. So, the purity of quantum mechanical truths must yield to the imperfections of practical quantum chemical calculations, or of even more approximate qualitative arguments based on molecular structure, polarity, and polarizability. The rest of the story of intermolecular interactions is a story on the one hand of the attempts to recast in at least partially classical terms all the quantum mechanical facts, and on the other hand of the attempts at partitioning the total electronic energies into recognizable and interpretable separate contributions [1–3].

Consider what happens when two closed shell molecules approach one another. When the distance between them is large, each electron density preserves its identity, and there is very little overlap. At this stage, the interaction between the two charge distributions can be appropriately described in classical terms by Coulomb's law. As the distance decreases, the electron density clouds begin to overlap. In quantum mechanical terms, a new wavefunction is needed to describe the properties of the dimer, and the electron density of the complex becomes different from a juxtaposition of the electron densities of the two monomers. The new wavefunction can be written as a combination of the wavefunctions of the monomers, but antisymmetrization is required, to comply with the Pauli principle. Furthermore, the electric field of one charge distribution acts upon the other, producing what is called in classical physics terms a polarization, that is a displacement of the elements of one charge distribution under the action of the electric field generated by the other. In molecular orbital terms, there is an orbital interaction step, in which the molecular orbitals of the two interacting fragments, including virtual unoccupied orbitals, are allowed to mix appropriately into the molecular orbitals of the complex.

Historically, all these classical and quantum mechanical facts have been considered under different perspectives and with different degrees of approximation [4]. A very important point is that coulombic and polarization electrostatic energies cannot explain all of the cohesive energies, and a further term is required to include the missing part of inter-electronic action. The rationalization in chemical terms of intermolecular interaction has led to the subdivision of intermolecular energies into coulombic, polarization-charge transfer, dispersion (the missing term), and repulsion contributions. This subdivision does not stem from firm physical principles and is by no means unique, so that the actual values of such partitioned energies may depend on the conceptual basis on which a particular method is framed, and also, within the framework of quantum chemical calculations, on the particular choice of the computational method and of the computational details. The first two terms may be classified as purely electrostatic terms, the third term depends on the interaction between variable dipoles ("moving" electrons), while the repulsion term is a consequence of spin, and hence may be considered separately.

4.2 The representation of the molecular charge distribution and of the electric potential

4.2.1 *Full electron density*

The distribution of electric charges in a molecule plays a central role in all discussions of intermolecular interactions. No wonder, a very large amount of conceptual and computational effort has been and is being dedicated to its expression and interpretation.

Consider molecule A identified by an electron density $\rho_A(\mathbf{r}_1)$, a charge distribution function $q_A(\mathbf{r}_1)$ and a number N_A of nuclei of charge $Z_k(A)$. The electron density is the number of electrons per cubic angstrom at point **r**: it can be readily obtained by a molecular orbital calculation (see equation (l) of Box 3.1), and quantum chemical programs produce it in the form of discrete points along a regular grid. The charge distribution is the number of electrons in a cube element whose center is at **r**; it is obtained from the electron density by multiplication by the volume of the elementary cube. This representation typically involves something like 10^6 points, even for a rather small molecule, and the information it carries is extremely detailed, but somewhat opaque in terms of common comprehension.

When two molecules approach, intermolecular forces drive them into contact, ideally to the geometry with the most favorable cohesion energy. In this process, longer-range forces act first. The coulombic potential decays as the first power of distance, so that coulombic forces decay as the inverse square power of distance. All other attractive effects act at shorter range, typically of the order of R^{-3} to R^{-6}. Repulsion sets in at very short range. So it is reasonable to expect that coulombic forces will act first, to be later corrected by adjustments due to polarization and dispersion, while repulsion prevents too large a molecular overlap in the final stages of the recognition process. This analysis applies both to molecular recognition during crystal nucleation and to the early stages of a chemical reaction.

Recalling the definition of the density matrix in Box 3.1, for a molecule with N_A nuclei of charge Z_k at locations $\mathbf{R}_k$ and whose wavefunction is expressed in the usual LCAO form 3.46, the quantum mechanical (QM) electric potential at point **r** is given by [5]

$$(4\pi\varepsilon°)V^{QM}(\mathbf{r}) = \sum_k Z_k (|\mathbf{r} - \mathbf{R}_k|)^{-1} - \sum_r \sum_s P_{rs} \int \chi_r(\mathbf{R})\chi_s(\mathbf{R}) \, (|\mathbf{r} - \mathbf{R}|)^{-1} \, d\mathbf{R} \quad (4.1)$$

where the first summation runs over nuclei, and the other two summations run on occupied molecular orbitals. Note that this is the electric potential, not to be confused with the coulombic potential energy, which is given by multiplication of the electric potential by the charge q that feels it, $E_{coul} = qV$. The molecular electrostatic potential is a quantum chemical observable, just as the electron density is, as can be guessed from the fact that equation 4.1 contains no approximation. $V(\mathbf{r})$ is in fact the expectation value of the r^{-1} operator.

4.2.2 Central multipoles

Using a rather far-fetched metaphor, tracing the molecular electrostatic potential is the computational equivalent of touching the "skin" of a molecule, and it is more than understandable that the visualization of the contours of the molecular electrostatic potential has attracted the chemists' attention from the very beginning [6]. Sophisticated graphics programs are nowadays available for that purpose. Yet, the collection of numbers that describe the electric potential, or even its contours drawn in beautiful graphics, are not immediately transferable into common chemical concepts. In many applications, much simpler schemes must be used. One possibility is suggested by a standard theorem of electrostatics that states that the potential due to a charge distribution, at any point at a distance R from the center of charges, can be expanded in a series of terms in the inverse powers of R, called multipoles [7]:

$$V(R) = A°/R + (1/R^2)(A_x p_x + A_y p_y + A_z p_z) + (1/R^3)(A_{z2}\, d_{z2} + A_{xy}\, d_{xy}$$
$$+ A_{xz}\, d_{xz} + A_{yz}\, d_{yz} + A_{x2-y2}\, d_{x2-y2}) + \text{higher terms over } 1/R^4,\ 1/R^5, \ldots \quad (4.2)$$

Quantum chemical calculations can nowadays give the components of the first few molecular multipoles for a rather cheap computational price. The first three terms in equation 4.2 are called monopole, dipole, and quadrupole terms, respectively. The A coefficients express the radial dependence of the potential, while the p and d functions are the spherical harmonics of Table 3.2. Their appearance is perhaps not surprising if one thinks that they are to represent a distribution in space of "pluses" and "minuses" with increasing number of nodes. If the molecule is uncharged (that is, not ionic) the monopole term is zero. The dipole is a vector (call it **D**) joining the center of negative charges and the center of positive charges. This vector is null, for example, if the molecule has a center of symmetry. The dipole moment, μ_D, is a vector defined as the product of the dipole charge Q (the sum of positive or negative charges) and the dipole vector:

$$\mu_D = Q\mathbf{D} \quad (4.3)$$

Equation 4.2 implies that if R is large, the potential may be taken in a first approximation as just the first non-zero term in the expansion, because further terms depend on increasing inverse powers of distance. So for a non-ionic molecule with non-zero dipole the monopole term is zero, while quadrupole and higher multipole terms are much smaller than the dipole term. At large intermolecular distance, therefore, the electric potential of a molecule is to a good approximation described by just the dipole term. For condensed phases, where molecules may come very close to one another, the dipolar approximation is unsatisfactory.

For an ensemble of dipolar molecules, the coulombic interaction is a sum of dipole–dipole terms, the energy of a dipole in the electric potential due to another dipole. More refined treatments of coulombic energies using multipole expansions may take into account all multipole–multipole interactions. However, note carefully that these are central multipoles, that is, they are formally centered at the molecular center of charges, and they should not be confused with the distributed multipoles that will be described in the next paragraph. Central multipoles may be appropriate for some very small molecules, but clearly the description of the complex charge distribution of a large organic molecule by a small number of such central multipoles is not recommendable and may generate more confusion than understanding. For example, it has been shown [8] that the multipole expansion of the electrostatic interaction between benzene and hexafluorobenzene is not convergent at the intermolecular distances of chemical interest below 4Å. Therefore, it would be a benefit to the chemical community if the concepts of central molecular dipoles and quadrupoles were to disappear from the scene.

4.2.3 *Distributed multipoles*

The direct calculation of the electric potential and of the coulombic potential energy using equation 4.1 requires the evaluation of a very large number of terms, is extremely time-consuming, and therefore is not applicable to large-scale molecular modeling. At another extreme, central multipoles are a very poor representation of the charge distribution. The idea of describing the charge distribution by a set of multipoles distributed at many molecular locations therefore seems appealing.

Consider again a molecular orbital in the LCAO form of equation 3.46. If the Gaussian basis functions are centered at points r_j, usually nuclear positions, the electron density at point r and the coulombic energy between the electron densities of molecules A and B can be written as follows:

$$\varphi_i = \sum_j c_{ij}\, \chi_j(\mathbf{r} - \mathbf{r}_j) \tag{4.4}$$

$$= (3.46)$$

$$\rho(\mathbf{r}) = \sum_r \sum_s P_{rs}\chi_r(\mathbf{r} - \mathbf{r}_r)\chi_s(\mathbf{r} - \mathbf{r}_s) \quad \text{(see Box 3.1)} \tag{4.5}$$

$$(4\pi\varepsilon^\circ)\, E_{coul} = \int\int \rho_A(\mathbf{r}_1)\rho_B(\mathbf{r}_2)/|\mathbf{r}_1 - \mathbf{r}_2|\, d^3\mathbf{r}_1\, d^3\mathbf{r}_2 \tag{4.6}$$

If equation 4.5 is substituted into 4.6, the resulting expression involves a quadruple summation of fourfold products of the basis functions. Terms for the nuclear contributions must then be added.

The basic ideas of the distributed multipole analysis (DMA) as developed by Stone [9] can be summarized as follows. Each product of two gaussian functions, like the ones appearing in 4.6, is again a gaussian function centered somewhere in between the two original ones. It can be shown [10] that this product in turn can be expressed as a series of multipoles centered at the new location. This analysis can be viewed [9] as similar to a Mulliken population analysis including higher moments of the overlap density to give, besides charge populations, also higher multipole populations. This adds flexibility to the model: for example, a lone pair may be better described by a dipole at the nucleus rather than by a point charge somewhere away from it. The choice of the location of these multipoles is not unique, but this is in a sense an advantage, because the distributed multipoles can be grouped together and strategically located. In a compromise between the requirements of convergence and economy, a reasonable choice is, for example, to distribute multipoles just at atomic nuclei and at the centers of the bonds. If so desired, the multipoles can all be transferred to the nuclear locations, thus allowing an atom–atom formulation of the entire procedure [11].

4.2.4 *Point charges*

Another, very popular alternative for the representation of the charge distribution in organic molecules is the point-charge model. In this approach, partial net charges (that is, positive or negative with respect to the neutral atom) are assigned to a number of

THE REPRESENTATION OF THE MOLECULAR CHARGE DISTRIBUTION

selected molecular locations, often the nuclei for lack of a better hypothesis. A rigorous definition of the amount of charge that pertains to a given atom in a molecule is given by the atoms-in-molecules (AIM) method (see Section 3.5.2), as the integral of the electron density on each atomic basin. However, the calculation of these AIM charges is complicated and this kind of partitioning does not lend itself easily to molecular simulation purposes. An alternative is offered by the so-called electrostatic potential-derived (ESP) partial charges. For a molecular object with a distribution of net point charges q_k at locations $\mathbf{R}_k$, the approximate electric potential is written as:

$$(4\pi\varepsilon°)V^{APP}(\mathbf{r}) = \sum_k q_k(|\mathbf{r} - \mathbf{R}_k|)^{-1} \quad (4.7)$$

The procedure [5] consists of evaluating the quantum mechanical potential at a grid of points in the molecular space, followed by a least-squares fit of the parameters, q_k, so that the difference $(V^{QM}(\mathbf{r}) - V^{APP}(\mathbf{r}))^2$ is minimized over the collection of grid points, subject to the condition of neutrality, that is, of a zero sum of the q_ks. These are located either at nuclei or at any other strategical molecular position, like lone pairs, or electron-rich bonds, etc. The physical meaning of the derived charges is indeed marginal, and in fact there exist many different sets of q_ks that reproduce the same electric potential. The results are sometimes not entirely free from numerical noise, including a breaking of the molecular symmetry if the grid is not judiciously constructed. Nevertheless these ESP charges are much more consistent than Mulliken population analysis charges (see Table 3.4) [12].

Another simple and elegant method for the determination of point charges is Hirshfeld's stockholder recipe [13]. In this approach, a pro-molecule is defined as the collection of the electron densities $\rho_{i,AT}(\mathbf{r})$ pertaining to spherically averaged, hypothetically isolated atoms located at the position they occupy in the actual molecule. The molecular electron density, $\rho_{mol}(\mathbf{r})$, which includes a deformation from the pro-molecule density because of intramolecular bonding, is then apportioned among atoms according to the proportion of the contribution of each atom to the pro-molecule: like a stockholder, each atom "receives" electrons in proportion to how many it has "invested" in the formation of the molecule. The simple equations are:

$$w_i(\mathbf{r}) = \rho_{i,AT}(\mathbf{r}) / \sum_i \rho_{i,AT}(\mathbf{r}) \quad (4.8)$$

$$\rho_{i,bound}(\mathbf{r}) = w_i(\mathbf{r})\,\rho_{mol}(\mathbf{r}) \quad (4.9)$$

where $w_i(\mathbf{r})$ is the stockholder's weight, and $\rho_{i,bound}(\mathbf{r})$ is the density assigned to atom i bound into the molecule. The deformation density is then the difference between the bound and free atomic densities, and the net charge on atom i is the sum of the deformation density over atom i:

$$\delta\rho_i(\mathbf{r}) = \rho_{i,bound}(\mathbf{r}) - \rho_{i,AT}(\mathbf{r}) \quad (4.10)$$

$$q_i = -\int \delta\rho_i(\mathbf{r})d\mathbf{r} \quad (4.11)$$

An operational way of obtaining point charge parameters in a quick and consistent way is the rescaled-EHT method [14]. An Extended Hückel calculation (Section 3.6), which requires a negligible amount of time for medium-size organic molecules, is carried out, and a Mulliken population analysis is performed on the resulting wavefunction. This is in fact a way of apportioning electrons among atoms roughly according to the atomic ionization potentials, the only parameters that appear in the EHT calculation: however, charge separations are usually too large, and a rescaling has to be applied, using a fictitious ionization potential of -10 eV for hydrogen, and dividing the resulting charges by a factor, S. Calling p_k the Mulliken gross atomic population on atom k with a number of electrons Z_k, the charge parameter q_k is given by

$$q_k = 1/S(Z_k - p_k) \tag{4.12}$$

When $S = 3$, the resulting net atomic charges are almost undistinguishable from the Mulliken charges that are obtained in an MP2/6-31G** calculation, which requires a 1000-fold longer time than the EHT calculation.

One disturbing feature of point-charge parameters is their scarce transferability even over quite similar chemical groups. Although compilations of standard repertories of point charges for use in molecular simulation have been attempted, the results have never been entirely satisfactory. It turns out that point charges [15] and distributed multipoles [16] do change even for different conformations of the same molecule. For accurate simulations, therefore, there seems to be little choice but to calculate a high quality ab initio wavefunction for each new molecule and, indeed, for each significantly different conformation of the same molecule. This is a major disadvantage.

4.3 Coulombic potential energy

The coulombic potential energy is stabilizing whenever opposite charges are involved, or destabilizing among charges of the same sign, and the corresponding forces are attractive or repulsive, respectively (since the Coulomb potential has no minima, stabilizing energies always correspond to attractive forces and destabilizing energies to repulsive forces).

If the full molecular electron density is considered, the total classical coulombic energy between the two molecules is:

$$(4\pi\varepsilon°) E_{coul} = \iint \rho_A(\mathbf{r}_1) \rho_B(\mathbf{r}_2)/|\mathbf{r}_1 - \mathbf{r}_2| \, d^3\mathbf{r}_1 d^3\mathbf{r}_2 + \sum_k \int Z_k(A) \rho_B(\mathbf{r}_2)/|\mathbf{r}_k - \mathbf{r}_2| d^3\mathbf{r}_2 + \sum_m \int Z_m(B) \rho_A(\mathbf{r}_1)/|\mathbf{r}_m - \mathbf{r}_1| d^3\mathbf{r}_1 + \sum_k \sum_m Z_k(A) Z_m(B)/|\mathbf{r}_m - \mathbf{r}_k| \tag{4.13}$$

Formidable as this equation may seem, it is composed only of one term for electron–electron repulsion, two terms for electron–nuclear attraction, and a last term for nuclear–nuclear repulsion. In terms of the corresponding charge distribution, the expression for the coulombic energy may be recast in the following form, in which the integrals are replaced by the corresponding summations:

$$(4\pi\varepsilon°) E_{coul} = \sum_k \sum_m q_{k,A}(\mathbf{r}_k) q_{m,B}(\mathbf{r}_m)/|\mathbf{r}_k - \mathbf{r}_m| + \sum_k Z_k(A)[\sum_m q_{m,B}(\mathbf{r}_m)/|\mathbf{r}_k - \mathbf{r}_m|] + \sum_k Z_k(B)[\sum_m q_{m,A}(\mathbf{r}_m)/|\mathbf{r}_k - \mathbf{r}_m|] + \sum_k \sum_m Z_k(A) Z_m(B)/|\mathbf{r}_m - \mathbf{r}_k| \quad (4.14)$$

The charge distribution can be readily obtained by a standard MO calculation, and equation 4.14 is a practical way of obtaining intermolecular coulombic energies (Section 12.2.2). The computational load when using this equation increases steeply with the number of discrete charge points, while accuracy depends on the fineness of the grid mesh. Use of a smaller grid step improves the accuracy of the representation of the charge density, especially in the regions close to nuclei where the electron density rises very steeply, but also increases the computational demand very quickly. Use of a wider step results in a smaller number of charge density units, but decreases the accuracy of the representation of the electron density and might even result in a significant loss in the integral number of electrons. A convenient computational alternative [17] is to use a fine mesh in the calculation of the charge distribution, but then to condense it by summing into one charge unit the contents of a super-cube made of $n \times n \times n$ original charge units. In this way the number of charge points is reduced by n^3 and the time required for the calculation of the first term in equation 4.14 is reduced by n^6.

In terms of central or distributed dipoles, the coulombic energy is calculated as a sum of moment k to moment m terms, where each moment is a monopole, dipole, quadrupole moment, etc. The exact forms of the terms involving higher dipoles quickly become very complicated, [18] and will not be given here in their tedious detail.

Dipole–dipole interaction energies may be stabilizing or destabilizing according to the relative orientation of the dipoles, and in the most stabilizing arrangement the dipolar energy decays as the inverse third power of distance. An average interaction energy over an ensemble of freely rotating dipoles is equal to zero. In liquids, molecules undergo fast reorientation, but contacts between like charges are on average less frequent than contacts between opposite charges, so that the sum-total is generally stabilizing, because the different orientations are weighted by a Boltzmann factor. The Boltzmann orientation-averaged result is [19]:

$$E_{coul} = -C(\mu_A \mu_B)^2/R^6 \quad (4.15)$$

where C is a positive constant. This is the expression for the expected cohesive energy for a molecule represented by just one central dipole. In crystals, where there is no orientational freedom, coulombic energies are usually stabilizing overall, but some particular molecule–molecule coulombic energies may also be destabilizing.

In terms of point-charge models, the total coulombic energy is simply a sum of two-body terms between charges i on molecule A and charges j on molecule B:

$$E_{coul} = 1/(4\pi\varepsilon°) \sum\sum q_{i,A}\, q_{j,B}\, /R_{ij} \qquad (4.16)$$

Equations 4.13 and 4.14 are rigorous expressions for the coulombic energy between two charge distributions. When the delocalized charge distributions are replaced by more localized models such as multipoles or atomic point charges, the short-range effects of the actual size of the electron cloud are to some extent lost. The correction to the coulombic energy for the finite size of a charge distribution is called the penetration energy, and the same term is often used to denote the difference between the coulombic energy calculated by a delocalized electron distribution model and that calculated by a localized model. Localized models may be adequate for the calculation of global coulombic energies in crystals, but when attempting a description or a rationalization of molecular orientations at short range, a typical crystal packing analysis task, the inclusion of penetration energies is essential. It can be easily shown that the coulombic potential energy between approaching closed-shell molecules is stabilizing to very short intermolecular separation, much too short to be of chemical significance, while a much less stabilizing or even destabilizing energy is calculated by localized models, due to the lack of the penetration term. Thus, coulombic inferences based on point-charge or central multipole approximate models may easily be wrong or even mistake an attraction for a repulsion. The true nature of intermolecular repulsion at short range (Section 4.6) is entirely independent of coulombic potential energies.

4.4 Polarization (electrostatic induction) energy

Consider the particle picture of a hydrogen atom, in which a static nuclear particle of charge $+e$ is encircled by a revolving electron particle of charge $-e$ at a distance R. When an external electric field ε acts on such a system, the perturbation can be described by a net displacement Δx of the electron under the action of the electric force of the field, $F_{displ} = e\varepsilon$ (Fig. 4.1). This process can be represented as the formation of a dipole induced by the field, whose dipole moment μ is assumed as a first approximation to be proportional to the intensity of the external field. The proportionality

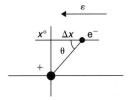

Fig. 4.1. An electron, denoted by e^-, orbiting at a distance R around a nucleus, denoted by (+), is displaced from its original position x_0 by an amount $\Delta x = R\cos\theta$ under the action of an electric field ε.

constant is called the polarizability, α so that $\mu = \alpha\varepsilon = e\Delta x$. The displaced electron is also subject to a restraining force given by the coulombic attraction to its nucleus, and at equilibrium the displacing and restraining forces balance out [20]:

$$F_{restr} = 1/(4\pi\varepsilon°)\, e^2/R^2 \, \cos\theta = 1/(4\pi\varepsilon°)\, e^2 \, \Delta x/R^3 = 1/(4\pi\varepsilon°)\, e/R^3 \, \mu = e\varepsilon \quad (4.17)$$

This simple model reveals that the polarizability has the dimensions of a volume, since $4\pi\varepsilon°$ is a dimensionless term:

$$\mu = 4\pi\varepsilon° R^3 \varepsilon = \alpha\varepsilon \quad (4.18)$$

$$\alpha = 4\pi\varepsilon° R^3 \quad (4.19)$$

The physical interpretation of polarizability is just in terms of its definition, that is, a measure of the propensity of a given electronic environment to yield under the action of an external electric field. Since the restraining force is given by coulombic attraction between the displaced charge and the nucleus, polarizability will be large when electrons are at a large distance from a weakly charged nucleus. Being a volume quantity, polarizability is an additive, extensive quantity, and tables of bond polarizabilities or even atomic polarizabilities have been prepared (Table 4.1) by a breakdown of experimental molecular polarizabilities into bond or atom terms [21].

The energy involved in the polarization process is always stabilizing because the induced dipole always points in the stabilizing direction (Fig. 4.2), being a sort of compliance of the polarized medium to the polarizing field. The polarization energy

Table 4.1 Atomic polarizabilities

Atom	$\alpha_{atom}(\text{Å}^3)$	Atom	$\alpha_{atom}(\text{Å}^3)$
C(alifatic)	1.06	F	0.30–0.50
C(unsaturated)	1.35	Cl	2.32
O	0.57–0.78	Br	3.01
N	0.85–1.09	I	5.42
H	0.39		
S	2.70–3.70		
Ar	1.70		

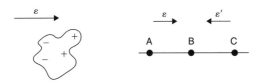

Fig. 4.2. Left: a sketch of the polarizing effect of an electric field ε on a charge distribution. Right: two opposite fields at point B cancel, polarization is not a two-body effect.

can be thought of as the energy required for "turning on" a field ε onto a dipole μ, and if the direction of the induced dipole is parallel to that of the inducing field, it is given by [22]

$$E_{\text{pol}} = -\int \mu \, d\varepsilon = -1/2 \, \alpha \, \varepsilon^2 \qquad (4.20)$$

The calculation of polarization energies by this formula is simple and straightforward, and requires only a calculation of the electric field, using one of the previously described charge distribution models, and some hypothesis about the polarizability of the molecular medium. Such a procedure has, however, several shortcomings. First, molecular electron densities are non-spherical three-dimensional objects, and the polarizability is far from being isotropic and uniform. In an anisotropic treatment, the polarizability is a tensor quantity and the corresponding expressions become orientation-dependent and consequently more complicated. Second, this linear polarization model breaks down under very strong fields or extreme polarization conditions when the induced moment is no longer proportional to the polarizing field. Third, this model accounts only for what may be called static polarization: one should more properly take into account the fact that the induced dipole generates a change in the field generated by the polarized location, and this change must in turn produce some effects at the polarizing site too. Polarization is mutual, and a more accurate treatment should include all these effects to self-consistency of the mutually polarized electron distributions.

The calculation of the electric field is quite straightforward if the electron charge distribution is given in the form of a discrete array of points of charge q_k. The electric potential at point P due to this charge distribution plus a number of nuclei of charge Z_j is given by

$$\Phi(P) = 1/(4\pi\varepsilon_0) \left(\sum_k q_k/R_{Pk} + \sum_j Z_j/R_{Pj} \right) \qquad (4.21)$$

The electrostatic potential energy of a charge q_P at point P is then given by $E(P) = q_P \Phi(P)$. The corresponding electrostatic field at point P, ε_P (the x component) is [23]

$$\varepsilon = -d\Phi/d\mathbf{R} \qquad (4.22a)$$

$$\varepsilon_{xP} = 1/(4\pi\varepsilon_0) \left\{ \sum_k [q_k \, (x_P - x_k)/R_{Pk}^3] + \sum_j [Z_j \, (x_P - x_j))/R_{Pj}^3] \right\} \qquad (4.22b)$$

This immediately shows that the induction energy, which depends on the square of the electric field, must have approximately an R^{-6} dependence on the distance between the polarizer and the polarized. In fact, when a molecule of isotropic polarizability α is represented by a central dipole μ and orientational Boltzmann averaging is carried out, the approximate expression for the induction energy is [24]:

$$E_{\text{pol}} = -\mu^2 \, \alpha (4\pi\varepsilon_0)^{-2} \, (R)^{-6} \qquad (4.23)$$

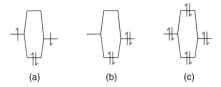

Fig. 4.3. Sketches of orbital interaction diagrams for two approaching molecular fragments: (a) one unpaired electron on each fragment, bond formation; (b) a couple of paired electrons on one fragment and an empty LUMO orbital on the other, donor–acceptor interaction, or charge transfer; (c) one couple of paired electrons on each fragment, non-bonding interaction. The zero-order energy is the sum of the occupied orbital energies, so (c) is net destabilizing.

Is there a quantum mechanical equivalent of polarization? As usual, this depends on how one wants to adapt a classical concept to a quantum mechanical reality. An analogy can be established between classical polarization and quantum mechanical molecular orbitals mixing, including virtual orbitals. In this respect, a special form of polarization is sometimes called charge-transfer energy. The name comes from a simple molecular orbital picture of the mixing, when an electron pair in a high lying occupied MO of one interacting fragment is "poured" into an empty low-lying molecular orbital of another interacting fragment as a result of the coupling (see Fig. 4.3(b)). It may be more convenient to include all these effects under the unique category of polarization effects.

4.5 Dispersion energy

Before quantum mechanics was developed, intermolecular interactions were treated in classical terms as the result of multipolar electrostatic interactions. An insurmountable difficulty was however encountered in the description of stabilizing energies in completely apolar systems: the paradigm of this challenge to the molecular physics of the beginning of the twentieth century was understanding the crystallization of argon [25]. No combination of multipole interactions could account completely for that energy, not even when oscillating multipoles were introduced. Only when the phenomenon is described as a coupling of quantum mechanical oscillators [26] does the final result amount to the correct stabilizing interaction.

As these energies depend on the simultaneous behavior of different electrons, they belong in what quantum chemistry calls electron correlation effects [27]. The introduction of dispersion energy is a way of re-casting the interelectronic phenomena that occur in a many-electron molecule in terms of a semi-classical model depending on simultaneous electron "displacements". The starting point is the assumption that an electron density is not static, but oscillating. The exact nature of these oscillations is difficult to describe; in semi-classical terms, one might think of an oscillation in time, as if the electrons were moving; alternatively, one may invoke fluctuations depending on the statistical nature of the quantum mechanical description of matter. In any case, when the electron density at a given point changes, a transient dipole is generated. This fluctuating dipole generates an induced dipole in a nearby electron distribution. The resulting quantum mechanical coupling in these dipole–dipole interactions is the basis of the stabilizing energetic effect called dispersion.

By writing the frequencies of two dipole oscillators, and using the zero-point energy expression for a harmonic oscillator with the same proper frequency (equation 3.24), London [28] was able to derive the following expression for the dispersion energy between two molecules of polarizability α at a distance R:

$$E_{disp} = [-3/4 \, (\alpha)^2 \, (h\nu)/(4\pi\varepsilon_0)^2](R)^{-6} = [-3/4 \, I_I \, (\alpha)^2/(4\pi\varepsilon_0)^2](R)^{-6} \quad (4.24)$$

This stabilization has to do with zero-point oscillator energies, so dispersion energy is a consequence of the uncertainty principle, like all zero-point energy effects (Section 3.1).

The energy in 4.24 is always stabilizing, as does polarization energy. No wonder, the expression contains the square of the polarizability and an R^{-6} term (compare with equation 4.23). The one novel quantity in this expression is the "oscillation frequency" of the electrons, ν, with its corresponding "oscillator strength", $h\nu$, analogous to a quantum mechanical vibrational energy. The "force constant" for the oscillation is also related to polarizability, and in a first approximation the quantity $h\nu$ is related to the potential energy of an electron in the molecule; hence, it was taken by London as equal to the first ionization potential of the molecule, I_I. Whatever the true justification for this assumption might have been, the calculations performed with London's formula gave the correct order of magnitude for the experimental interaction energies in spherical apolar molecules and the success of the model was instantaneously acknowledged. The simple expression 4.24 gives acceptable values for the cohesive energies in condensed noble gases and explains at a glance some simple and obvious experimental facts such as, for example, the increase in boiling point in similar molecules with substituents of increasing size and hence of increasing polarizability. Of course, in complex molecules the "oscillator strengths" of the electrons cannot all be represented by just one ionization potential but may depend on the atomic and molecular environment. This is the reason why all practical dispersion models must use different coefficients for different parts of the molecule, typically for different atoms in atom–atom models (Section 4.9).

In further refinements of the model, higher instantaneous multipoles can also be considered, and lesser contributions to dispersion energies are found to depend on R^{-8}, R^{-10}, and so on. In the formulation of equation 4.24 the dispersion energy is pairwise additive over interacting centers, but it might be expected that, being related to polarization, dispersion may have some many-body character, and indeed the cohesive energies obtained by accurate calculations using only two-body terms are slightly underestimated.

The reader will have noticed that there is no attempt to relate dispersion energies in the London model with some parts of the quantum mechanical expectation energies. It is not entirely clear, or at least it is not easily explained, if and how London forces correspond to some molecular orbital property, or why coupled quantum mechanical oscillators should represent some if not all of the higher-order interelectronic interactions in a complex plurimolecular system.

4.6 Pauli (exchange) repulsion energy

The above analysis of intermolecular stabilizing interactions with its breakdown into coulombic, polarization, and dispersion attractive terms has the advantage of clarifying, at least to some extent, the nature of these attractions in terms of chemically understandable interactions between charge densities. The price that has to be paid is that for a complete description of the interaction, some kind of repulsive effect must also be partitioned out of total energies. As already mentioned, repulsion must account for the scarce compressibility of condensed media – and, indeed, for the very fact that molecules just do not collapse onto one another, and macroscopic bodies have a finite spatial size.

The origin of short-range destabilization and repulsion is quantum mechanical, and may be ascribed to the effect of antisymmetrization of the wavefunction, so that it has no classical analogy. The basic concepts can be illustrated using a prototypical system consisting of just two 1s atomic orbitals, a and b, with electrons 1 and 2 in them with unpaired spins (Stone [29] gives a particularly clear description of the case). When the two atoms are at large distance, their electron densities do not overlap, hence spin prescriptions need not apply because the electrons do not occupy the same space, and the total wavefunction, $\Phi°$, can be just the product of the two atomic orbitals. In the short-range regime, where the two electron densities start overlapping, one must use a properly antisymmetrized wavefunction, Φ_{as} (Section 3.4), in the determinant form of equation 3.45. The two wavefunctions are, respectively:

$$\Phi° = a(1)b(2) \quad (4.25a)$$

$$\Phi_{as} = (1 - S^2)^{-1}(1/2)^{1/2}[a(1)b(2) - a(2)b(1)] \quad (4.25b)$$

where $a(1)$ means orbital a with electron 1 in it, as usual, and S is the overlap integral (equation (b) in Box 3.1). It can be shown that the difference between the expectation energies (equation 3.43) calculated with each of the two wavefunctions is a term in the form

$$E(\Phi_{as}) - E(\Phi°) = E_{as} = (1 - S^2)^{-1}(\Delta E' - K_{ab}) \quad (4.26)$$

where $\Delta E'$ includes a number of corrections to the total electronic energy, while K_{ab} is the $ab|ba$ exchange integral (equation (d), Box 3.1) [30]. E_{as}, the energy arising from the antisymmetrization required by the overlap, is called the exchange-repulsion energy because it is overall destabilizing; it vanishes when $S = 0$ [31]. Some confusion may arise in the terminology because the exchange integral gives a stabilizing contribution to the energy, so that one can speak of exchange stabilization in many-electron systems, but the exchange integral also gives the largest contribution to equation 4.26, where it is taken with a negative sign and is responsible for the main part of the destabilization. Confusion is not a rare occurrence when talking of intermolecular interactions in semi-classical terms versus approximate quantum chemical terms.

The electron density of the non-interacting wavefunctions is just the sum of the original densities, while the electron density from the antisymmetrized wavefunction contains supplementary terms that depend on the overlap integral. It can be readily shown that these extra terms are such that the electron density in between the nuclei is less than the sum of the separate densities by a factor of $1/(1 + S)$, while the density on the farther side of each nucleus is larger than for the sum of the separate densities by a factor of $1/(1 - S^2)$ (S is always > 0 in this system made of $1s$ orbitals). In approximate words along these lines, the ultimate physical representation of the repulsion results in one saying that the exclusion principle drives electrons with same spins away from the region of intermolecular contact, so that a net force, from the electrostatic Hellman–Feynman theorem (Chapter 1) pulls the nuclei apart. This is what chemists call steric repulsion: the same reasoning, here presented for what would formally be a triplet state of the hydrogen molecule, holds for any closed shell systems where electrons pushed in the same region of space cannot pair their spins. Note that this explanation is quite different from saying that repulsion arises from coulombic interaction between nuclear charges or between electron clouds; also, this language has no equivalent in the language of AIM theory (Section 3.5) in which steric repulsion has no place.

In terms of energy, alternative explanations may be offered [32]. When electrons are driven away from the space between the nuclei towards locations that are closer to the nuclei, the change in coulombic energy is actually stabilizing, because the regions closer to the nuclei are deep coulombic potential energy wells; this picture contrasts to the popular view of chemical bonding as a coulombic effect due to the bond electron pair. The destabilizing contribution in this picture comes from the rise in kinetic energy, due to the enhanced space restriction of the electrons, or, in other words, to the higher gradient of the wavefunction. In molecular orbital terms, two MOs are formed from the two starting atomic orbitals, one being $a + b$ and the other $a - b$; the antibonding orbital has a nodal plane, and hence a larger gradient. In these terms, repulsion is seen as an effect of variations of the kinetic energy of electrons.

Semi-classical definitions of intermolecular forces are always partial or incomplete and leave something to be desired. In the present case, the unadorned truth is that some sort of short range, steadily destabilizing potential with corresponding repulsive forces must be introduced in the practical modeling of intermolecular forces, and there is little alternative to taking these potentials and forces in the form of empirical functions of intermolecular distance. A good repulsion function is then anything that rises steeply at short distances and falls off very quickly with intermolecular separation. In searching for quantum mechanical quantities that can be used in empirical correlations with repulsion energies, a good opportunity is offered by the total overlap integral, S_{AB}, between the undistorted electron densities of the two separate approaching molecules A and B, because, as the above discussion suggests, the influence of antisymmetrization requirements is larger as the overlap increases [33]. There are two ways of computing this integral: over an electron density known in analytical form or over a charge distribution given in the discrete form of a grid of volume elements V, centered at $\mathbf{r}_k$:

$$S_{AB} = \int \rho_A(\mathbf{r})\rho_B(\mathbf{r})d^3\mathbf{r} \qquad (4.27)$$

$$S_{AB} = V \sum_k q_A(\mathbf{r}_k)q_B(\mathbf{r}_k) \qquad (4.28)$$

In the second expression, the summation runs on all charge elements that represent the molecule, and accuracy depends on a number of numerical factors, the most important of which is obviously the resolution of the charge distribution grid.

Popular empirical expressions for the repulsion energy are then:

$$E_{rep} = KR^{-n} \quad \text{with } n \approx 12 \qquad (4.29)$$

$$E_{rep} = K_1 \exp(-K_2 R) \qquad (4.30)$$

$$E_{rep} = K(S_{AB})^n \quad \text{with } n \approx 1 \qquad (4.31)$$

where R is a distance between "repulsion centres" to be specified, and the Ks are free game for empirical calibration. Forms 4.29 and 4.30 readily lend themselves to atom–atom formulations, while form 4.31 is in principle a molecule–molecule formulation, but can be turned into an atom–atom formulation if the electron density can be subdivided into atomic basins and the total overlap into atomic contributions. All forms are intrinsically two-body forms and there is little or no room for many-body effects.

4.7 Total energies versus partitioned energies

4.7.1 *Pairwise additivity*

Coulombic energies, dispersion energies from equation 4.24, and repulsion energies from equation 4.29–4.31 result from integrals or summations whose terms involve only the distance between two points, and hence can be said to be pairwise-additive: the total energy in a system of N molecules is the sum of the energies between molecules 1 and 2, 1 and 3,... $N-1$ and N. Consider three centers, A, B, and C; the total coulombic energy is

$$E = q_A q_B / R_{AB} + q_A q_C / R_{AC} + q_B q_C / R_{BC} = E_{AB} + E_{AC} + E_{BC} \qquad (4.32)$$

and $E_{AB} = E_{BA}$ because $\Phi_A q_B = \Phi_B q_A$. The total sum can be unequivocally subdivided into center-to-center terms.

Equations 4.20 and 4.22 show instead that polarization energies depend on an intrinsically many-body effect and are not additive over centers. Consider three centers, A, B, and C along a direction x (Fig. 4.2). The field at point B is given by:

$$\varepsilon_B = q_A/(x_B - x_A)/R_{AB}^3 + q_C/(x_B - x_C)/R_{BC}^3 \qquad (4.33)$$

If the distances and the charges are equal, the total field at B is zero because $(x_B - x_A) = -(x_B - x_C)$. Moreover, in the general case, the energy of a molecule B polarizing a molecule A is different from the energy of molecule A polarizing molecule B because the two fields are different and $\alpha_A \neq \alpha_B$. Therefore, in lattice energy calculations the polarization term must not be halved when comparing with the sublimation energy (see Section 8.7.1).

4.7.2 Interpretation

As a result of the above partitioning, total intermolecular energies can be written as

$$E_{\text{tot}} = E_{\text{coul}} + E_{\text{pol}} + E_{\text{disp}} + E_{\text{rep}} \tag{4.34}$$

These partitioned energies lend themselves to chemical interpretations; for example, coulombic end polarization energies will be large in molecules with a permanent polarization [34]; dispersion energies will be large for molecules with highly polarizable moieties, like aromatic electron clouds.

A separate evaluation of the terms appearing in 4.34 is possible using semiempirical methods, but some terms can also be obtained by partitioning the total electronic energy in a molecular orbital calculation into the expectation energies from the various terms of the hamiltonian, using appropriate dissections, or by the application of perturbation theory [35]. A detailed analysis of these methods is beyond the scope of this book, also because these formulations are applicable only to small oligomers and cannot be used for the calculation of the actual cohesive energies in liquids or in crystals. The Morokuma analysis is however simple enough and instructive. Consider two molecules, A and B, with Hamiltonians H_A and H_B, for which the Hartree–Fock equations have been solved separately yielding the reference energy (equation 3.8) and the wavefunctions in terms of molecular orbitals ϕ_{kA} and ϕ_{mB}:

$$E^\circ = \int \psi_A H_A \psi_A d\tau + \int \psi_B H_B \psi_B d\tau \tag{4.35}$$

The MOs of the AB complex are constructed using the MOs of the separate fragments as basis set:

$$\psi_i = \sum_k C_{ik} \phi_{kA} + \sum_m C_{im} \phi_{mB} \tag{4.36}$$

In this combination, both occupied and vacant MOs are used. This leads to the secular equation (see Box 3.1):

$$(\mathbf{F} - \varepsilon \mathbf{S})\mathbf{C} = (\mathbf{F}^\circ - \varepsilon \mathbf{I}) + \Sigma = 0 \tag{4.37}$$

where $\mathbf{F}$ is the Fock matrix, $\mathbf{F}^\circ$ is the corresponding matrix at infinite separation, $\mathbf{S}$ is the overlap matrix and $\mathbf{I}$ is the identity matrix. Σ is the molecular interaction matrix, which collects all the integrals pertaining to the joint complex. This matrix, thanks to the definition 4.36, can easily be subdivided into blocks corresponding to the interaction between free and occupied MOs of the two fragments. The Fock equations are then solved in turn for each of the selected components of the interaction matrix, yielding partitioned energies, as follows: (1) electrostatic; interaction between occupied MOs without any mixing; (2) polarization; mixing of occupied and vacant orbitals within each fragment; (3) exchange; interaction between occupied MOs with electron exchange (delocalization) between fragments; (4) charge transfer: mixing of occupied MOs on one molecule with vacant MOs of the other molecule. The whole scheme amounts to a subdivision of the total interaction energy by a subdivision of the integrals into occupied–occupied and occupied–vacant blocks.

Of course, in the end, the best estimate of an interaction energy is given by a very accurate quantum mechanical calculation. For dispersion-dominated organic molecules, the required level of accuracy is however such that, again, only small oligomers of rather small molecules can be safely handled [36]: the benzene dimer is already a limiting case, and the evaluation of the relative energies of parallel, parallel offset, and perpendicular benzene dimers has been one of the most popular issues (and stumbling points) in the history of such calculations [37].

Density functional theory (DFT) calculations are problematic because electron correlation is only partially represented in the functionals, or because any proposed functional must represent all of the dispersion contribution and nothing else but the dispersion contribution, making sure that other effects are not counted twice [38]. In a different strategy, once it has been assured that the missing terms are dispersion terms only, an empirical correction in the form of a supplementary R^{-6} contribution can be applied [39]. The obvious objection to the latter strategy is that a method is as accurate and as rigorous as the less accurate and rigorous of its parts, and a long and expensive DFT calculation would only be contaminated by an empirical addition.

A final remark concerns the units in which energies are expressed. SI energy units (kJ/mole) are adopted here, but the conversions from the more familiar units of electrostatics and of theoretical chemistry can be a nightmare [40].

4.8 Intermolecular hydrogen bonding

So far, an implicit distinction has been made between covalent bonds and intermolecular bonding between closed-shell units. There is, apparently, something in between. The experimental sublimation enthalpies of toluene and benzoic acid are 45 and 89 kJ mol^{-1}, respectively. Similar differences are consistently found when comparing the sublimation enthalpies of other hydrocarbons with those of the corresponding carboxylic acid of comparable molecular size (propane 29, propionic acid 74 kJ mol^{-1}). Each atom in the toluene molecule contributes on average 45/15= 3 kJ mol^{-1} to the crystal cohesive energy. In the crystal structure of benzoic acid, the carboxylic proton is seen to be in close contact with the carboxylic oxygen of a neighbor molecule in a cyclic pattern, with a clear directional preference expressed in a linear C=O $\cdots$ O—H alignment, and with two O $\cdots$ H distances of about 1.7 Å, incomparably shorter than the sum of the average intermolecular contact radii in crystals listed in Table 1.2. This structural fact and the large energy difference in cohesive energy can be taken together to signify that some special bonding and stabilizing interaction is taking place between the carboxylic O and H atoms, called a hydrogen bond. A crude but not unrealistic estimate of the strength of the carboxylic acid hydrogen bond is $(89 - 45)/2 = 22$ kJ mol^{-1}. Therefore, what happens between the O and H atoms has an energetic relevance that is an order of magnitude larger than that of an atom in toluene, and this is undisputable evidence that the hydrogen bond is indeed a special form of intermolecular interaction.

These facts force the development of a separate interpretation, because that interaction is repeatedly and predictably observed in different compounds and in different chemical contexts. The intramolecular hydrogen O—H ... O bond and hydrogen bonds between charged species have been measured to be worth up to 80 kJ mol^{-1} [41]. For hydrogen bonding, the discussion of intermolecular attraction and stabilization in terms of coulomb-polarization and dispersion terms between separate electron densities runs into conceptual difficulties because the wavefunctions of the interacting fragments overlap to a large extent, and the liaison has some character of a covalent bond. If however one insists in pushing the partitioning model slightly beyond its natural limits, it is found that coulomb-polarization terms give the largest contribution to the stabilization, as might have been expected, supporting hasty explanations of the hydrogen bond as a H(+) $\cdots$ O(−) coulombic interaction. Theories of the X—H $\cdots$ Y (X,Y=O,N) hydrogen bond have been presented, but their detailed analysis is outside the present scope. A complete discussion of the many geometrical, structural, and energetic aspects of hydrogen bonding would require a separate monography, and indeed several books and essays have been devoted to the subject [42].

The difference in sublimation energy between xylene and benzoquinone is 20 kJ mol^{-1}, and that between dimethylanthracene and anthraquinone is only 10 kJ mol^{-1}. The distances between some aromatic hydrogens and some carbonyl oxygens in the quinone crystals are less than average distances estimated from the radii in Table 1.2, but there are many of them, sometimes bifurcated, and the directional preference is not so clear, also because in a hypothetical O $\cdots$ H hydrogen bond in these crystals, the oxygen atom may choose among many different partners. The obvious question is, do these facts force the definition of a hydrogen bond for the O $\cdots$ H—C interaction, or is it sufficient to describe the structural motifs found in these crystals as the result of coulombic, polarization, and dispersion interactions? There is no answer to this question in terms of principles, and the choice depends on a matter of convenience. Again quoting Pauling, is the presumed C—H $\cdots$ O bond "an aggregate with sufficient stability to make it convenient for the chemist to consider it as an independent chemical species"?

4.9 Simulation methods

4.9.1 *The intermolecular atom–atom model for organic crystals*

In the early 1970s, the availability of an ever increasing amount of crystal structure data for organic compounds prompted the development of empirical schemes for the calculation of lattice energies. As discussed in the preceding sections, the total lattice energy of an organic crystal is a sum of coulombic, polarization, dispersion, and repulsion terms, which all depend on intermolecular distances. The problem arose of finding the appropriate reference centers for the calculation of these distances: just as a few central multipoles are inadequate for the description of the charge distributions, so just one distance between centers of mass or of charge is not sufficient for the description of the diffuse and shape-dependent interaction between complex organic molecules. Since the results of X-ray diffraction experiments came in the form of a list of coordinates for the positions of atomic nuclei, the use of these centers was an obvious option. The lattice energy sums then became sums over atom–atom terms and the method came to be known as the atom–atom method. Pioneers in this field were the members of the Russian school grouped around Alexander Kitaigorodski [43]. Decisive steps forward were made by the classical, systematic studies in parameter optimization presented by the Williams school in the years from 1970 to 1990 [44]. The history of the development of the atom–atom method for organic crystals has been told in an excellent book [45].

Once again, there is no compelling physical reason for using atomic centers; it is a matter of computational convenience, because the terms in the summation are relatively few, that is $(N_{atom})^2$ (N_{mol}) for a crystal made on N_{mol} molecules each of which is made of N_{atom} atoms. The lattice summation for a crystal model composed of 100 naphthalene molecules has thus 32,400 terms, an affordable number even for the computers of 35 years ago. The price to pay is of course a drastic approximation to the already approximated formulas for the various energy contributions. Attractive terms, including dispersion and polarization, are lumped together in R^{-6} terms in the atom–atom distance R, while repulsion terms are approximated by R^{-12} or exponential terms in the same distance. Sometimes, even coulombic energies are incorporated in R^{-6} terms, under the more or less explicit assumption that in scarcely polar molecules the leading multipolar energy terms are the dipole and quadrupole terms, which fall off as R^{-n} with $n \approx 6$. The resulting atom–atom intermolecular interaction function then has the same form as the functions that are used for the representation of intramolecular non-bonded energies (Section 2.2):

$$E_{ij} = A(R_{ij})^{-12} - B(R_{ij})^{-6} \tag{4.38}$$

$$E_{ij} = A\exp(-BR_{ij}) - C(R_{ij})^{-6} \tag{4.39}$$

$$E_{ij} = A + \exp(-BR_{ij}) - C(R_{ij})^{-6} + 1/(4\pi\varepsilon°)q_iq_j/R_{ij} \tag{4.40}$$

These are the so-called 12-6, 6-exp, and 6-exp-1 forms. The numbers $A, B, C \ldots$ are disposable parameters, and for equation 4.40 some quick method for the determination of point charges is needed. Figure 4.4 shows the typical shape of these functions. The curve has a destabilizing repulsive branch, a stabilizing repulsive branch, a minimum at a given separation with a given depth, and a stabilizing and attractive branch vanishing asymptotically at infinite separation. Another useful feature of these simple representations is that the analytical derivatives of the potential are easily accessible, so that the optimization of crystal structures and the calculation of some mechanical

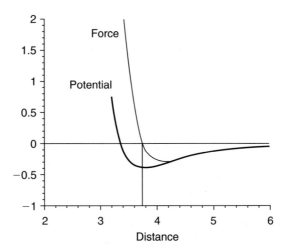

Fig. 4.4. The potential and the force for the interaction between two particles as a function of their separation. The region where the potential is < 0 is stabilizing, the region where the force is < 0 is attractive. The concepts of stabilization and attraction are different and should not be confused. In a small region at the left of the minimum, the potential is still stabilizing but the force is already repulsive.

properties and of lattice vibrational frequencies in crystals are easily carried out (Section 6.3).

The physical interpretation of the necessary parameters is difficult or impossible, if one considers how difficult it must be to reconcile the simple expressions 4.38–4.40 with the intricate electronic effects that have been sketched in the preceding sections. A recommendable attitude [46] is then to consider the atom–atom model as a useful numerical machinery for the calculation of lattice energies, without attaching too much physical significance to it. Thus, R^{-6} terms are attractive terms but cannot be considered to represent dispersion or polarization, R^{-12} or exponential terms are repulsive terms but do not necessarily represent Pauli repulsion, and even the R^{-1} terms are of a coulomb-law type but, as discussed in former sections of this chapter, are but a poor representation of the true coulombic potential energy.

The calibration of the parameters in an atom–atom interaction energy curve is better carried out in terms of the position of the minimum, $R°$, the well depth, ε, and the curvature parameter, λ [47]. An atom–atom potential can be written as:

$$E(r) = \varepsilon(n+\lambda)/(n+\lambda-6)[-(R°/r)^6 + 6/(n+\lambda)(R°/r)^n \exp(\lambda(1-r/R°))] \tag{4.41}$$

where n is an integer and r is the actual interatomic distance. If $n = 0$, the 6-exp form 4.39 obtains:

$$A = 6\varepsilon e^\lambda/(\lambda-6); \quad B = \lambda/R°; \quad C = \varepsilon\lambda(R°)^6/(\lambda-6) \tag{4.42}$$

If $n = 12$ and $\lambda = 0$, the 6–12 form 4.38 obtains:

$$A = (\varepsilon R°)^{12}; \quad B = 2\varepsilon(R°)^6 \tag{4.43}$$

The meaning of the λ parameter can be understood by writing the curvature of the potential at the minimum ($E^* = E/\varepsilon, z = r/R°$):

$$(dE^*/dz) = 6\lambda(\lambda - 7)/(\lambda - 6) \tag{4.44}$$

For a 12–6 potential curve, $(dE^*/dz) = 72$ (a constant); by comparison with equation 4.44, a 12–6 curve corresponds to a 6-exp curve with $\lambda = 13.8$.

Each kind of atom–atom interaction requires two parameters for form 4.38 and three parameters for form 4.39 so that the calculation for a molecule including N_{sp} different atomic species requires $[(N_{sp}(N_{sp} + 1)]$ or $3/2\ [(N_{sp}(N_{sp} + 1)]$ parameters, respectively. In principle, even the same atomic species may require different parameters for different chemical environments, e.g. an aliphatic carbon may be treated differently from an aromatic ring carbon, etc. The efficiency and significance of the method decrease with increasing number of parameters, but the flexibility and accuracy increase with increasing number of parameters. Some expedients have often been adopted for parameter economy, like, for example, taking the parameters of the cross i–j interaction as arithmetic or geometrical means of the corresponding i–i and j–j parameters.

Ultimately, all parameters are to be determined by comparison with some kind of experimental data. These data are of three main types.

1. Thermodynamic data: the total lattice energy can be compared with the sublimation enthalpy of the crystal, a requirement that is met mainly by adjusting the depths of the atom–atom potential energy wells.
2. Structural data: the model must reproduce the observed crystal structure, meaning that a relaxation of the observed crystal structure under the action of the postulated potential must be small. This requirement is mainly met by adjusting the equilibrium distances in the atom–atom potential energy curves.
3. Vibrational and mechanical data: the potential should reproduce the measured lattice vibrational frequencies and the crystal strain and stress tensors. These requirements are much more difficult to satisfy, because they require a careful calibration of the curvature at the minimum and of the steepness of the repulsive branch.

Success in the optimization of the empirical potential parameters depends also on the accuracy of the experimental data. Accurate experimental measurements of sublimation enthalpies (Section 7.6) are time-consuming, and require comparatively large amounts of material and the measurement of vapor pressures of the solids, which can be as small as 10^{-7} atm, and correspond to very small amounts of sublimated matter. Even a minor decomposition of the material during the heating cycle may significantly alter the results. For that reason, experimental sublimation enthalpies should always be regarded with a critical attitude. Structural data are abundant and reliable, thanks to the accuracy of X-ray diffraction experiments, but the comparison with calculated equilibrium structures must take into account the fact that the calculation neglects the effects of kinetic energy, so that the relaxed crystal structure will usually have a higher density than the experimental one; as a rule of thumb, the percent density variation is close to 0.01–0.02 times the temperature of the crystal structure determination. Vibrational and mechanical experimental data are scarce and relatively inaccurate for organic crystals, and hence have seldom been successfully exploited in the calibration of empirical force fields. Lattice-vibrational frequencies are few in number and restricted to the narrow range between 30 and 150 cm^{-1} (section 6.3), so that fitting procedures are seldom significant. A more useful test is provided by checking that, at least, all calculated frequencies are real (a positive-definite hessian) so that the structure predicted by the force field is a true minimum in the potential energy hypersurface.

The weakest spot of empirical intermolecular potentials is perhaps the repulsive branch, which is always ill-determined since the repulsive regime is seldom accessed in equilibrium crystal structures at room pressure. The case is different in liquids, where the high molecular mobility occasionally brings molecules into collision. Atom–atom intermolecular energy curves for liquids are usually calibrated separately from those for crystals.

4.9.2 Distributed multipole methods

If the charge distribution is described by a set of distributed multipoles, as described in Section 4.2.3, the coulombic contributions to the intermolecular potential energy are calculated as multipole–multipole terms. The main disadvantage of even a rigorous distributed multipole model is that such a representation is still very localized, so that coulombic energies miss a large part of the penetration contribution. For use in a complete representation of intermolecular interactions, the dispersion, polarization, and repulsion terms must be evaluated separately by some semi-empirical or fully empirical method, for example the approximate atom–atom formulations of equations 4.38–4.39. This approach has been extensively exploited by S. L. Price and coworkers over the years, in applications to molecular crystals [48].

A conceptually similar approach based on a completely different computational scheme has been developed by Leiserowitz and Berkovitch-Yellin [49] using experimental electron densities obtained from experimental electron density distributions from accurate X-ray experiments. The Leiserowitz–Yellin potential is in the following atom–atom form:

$$V(r_{ij}) = A(r_{ij})^{-9} - B(r_{ij})^{-6} + \iint \delta\rho_i(\mathbf{r}_1)\delta\rho_j(\mathbf{r}_2)/|\mathbf{r}_1 - \mathbf{r}_2| d^3\mathbf{r}_1 d^3\mathbf{r}_2 \quad (4.45)$$

where r_{ij} is an internuclear distance, and $\delta\rho$ is the atomic deformation density, the difference between the electron density of an atom in the molecule and the electron density of the free atom; this requires a partitioning of the total density into atomic contributions, done according to Hirshfeld's recipe [14].

In a related approach, Spackman has developed a force field [50] that includes an empirically derived part for the dispersion and repulsion terms, while coulombic energy terms are treated in the following manner. The molecular electron distribution is divided into a promolecule term and a deformation term, as in the Hirshfeld definition. The product of the distributions in two interacting monomers A and B, as required in the evaluation of the coulombic energy, is then expanded as:

$$\rho_A \rho_B = \rho_{AT,A}\, \rho_{AT,B} + \rho_{AT,A}\, \delta\rho_B + \rho_{AT,B}\, \delta\rho_A + \delta\rho_A\, \delta\rho_B \quad (4.46)$$

where ρ_{AT} and $\delta\rho$ are the promolecule and deformation parts of the distribution, respectively. The first term in this expression is the promolecule coulombic energy. The next two terms represent the interaction of the deformation densities with promolecule atoms, and are called penetration terms (the meaning of the word is here different from the meaning it was given in Section 4.3). The last term is the net coulombic energy between charged atoms bound in the molecule. The actual energy calculations are done by expanding the distributions in terms of monopole, dipole, and quadrupole terms.

In the first applications the electron distributions were derived from ab initio molecular orbital calculations. In an interesting development, in further work [51] the experimental electron densities as obtained by X-ray diffraction work were used. In this sense the method is similar to the Leiserowitz–Yellin approach, but the analytical derivation is somewhat different. The electron density is written as

$$\rho(r,\theta,\phi) = \rho_{AT}(r) + \sum_{l,m} C_{lm} R_{nl} Y_{lm} \quad (4.47)$$

where R_{nl} are radial functions and Y_{lm} are spherical harmonics. The C_{lm} coefficients are determined by least-squares fit to the diffraction data; this fitting is the equivalent of the more usual fitting procedure between observed and calculated structure factors (see Section 5.4.4).

Equations 4.46 and 4.47 open the way to a calculation of coulombic energies in crystals directly from experimental electron densities, and represent a connection between X-ray diffraction and energies, a step forward from the usual interpretation of diffraction data just in terms of molecular structure. A rich literature has developed along these lines [52].

4.9.3 Other density-based methods

The availability of cheap and accurate charge distributions in numerical form for medium-size and large molecules has prompted the idea of using these distributions for a calculation of the first three terms in equation 4.34 by numerical integration. Just as coulombic energies are obtained by integration over charge units, polarization energies are obtained over electric field and polarization units, and dispersion energies over London-type dispersion units. Repulsion is treated as proportional to density overlap, also calculated by numerical integration. The limiting factor is that the summations include N^2 distance terms between charge units, where for an organic molecule like benzene N can be of the order of 20,000. This approach, called the semi-classical density sums (SCDS) approach [17], and more familiarly known as the "Pixel approach", is particularly suited for the analysis of molecular packing in crystals. The point will be taken up again in Section 12.2.

4.10 Ad hoc or transferable? Force field fitting from ab initio calculations

The value of a scientific model that includes some disposable numerical parameters increases with the ratio of the number of explained facts to the number of necessary parameters. In this typically academic perspective, transferability is the keyword and the ideal scientist explains all the universe from first principles as the number of parameters tends to zero. Even in this hyperuranean view, one may ask where is the divide between first principles and adaptable parameters: for example, the numbers, 6 and 12, in the atom–atom R^{-6} and R^{-12} potentials of equation 4.38 are of a clear empirical origin, but the R^{-6} term in London's dispersion energy seems to come from quantum chemical laws; eventually, one may ask if the distance exponents in Coulomb's law and Newton's gravitational law are first principles or parameters. Experiment tells that they do not differ from 1 at least at the present level of accuracy, and yet the question is a legitimate one. In a different attitude, a commercial company might wish to optimize the production of a given chemical by predicting its crystalline properties. The knowledge of a certain property may help in the design of a batch reactor, with a saving or increased profit of many million dollars a year. In this case, anything goes and one should not be too fussy about the philosophy, provided that the results are good.

Clearly, the issues of first principles and transferability are more pertinent to an essentially academic book such as the present one, while production details are more aptly included in internal technical manuals (heaven knows how much vital information is secreted in confidential company reports and thus hidden from the scientific community). The matter can be reformulated as follows: the value of a parametric theory is directly proportional to the extension of the factual landscape to which it applies, and inversely proportional to the effort made in its development. Atom–atom formulations cannot yet be beaten in this respect. They apply with success to a wide scope of different problems and can give reliable theoretical estimates of crystal sublimation enthalpies and liquid evaporation enthalpies. And yet they are limited by their intrinsically scarce adherence to first principles, with a lack of contact between parameters and the implied physics. For these reasons, the future is not in their direction, and different paths must be sought, presumably in the direction of closer relationships between empirical force field parameters and quantum chemical data, considering that many molecular properties are nowadays more cheaply and reliably calculated by ab initio quantum chemistry than measured in expensive and awkward experiments.

In a typical example [53], high quality ab initio calculations were carried out for 111 different geometries of approach in methanol dimers and trimers, and the resulting energies were partitioned over well defined coulombic, polarization, dispersion, and repulsion terms. The electrostatic potentials were fitted by atomic multipole moments. All these data were converged into an empirically fitted intermolecular force field, in which the separate terms could be assigned a definite physical meaning according to the partitioned energies from which they were derived. The computational effort in the derivation of the force field is, however, massive.

References and Notes to Chapter 4

[1] Stone, A. J. *The Theory of Intermolecular Forces*, 1996, Clarendon Press, Oxford (reprinted with corrections, 2000).
[2] Rigby, M.; Smith, E. B.; Wakeham, W. A.; Maitland, G. C. *The Forces between Molecules*, 1986, Clarendon Press, Oxford.
[3] Israelachvili, J. N. *Intermolecular and Surface Forces*, 1985, Academic Press, London.
[4] An excellent historical overview of the subject of intermolecular forces is in Rowlinson, J. S. *Cohesion, A Scientific History of Intermolecular Forces*, 2002, Cambridge University Press, Cambridge.
[5] Williams, D. E. Net atomic charge and multipole models for the ab initio molecular electric potential, in *Reviews in Computational Chemistry*, edited by K. B. Lipkowitz and D. B. Boyd, Vol. 2, 1991, VCH, New York pp. 219–271.
[6] For early work in this direction see e.g. Weiner, P. K.; Langridge, R.; Blaney, J. M.; Schaeffer R.; Kollman, P. A. Electrostatic potential molecular surfaces, *Proc. Natl. Acad. Sci. USA* 1982, **79**, 709–713; Purvis, G. D.; Culberson, C. On the graphical display of molecular electrostatic force-fields and gradients of the electron density, *J. Mol. Graphics* 1986, **4**, 88–92; Hermsmeier, M. A.; Gund, T. M. A graphical representation of the electrostatic potential and electric field on a molecular surface, *J. Mol. Graphics* 1989, **7**, 150–152; Honig, B.; Nicholls, A. Classical electrostatics in biology and chemistry, *Science* 1995, **268**, 1144–1149.
[7] Kauzmann, W. *Quantum Chemistry. An Introduction*, 1957, Academic Press, New York, pp. 95–96. This form is the general solution of Poisson's equation outside a cluster of positively and negatively charged points. See a thorough discussion of the definition of the various moments in electrostatic expansions in an excellent review: Dykstra, C. E. Electrostatic interaction potentials in molecular force fields, *Chem. Rev.* 1993, **93**, 2339–2353.

[8] Ref [1], p.106; Fowler, P. W.; Buckingham, A. D. Central or distributed multipole moments? Electrostatic models of aromatic dimers, *Chem. Phys. Letters* 1991, **176**, 11–18.

[9] Stone, A. J., Distributed multipole analysis, or how to describe a molecular charge distribution, *Chem. Phys. Letters* 1981, **83**, 233–239; Stone, A. J.; Alderton, M. Distributed multipole analysis. Methods and applications, *Mol. Phys.* 1985, **56**, 1047–1064.

[10] Ref. [1], p.108

[11] Faerman, C. H.; Price, S. L. A transferable distributed multipole model for the electrostatic interactions of peptides and amides, *J. Am. Chem. Soc.* 1990, **112**, 4915–4926.

[12] A critical evaluation of the various methods for the derivation of point charges from electrostatic potentials is in Breneman, C. M.; Wiberg, K. B. Determining atom-centered monopoles from molecular electrostatic potentials. The need for high sampling density in formamide conformational analysis, *J. Comp. Chem.* 1990, **11**, 361–373.

[13] Hirshfeld, F. L. Bonded-atom fragments for describing molecular charge densities, *Theor. Chim. Acta* 1977, **44**, 129–138.

[14] Gavezzotti, A.; Filippini, G. Geometry of the intermolecular X—H···Y (X,Y=N,O) hydrogen bond and the calibration of empirical hydrogen-bond potentials, *J. Phys. Chem.* 1994, **98**, 4831–4837.

[15] Stouch, T. R.; Williams, D. E. Conformational dependence of electrostatic potential-derived charges: studies of the fitting procedure, *J. Comp. Chem.* 1993, **14**, 858–866.

[16] Koch, U.; Stone, A. J. Conformational dependence of the molecular charge distribution and its influence on intermolecular interactions, *J. Chem. Soc. Faraday Trans.* 1996, **92**, 1701–1708.

[17] Gavezzotti, A. Calculation of intermolecular interaction energies by direct numerical integration over electron densities. I. Electrostatic and polarization energies in molecular crystals, *J. Phys. Chem.* 2002, **106**, 4145–4154.

[18] See e.g. ref. [1], pp. 39–40.

[19] Ref [2], p. 8.

[20] Ref. [3], p.53.

[21] Miller, K. J. Additivity methods in molecular polarizability, *J. Am. Chem. Soc.* 1990, **112**, 8533–8542.

[22] Ref. [3], pp. 56–57.

[23] A standard theorem of electrostatics states that the field due to several charges is the vector sum of the field due to each of the contributing charges: Feynman, R. P. *The Feynman Lectures on Physics, Commemorative Issue*, Volume II, Chapter 11, 1989, Addison-Wesley, Reading, Mass.

[24] Ref. [2], p.9.

[25] Ref. [4], pp. 245ff.

[26] A simple demonstration that the quantum mechanical coupling of oscillating charges produces a non-zero stabilizing term is the Drude model; see ref. [2], pp. 32–34. That some physical facts could only be explained by quantum mechanics (quantal "fiat", see ref. [4], p. 200) was an epistemological scandal to many physicists of the early twentieth century.

[27] "Correlation energy" is ultimately a convenient container for all that escapes from first-order coulombic and polarization terms. See Bickelhaupt, F. M.; Baerends, E. J. Kohn-Sham density functional theory: predicting and understanding chemistry, *Reviews in Computational Chemistry*, Vol. 15, edited by K. B. Lipkowitz and D. B. Boyd, 2000, Wiley-VCH, New York p.11: "Is electron correlation (not defined in a statistical sense, but according to either the quantum chemical or the DFT definition) a true physical phenomenon? It is a man-made concept, related to the introduction of a convenient trial wavefunction, that is useful for our communication and understanding."

[28] London, F. The general theory of molecular forces, *Trans.Farady Soc.* 1937, **33**, 8–26: "These very quickly varying dipoles, represented by the zero-point motion of a molecule, produce an electric field and act upon the polarisability of the other molecule and produce there induced dipoles, which are in phase and in interaction with the instantaneous dipoles producing them ... we may imagine a molecule in a state k as represented by an orchestra of periodic dipoles μ_{kl} which correspond with the frequencies $v_{kl} = (E_l - E_k)/h$ of (not forbidden) transitions to the states l. These "oscillator strengths", μ_{kl}, are the same quantities which appear in the "dispersion formula" which gives the polarisability of the molecule in the state k when acted on by an alternating field of the frequency v." See also Maitland, G. C.; Rigby, M.; Smith, E. B.; Wakeham, W. A. *Intermolecular Forces: Their Origin and Determination*, 1981, Clarendon Press, Oxford.

[29] Ref. [1], pp. 79–82.

[30] Exchange is not a real physical process. Exchange energies occur because the molecular wavefunction is written in terms of a determinantal combination of atomic wavefunctions, like equation 3.45.

[31] The exchange-repulsion term vanishes when electrons have opposite spin, because the overlap integral is zero over spin-orbitals (see Section 3.4, the discussion around equation 3.38). This corresponds to the formation of a chemical bond.

[32] Bickelhaupt, F. M.; Baerends, E. J. Kohn–Sham density functional theory: predicting and understanding chemistry, in *Reviews in Computational Chemistry*, Vol. 15, edited by K. B. Lipkowitz and D. B. Boyd, 2000, Wiley-VCH, New York, pp. 14–19.

[33] Nobeli, I.; Price, S. L.; Wheatley, R. J.; Use of molecular overlap to predict intermolecular repulsion in N ... H—O hydrogen bonds, *Mol. Phys.* 1998, **95**, 525–537. The relations between repulsion and overlap can be probed by molecular beam experiments; see e.g. Kita, S.; Noda, K.; Inouye, H. Repulsive potentials

for Cl–R and Br–R (R = He, Ne and Ar) derived from beam experiments, *J. Chem. Phys.* 1976, **64**, 3446–3449.

[34] Quoting from Hirshfeld [13]: "Chemists have acquired, with experience, a reasonably serviceable notion of the charge distributions in the molecules they work with. In many instances a look at the structural formula of a molecule is enough to suggest which regions are electron-rich and so vulnerable to electrophilic attack and which are more likely to attract nucleophilic reagents." This statement reveals the analogy between coulombic interactions in crystals and in reaction pathways.

[35] For Inter Molecular Perturbation Theory (IMPT) see Hayes, I. C.; Stone, A. J. An intermolecular perturbation theory for the region of moderate overlap, *Mol. Phys.* 1984, **53**, 83–105; papers of this kind, however, contain a large amount of theoretical and mathematical detail and are not transparent to the uninitiated. For Symmetry-Adapted Perturbation Theory (SAPT) see e.g. Bukowski, R.; Szalewicz, K.; Chabalovski, C. F. Ab initio interaction potentials for simulations of dinitramine solutions in supercritical carbon dioxide with cosolvents, *J. Phys. Chem.* 1999, **A103**, 7322–7340, and references therein. The Morokuma decomposition scheme is described in Kitaura, K.; Morokuma, K. A new energy decomposition scheme for molecular interactions within the Hartree–Fock approximation, *Int. J. Quantum Chem.* 1976, **10**, 325–340.

[36] For some contemporary examples of high level quantum chemical calculations on molecular dimers, see the literature cited in Gavezzotti, A. Quantitative ranking of crystal packing modes by systematic calculations on potential energies and vibrational amplitudes of molecular dimers, *J. Chem. Theor. Comp.* 2005, **1**, 834–840.

[37] Sinnokrot, M. O.; Valeev, E. F.; Sherrill, C. D. Estimates of the ab initio limit for π–π interactions: the benzene dimer, *J. Am. Chem. Soc.* 2002, **124**, 10887–10893, and references therein.

[38] Wu, X.; Vargas, M. C.; Nayak, S.; Lotrich, V.; Scoles, G. Towards extending the applicability of density functional theory to weakly bound systems, *J. Chem. Phys.* 2001, **115**, 8748–8757. Byrd, E. F. C.; Scuseria, G. E.; Chabalowski, C. F An ab initio study of solid nitromethane, HMX, RDX, and CL20: successes and failures of DFT, *J. Phys, Chem. B* 2004, **108**, 13100–13106. It is clearly said that DFT functionals are unsuitable for the prediction of the lattice parameters of organic compounds.

[39] Grimme, S. Accurate description of van der Waals complexes by density functional theory including empirical corrections, *J. Comput. Chem.* 2004, **25**, 1463–1473.

[40] Chemists like to use distances in Å and charges in electrons. The following is a list of factors for the conversion of results of electrostatic calculations to SI units: (a) coulombic energies: the conversion factor is 1389.355 (two 1-electron charges 1 Å apart interact with an energy of 1389.355 kJ mol^{-1}). (b) Electric field: the conversion factor is 1.439965×10^{11} (a 1-electron charge generates

at a distance of 1 Å an electric field $\varepsilon = 1.439965 \times 10^{11}$ V m^{-1}). (c) Polarization energies: with volume polarizabilities in Å^3, and $E_{POL} = -1/2(4\pi\varepsilon°)\alpha\varepsilon^2$, the overall conversion factor is $8.60003 \times 10^{-18}/(4\pi\varepsilon°)^2$, or 694.676 (a 1-electron charge at a distance of 1 Å from a point whose polarizability is 1 in $4\pi\varepsilon°$Å^3 units generates a polarization energy of -694.676 kJ mol^{-1}). (d) Dispersion energies: using mixed units of distances in Å, polarizabilities in Å^3, and the ionization potential in hartree as it comes out of a MO calculation, the proper conversion factor is E_{DISP} (kJ mol^{-1}) = 1969.126 E_{DISP} (mixed units).

1 a.u. of energy = 1 hartree = $4.3597482 \times 10^{-18}$ J; 1 a.u. of distance = the Bohr radius = 5.29177 10^{-11} m; Avogadro's constant is 6.02214×10^{23} mol^{-1}.

[41] Meot-Ner, M. (Mautner), The ionic hydrogen bond and ion solvation. *J. Am. Chem. Soc.* 1984, **106**, 1257–1272.

[42] Pimentel, G. C.; McClellan, A. L. *The Hydrogen Bond*, Freeman, San Francisco 1960; Gilli, G.; Gilli, P. Towards an unified hydrogen-bond theory, *J. Mol. Struct.* 2000, **522**, 1–15, and references therein; Jeffrey, G. A.; Saenger, W. *Hydrogen Bonding in Biological Structures*, Springer-Verlag, Berlin 1991; Steiner, Th. The hydrogen bond in the solid state, *Angew. Chem. Int. Ed. Eng.* 2002, **41**, 48–76; Taylor, R.; Kennard, O. Hydrogen-bond geometry in organic crystals, *Acc. Chem. Res.* 1984, **17**, 320–326; Etter, M. C. Encoding and decoding hydrogen-bond patterns of organic compounds, *Acc. Chem. Res.* 1990, **23**, 120–126; Gilli, P.; Bertolasi, V.; Ferretti, V.; Gilli, G. Covalent nature of the strong homonuclear hydrogen bond. Study of the O—H ··· O system by crystal structure correlation methods, *J. Am. Chem. Soc.* 1994, **116**, 909–915.

[43] See for example Mirsky, K. Interatomic potential functions for hydrocarbons from crystal data: transferability of the empirical parameters, *Acta Cryst.* 1976, **A32**, 199–207 and references therein; Kitaigorodski, A. I. in *Advances in Structure Research by Diffraction Methods*, Vol.3, edited by R. Brill and R. Mason, 1970, Pergamon Press, Oxford, pp. 173–247.

[44] Williams, D. E. *J. Chem. Phys.* 1967, **47**, 4680–4684; Williams, D. E.; Starr, T. L. Calculation of the crystal structures of hydrocarbons by molecular packing analysis, *Comp. Chem.* 1977, **1**, 173–177; Hsu, L.-Y.; Williams, D. E. Intermolecular potential-function models for crystalline perchlorohydrocarbons, *Acta Cryst.* 1980, **A36**, 277–281; Cox, S. R.; Hsu, L.-Y.; Williams, D. E. Nonbonded potential function models for crystalline oxohydrocarbons, *Acta Cryst.* 1981, **A37**, 293–301; Williams, D. E.; Cox, S. R. Nonbonded potentials for azahydrocarbons: the importance of coulombic interactions, *Acta Cryst.* 1984, **B40**, 404–417; Williams, D. E.; Houpt, D. J. Fluorine nonbonded potentials derived from crystalline perfluorocarbons, *Acta Cryst.* 1986, **B42**, 286–295. The Williams force field requires the calculation of atomic point and site charges for each molecule. The hydrocarbon force field was later improved (Williams, D. E. Improved intermolecular force field for crystalline hydrocarbons containing four- or three-coordinates carbon, *J. Mol. Struct.* 1999, 485–486, 321–347)

using different parameters for carbon atoms in different environments, and site charges at methylene groups or aromatic ring centers. The original formulation did not consider hydrogen bonding, which was included in further developments (Williams, D. E. Improved intermolecular force field for crystalline oxohydrocarbons including O—H···O hydrogen bonding, *J. Comp. Chem.* 2001, **22**, 1–20; Williams, D. E. Improved intermolecular force field for molecules contaning H, C, N and O atoms, with application to nucleoside and peptide crystals, *J. Comp. Chem.* 2001, **22**, 1154–1166). These latter papers propose a more flexible force field formulation, at the expense of an increase in number of parameters.

[45] Pertsin, A. J.; Kitaigorodski, A. I. *The Atom–Atom Potential Method*, 1987, Springer-Verlag, Berlin. See also Kitaigorodski, A. I. *Molecular Crystals and Molecules,* 1973, Academic Press, New York.

[46] Ref [45], Chapter 3: "...the atom–atom potential method represents a variational treatment using...A's, B's and C's as variational parameters. Clearly, there is no reason to attach any physical meaning to the parameters ... there is no reason to regard the sum of the $-AR^{-6}$ terms as the dispersion energy ... and the sum of $qq'r^{-1}$ terms as the electrostatic energy".

[47] Gavezzotti, A.; Filippini, G. Energetic aspects of crystal packing: experiment and computer simulations, in *Theoretical Aspects and Computer Modelling of the Molecular Solid State*, edited by A. Gavezzotti, Wiley & Sons, Chichester 1997.

[48] Willock, D. J.; Price, S. L.; Leslie, M.; Catlow, C. R. A. The relaxation of molecular crystal structures using a distributed multipole electrostatic model, *J. Comp. Chem.* 1995, **16**, 628–647; Price, S. L. Applications of realistic electrostatic modeling to molecules in complexes, solids and proteins, *J. Chem. Soc. Faraday Trans.* 1996, **92**, 2997–3008; Beyer, T.; Day, G. M.; Price, S. L. The prediction, morphology, and mechanical properties of the polymorphs of paracetamol, *J. Am. Chem. Soc.* 2001, **123**, 5086–5094.

[49] Berkovitch-Yellin, Z.; Leiserowitz, L. Atom–atom potential analysis of the packing characteristics of carboxylic acids. A study based on experimental electron density distributions, *J. Am. Chem. Soc.* 1982, **104**, 4052–4064.

[50] Spackman, M. A. Atom–atom potentials via electron gas theory, *J. Chem. Phys.* 1986, **85**, 6579–6586; Spackman, M. A. A simple quantitative model of hydrogen bonding, *J. Chem. Phys.* 1986, **85**, 6587–6601.

[51] Spackman, M. A.; Weber, H. P.; Craven, B. M. Energies of molecular interactions from Bragg diffraction data, *J. Am. Chem. Soc.* 1988, **110**, 775–782.

[52] Koritsanszky, T. S.; Coppens, P. Chemical applications of X-ray charge-density analysis, *Chem. Revs.* 2001, **101**, 1583–1627; Volkov, A.; Coppens, P. Calculation of electrostatic interaction energies in molecular dimers from atomic multipole moments obtained by different methods of electron density partitioning, *J. Comp. Chem.* 2004, **25**, 921–934; Suponitsky, K. Y.; Tsirelson, V. G.; Feil, D. Electron-density-based calculations of intermolecular energy: the case of urea, *Acta Cryst.* 1999, **A55**, 821–827.

[53] Mooij, W. T. M.; van Duijneveldt, F. B.; van Duijneveldt-van de Rijdt, J. G. C. M.; van Eijck, B. P. Transferable ab initio intermolecular potentials. 1. Derivation from methanol dimer and trimer calculations, *J. Phys. Chem.* A 1999, **103**, 9872–9882; Moij, W. T. M.; van Eijck, B. P.; Kroon, J. Transferable ab initio intermolecular potentials. 2. Validation and application to crystal structure prediction, *J. Phys. Chem. A* 1999, **103**, 9883–9890.

5

Crystal symmetry and X-ray diffraction

When electromagnetic radiation passes through a crystal (or matter in general), the electrons are perturbed by the rapidly oscillating electric field and are set into oscillation about their nuclei, with a frequency identical to that of the incident radiation. According to electromagnetic theory an oscillating dipole acts as a source of an electromagnetic wave. Thus, each electron in the medium acts as a source of a radiation that travels outwards with a spherical wave front. In other words, the incident radiation is scattered by the medium without alteration in its frequency. This is coherent scattering.

> Dunitz, J.D. *X-Ray analysis and the Structure of Organic Molecules*, 1995, 2nd corrected reprint, Verlag Helvetica Chimica Acta, Basel, p. 25.

Scatter, v.t. & i.. 1. send, go, in different directions ... **2.** throw or put in various directions ...

> Hornby, A.S.; Gatenby, E.V.; Wakefield, H.
> *The Advanced Learner's Dictionary of Current English*, 1973,
> Oxford University Press, London.

5.1 A structural view of crystal symmetry: bottom-up crystallography

The study of crystal symmetry started as a study of the symmetry of macroscopic crystalline objects, mostly minerals. When it was realized that external symmetry reflected the internal symmetry of the disposition of atoms and molecules within the crystal, mathematical methods were applied to the classification of all possible combinations of symmetry operations under the condition of periodic translational repetition, within the framework of group theory. As a consequence of this historical development, the teaching of crystallography has always proceeded – and continues to proceed – through a study of crystal systems, of Bravais lattices and of space group theory. The proceedings are thus rigorous but somewhat abstract, and chemistry students understandably do find crystallography a dull if not a repulsive subject.

The study of molecular crystals from a structural point of view requires neither abstract lattices nor group theory. A molecular crystal in flesh and bones is an aggregate of molecules that, at least within finite domains of size much larger than the size of one molecule, arrange themselves according to a very small number of periodic symmetry conditions under the action of an intermolecular potential. An organic crystal can therefore be studied by defining just a handful of such symmetry conditions and analyzing how they act and combine into the ten to fifteen periodic symmetries that are relevant to organic crystal chemistry.

Consider a molecular object composed of a certain number of atoms bound together into a known molecular geometry. As discussed in Section 1.2, the molecule has a surface envelope that contains all its nuclei and nearly all of its electrons. According to the definition of symmetry given in Chapter 1, a symmetry operation is a geometrical operation that transforms an object into itself without distortion. In an organic crystal, symmetry operations transform a reference molecule into many surrounding molecules, and all these operations must be such that the overlap between the molecular envelopes of the original object and of the symmetry-related objects be null or very small. Besides, a crystalline system is a periodic system because its structural motifs repeat themselves in space by translations with given periods. We will see that these two conditions are enough to derive most of the principles of lattice and space group theories, that is, to build a bottom-up crystallography that starts from molecules to arrive at space groups.

Since translation is so important in crystals, consider first what happens if a molecular object is translated in two directions in space to form a periodic structure defined by two translational periods (Fig. 5.1). There are many different choices of the unit cell; the cell defined by periods a and b is preferable over the cell defined by a' and b because the cell angle ($a0b$) is closer to 90° than the cell angle ($a'0b$). The unit cell defined by a' and b' is called a non-primitive (centered) unit cell, because it encloses one molecule related by pure translation by a sub-multiple of the cell periodicity. The choice of such a cell introduces a quite unnecessary complication in this case.

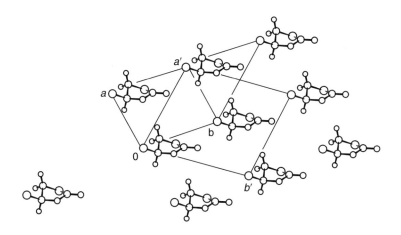

Fig. 5.1. A molecular object (left) packs in a layer (right). Different choices of the unit cell are shown in this two-dimensional example: ($0ab$), primitive; ($0a'b$), more oblique primitive; ($0a'b'$), centered. The choice of the origin is arbitrary.

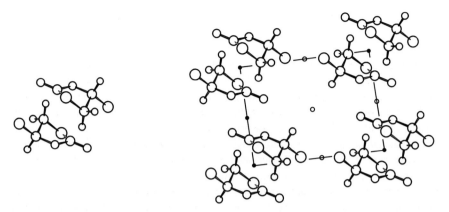

Fig. 5.2. The same scheme as in Fig. 5.1, except that the repeating unit is not a single molecule but a pair of molecules related by an inversion center. In the figure at the right, the small circles denote the positions of the centers of symmetry in the crystal cell. It is strongly advisable to choose the cell origin in the center of symmetry.

Extension to the third dimension, with periods a, b, and c, is obvious: the cell angles are called α (the bc angle), β (the ac angle), and γ (the ab angle). The collection of points at the corners of the unit cell parallelepiped, repeated by translation, form what is called the crystal lattice. There is no preferred location and there are no restrictions on the position of the origin; there are no restrictions to the length of the three translational periods, nor to the angles between the unit cell translation vectors. Such a crystal system is called triclinic.

Consider now a molecular pair formed over a center of symmetry, symbolized here by the capital letter I, and apply the same translation procedure to the molecular pair (Fig. 5.2). It is now intuitively obvious that the origin of the cell system must be chosen in the one very unique point in the whole structure, the center of symmetry. It turns out that with this choice not only does the visual picture of the structure becomes clearer and simpler, but also all the mathematical manipulations pertaining to crystal structure analysis become much easier. The space-group symbols for the pure translation of a single object or of a centrosymmetric pair are $P1$ and $P\bar{1}$ respectively, where the letter P stands for "primitive cell" and the symbol $\bar{1}$ is the traditional crystallographic symbol for the inversion operator. The contents of the unit cell is the portion of crystalline matter that is repeated in space by pure translation.

The $P1$ unit cell contains one molecule, while the $P\bar{1}$ unit cell contains two molecules altogether, although not necessarily one or two connected molecules, because the cell borders may cut through some molecular units. If the molecular object itself has a center of symmetry, that is, the inversion center is a point-group symmetry element, translation of a centrosymmetric molecule generates a $P\bar{1}$ space group where the cell contents is one molecule while the symmetry-independent part, the asymmetric unit, is half a molecule. The number of molecules in the unit cell is called Z, the number of molecules in the asymmetric unit is called Z', so the two cases of $P\bar{1}$ space group are denoted by $P\bar{1}, Z = 2, Z' = 1$, and by $P\bar{1}, Z = 1, Z' = 1/2$, respectively.

Another notable fact in Fig. 5.2 is that at the middle of each of the translational periods, as well as at the center of the faces and of the body of the unit cell, new inversion centers are generated by the translation operation. Thus, the symmetry of the structure repeats itself periodically throughout the crystal, i.e. the overall structure is translationally periodical as a true crystal should be.

Consider now the molecular dimers formed over a twofold rotation axis or a mirror plane (Fig. 5.3). One could follow the same line of reasoning as for the dimer over the inversion center, and build a layer or a three-dimensional structure using pure translation. This leads to arrays in which large empty spaces are left between molecular envelopes, not a favourable condition from the point of view of cohesive energies, because dispersion forces (section 4.5) are short-range, so that molecules must be as close as possible to one another. If the twofold axis or mirror plane operations are followed by a translation of the rotated object, these translation-corrected symmetry operations cause the molecular envelopes to interlock in a much more compact manner. They are called a twofold screw axis, symbol 2_1 (briefly, in the following, S) and a glide plane, whose symbol is the letter corresponding to the cell direction to which the mirror plane is parallel, a, b, or c (in the following, for coherence with previously defined symbols, the unique capital symbol G will be used). These symmetry operations generate infinite molecular ribbons along the screw or

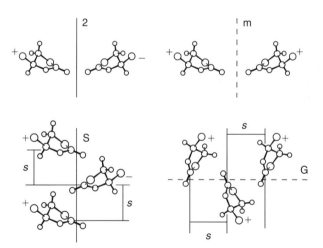

Fig. 5.3. Top to bottom, left to right: a pair of molecules related by a twofold axis (2), or by a mirror plane (m) perpendicular to the plane of the sheet, whose trace is denoted by a broken line. Strings of molecules along a screw axis (S) or along a glide plane (G) whose trace is denoted by a broken line. + and − denote symmetry-related atoms that point towards and away from the observer, respectively. s is the screw or glide displacement.

glide displacement direction: if the length of the screw or glide displacement is s, since double application of the twofold or mirror symmetry operators brings back the molecule to its original orientation, every third molecule in the ribbon is related by a pure translation of period $2s$. Thus, the crystal that contains a screw or a glide operator must also contain a cell periodicity equal to twice the intrinsic screw or glide displacement.

A screw ribbon along y can be translated, say along z, to give an array of neighboring ribbons, or a layer. However, as Fig. 5.4(a) shows, the direction of translation can only be perpendicular to the direction of the screw axis, because only in this case does a new screw axis S′ survive at half translation and the symmetry becomes translationally periodic. The yz, or α angle must be equal to 90°. A translation of the layer (along x) produces a three-dimensional crystal structure: the direction of this second translation must again be perpendicular to y, the γ angle must be 90°, but the β angle has no restrictions (Fig. 5.4(b)). The introduction of twofold or mirror symmetry changes the crystal system to monoclinic, because two of the three angles between cell translation vectors must be equal to 90°. No explanation is needed to understand why the space groups that contain only one screw axis or one glide plane are called $P2_1$ and Pc, respectively. In $P2_1$, the origin is located at any point along the screw axis, for the

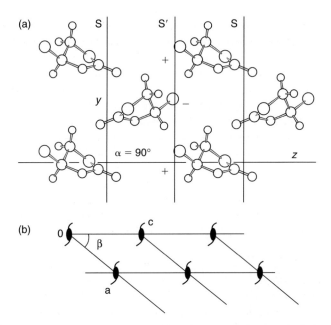

Fig. 5.4. (a) The S string in Fig. 5.3 translated in a direction (z) perpendicular to the direction of the screw axis. Angle α must be 90° otherwise the screw axis S′ disappears and the symmetry is no longer periodic. + and − denote symmetry-related atoms that point towards and away from the observer, respectively. (b) A view perpendicular to that in (a): the black helix symbols denote the traces of the screw axes.

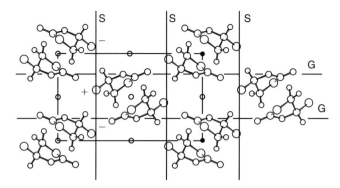

Fig. 5.5. The same scheme as in Fig. 5.4(a), except that the repeating unit is not one molecule but a pair of molecules related by a center of symmetry. Small circles denote inversion centers, S denotes screw axes, G denotes the trace of glide planes perpendicular to the plane of the page. + and − denote symmetry-related atoms that point towards and away from the observer, respectively.

same reasons of convenience that have been invoked for space group $P\bar{1}$. This leaves one degree of indeterminacy for the location of the origin.

Consider now again the I dimer of Fig. 5.2. If the whole dimer is carried through the screw symmetry generation and translation procedures that have been described above for a single molecule, the resulting layer (Fig. 5.5) has both I and S symmetry. The combination of these two operators generates a third symmetry operator, a glide plane perpendicular to the direction of the screw axis. The b periodicity is fixed as twice the intrinsic screw displacement, while the c periodicity is fixed as twice the intrinsic glide displacement, or four times the distance between the center of symmetry and the screw axis. As before, the third translation, the one along a that involves the whole bc layer and generates the full crystal structure, is perpendicular to the b direction. Angle β can take any value, as before, and this space group still belongs in the monoclinic system. This space group is called $P2_1/c$, the slash being shorthand to indicate that the 2_1 axis is perpendicular to the glide plane. The number of symmetry-related molecular objects in the unit cell (the number of equivalent positions) is 4, and the number of molecules in the cell is $Z = 4$, or $Z = 2$ if the molecule has a point-group inversion center.

At this point the reader may have noticed that the backbone of Fig. 5.5, with the indication and reciprocal location of the symmetry operators, is quite similar to the scheme that appears in the International Tables for Crystallography for space group $P2_1/c$. These schemes usually look rather obscure to the beginner, and generally have a discouraging aspect. The above derivation, however, should disclose a more viable approach and should explain the reasons why that scheme must be what it is.

Figure 5.6 shows what happens if the ribbon generated by a screw axis (Fig. 5.3) is subjected to a second screw operation along an axis perpendicular to the first one. The result is the generation of a third axis in the direction perpendicular to the first

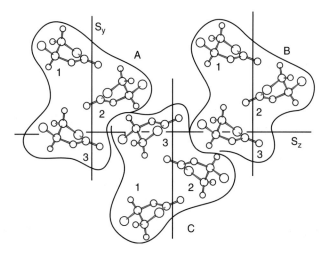

Fig. 5.6. The S repetition in Fig. 5.3, along z, for a triplet of molecules related by a screw operation along y. Triplets are denoted by A, B, and C, and molecules within each triplet by 1, 2, and 3. The product of these two screw operations is a third screw axis, S_x, whose trace is denoted by the helix symbol between the A2 and C3 molecules. Each symmetry operation can be denoted as $Xp(M)Yq$, where Xp is the starting molecule, M is the operator and Yq is the symmetry-related molecule. The following relationships hold: $A3(S_y)A2(S_y)A1$; $A(S_z)C(S_z)B$; $A2(S_x)C3$; $A1(S_x)C2$; $A1(S_yS_z)C2$.

two, and the resulting space group is called $P2_12_12_1$, with four equivalent positions. As three-dimensional propagation of screw ribbons must always be in a direction perpendicular to the direction of the screw axis, in this space group the angle between any two cell axes must be 90°. The crystal system is called orthorhombic. If the same procedure is applied to an I pair, three glide planes result as well, and the space group is called *Pbca*, with eight equivalent positions. At this point a figure with a representation of the whole set of symmetry operators becomes rather complicated. If the inversion center is a point-group operation, the result is $Pbca, Z = 4, Z' = 1/2$, which is the equivalent of the $P2_12_12_1$ space group for a centrosymmetric molecule. The extension of these same concepts to other combinations of T, I, S, and G operators generates all other space groups in the monoclinic and orthorhombic systems. Calling O the total number of symmetry operators, the number of equivalent positions in the space group, then $Z' = Z/O$.

A further step along these lines is to use some higher symmetry operators, such as threefold or fourfold rotation axes, and try the same combination of rotation and translation, preserving the periodic repeat of symmetry operators. With a fourfold axis, two cell parameters must be equal, and all cell angles must be 90°; this is called the tetragonal system. With a threefold axis, one cell angle must be 120°, and this is the trigonal system. When still higher symmetry operations are present one gets the cubic system, where all cell parameters must be equal. However, since close packing

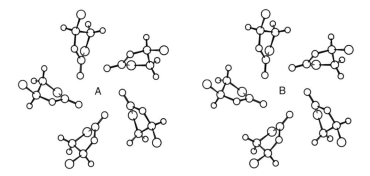

Fig. 5.7. An attempt at constructing a translationally periodic structure from a quintet of molecules related by a fivefold rotation axis located at points A and B. There is no symmetry relationship between molecules of quintet A and those of quintet B. The structure is not periodic and cannot be made periodic by any translational displacement, not even to the corners of a pentagon.

(Section 8.6.2) is easily achieved in systems up to orthorhombic, organic molecules very seldom if ever choose to pack in higher crystal systems.

Figure 5.7 attempts a visualization of what happens if one tries to use a fivefold rotation axis to build a crystal. There is no way in which different molecules may be related by a fivefold rotation in a real crystal while preserving the structural periodicity of a proper crystal. The same is true for seven-fold rotation. This is why no space group may contain these symmetry operators. Of course molecules with fivefold symmetry, like ferrocene, do crystallize, but the fivefold symmetry is not crystallographic. The analysis of all the possible and independent combinations of symmetry operators was carried out a century ago and has led to the definition of the 230 space groups.

5.2 Space group symmetry and its mathematical representation

In X-ray crystallography atomic coordinates are usually given as fractions of the unit cell parameters in a reference system that coincides with the cell axes system. This choice has the disadvantage of accepting an oblique reference frame, at least in monoclinic and triclinic systems. The advantages will soon become clear.

Given the position of an atom as $\mathbf{x} = (x, y, z)$, the translation of one period along the cell axis a can be written as

$$(x, y, z) \rightarrow (x + 1, y, z) \tag{5.1}$$

and the entire set of translation-repeated atoms in the infinite periodic crystal can be written as

$$\mathbf{x} = (x, y, z) \rightarrow \mathbf{x}' = (x + n, y + m, z + p) \tag{5.2}$$

with n, m, p integers, $-\infty < n, m, p < +\infty$. If the origin of the reference frame is located in a center of symmetry, the inversion center operation and successive translation can be schematized as:

$$\mathbf{x} = (x, y, z) \rightarrow \mathbf{x}' = (-x, -y, -z), (x+n, y+m, z+p), (-x+n, -y+m, -z+p) \tag{5.3}$$

For a screw axis operation, if the origin is placed on the screw axis running in the y direction, the corresponding coordinate transformation for a single and a double application of the operation are:

$$\mathbf{x} = (x, y, z) \rightarrow \mathbf{x}' = (-x, y+1/2, -z) \rightarrow x, y+1, z \tag{5.4}$$

The double application of a screw axis operation is equivalent to one cell translation along the screw axis direction.

Consider now the joint application of a center of symmetry and of a screw operation (see Fig. 5.8). The origin is kept in the center of symmetry, while the screw axis is at a distance s along the z axis. The two operations can be written as, respectively:

$$\mathbf{x} = (x, y, z) \rightarrow \mathbf{x}' = (-x, -y, -z) \tag{5.5}$$

$$\mathbf{x} = (x, y, z) \rightarrow \mathbf{x}' = (-x, y+1/2, -z+2s) \tag{5.6}$$

where the difference between 5.4 and 5.6 ($-z$ instead of $-z+2s$) is due to the fact that the screw axis is at a distance s from the origin. If transformation 5.6 is performed on 5.5, one gets:

$$-x, -y, -z \rightarrow x, -y+1/2, z+2s \tag{5.7}$$

This operation corresponds to a mirror plane perpendicular to y, at a distance of 1/4 fractional from the origin, and with a glide displacement of $2s$ along z, and hence implies a cell periodicity of $4s$ along the same direction (recall Fig. 5.5). Thus, in fractional coordinates $s = 1/4$ and the four complete transformations of space group $P2_1/c$ are written as:

$$x, y, z; \quad -x, -y, -z; \quad -x, 1/2+y, 1/2-z; \quad x, 1/2-y, 1/2+z \tag{5.8}$$

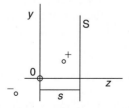

Fig. 5.8. The essential geometrical elements in the building of space group $P2_1/c$. An inversion center is in the origin (0) and relates points (+) at x, y, z with point (−) at $-x, -y, -z$. A screw axis runs along y at a distance s from the origin.

The same result would have been obtained starting with I and G to derive S, or with G and S to derive I. A space group is a closed ensemble of symmetry operations, such that the product of any two operations is still a group operation (not counting integer numbers of cell translations). In $P2_1/c$, calling E the identity operation (no operation or pure translation), it is easy to see that one ends up with the following multiplication table:

	E	I	S	G
E	E	I	S	G
I	I	E	G	S
S	S	G	E	I
G	G	S	I	E

The diagonal of this table consists of E operations only, since double application of any operation results in no operation (for the inversion center) or pure translation (for the other two operations). As required, the successive application of any two operations results in another operation of the group. Note also that since the choice of the axes is arbitrary, x, y and z (a, b, and c) can be interchanged, and the same space group can be written as $P2_1/a$ (glide displacement along x), or $P2_1/b$ (screw axis along x, glide displacement along y), and so on.

A more compact notation that is extremely practical especially for computer applications, uses matrix algebra. In this notation any space group symmetry operation is written as

$$\mathbf{x}' = \mathbf{M}\ \mathbf{x} + \mathbf{t} \tag{5.9}$$

where $\mathbf{x}$, $\mathbf{x}'$ are column vectors of the original and transformed coordinates, and $\mathbf{t}$ is a column vector with the translation, which can be either a screw or glide displacement or a full cell translation. Space group $P2_1/c$ in matrix notation is written as four pairs of $\mathbf{M}$ matrices and $\mathbf{t}$ vectors:

$$\begin{bmatrix} 1 & 0 & 0 \\ 0 & 1 & 0 \\ 0 & 0 & 1 \end{bmatrix} \begin{bmatrix} 0 \\ 0 \\ 0 \end{bmatrix} ; \begin{bmatrix} -1 & 0 & 0 \\ 0 & -1 & 0 \\ 0 & 0 & -1 \end{bmatrix} \begin{bmatrix} 0 \\ 0 \\ 0 \end{bmatrix} ; \begin{bmatrix} -1 & 0 & 0 \\ 0 & 1 & 0 \\ 0 & 0 & -1 \end{bmatrix} \begin{bmatrix} 0 \\ 1/2 \\ 1/2 \end{bmatrix} ; \begin{bmatrix} 1 & 0 & 0 \\ 0 & -1 & 0 \\ 0 & 0 & 1 \end{bmatrix} \begin{bmatrix} 0 \\ 1/2 \\ 1/2 \end{bmatrix} \tag{5.10}$$

An integer number of translations can be added at will to any of the three components of the $\mathbf{t}$ vectors. Thus, for example, a computational box containing a large number of unit cells of a given crystal can be obtained by starting from the atomic coordinates of a reference molecule and applying 0, 1, 2, 3, and 4 unit translations to each component of the four $\mathbf{t}$ vectors. The box will thus contain $5 \times 5 \times 5 = 125$ unit cells and 500 molecules. If cartesian coordinates are required, an operation called orthogonalization can be performed [1]. The calculation of the cartesian coordinates of all atoms within the box can be easily programmed for a computer. This is how packing diagrams are generated.

The fundamental concept in the study of crystals is that any scalar property calculated or measured at point x will repeat unchanged at point x'.

5.3   von Laue's idea, 1912

The terse statement that constitutes the first epigraph to this chapter summarizes about all there is to say about the phenomenon of scattering of radiation by matter, a phenomenon that is crucial to this book and to a large part of structural science. All one needs to know in order to appropriate the information carried by that statement are the answers to: what is electromagnetic radiation; what is an atom; why a moving charge generates an electromagnetic wave? These concepts are normally supplied to students in any high school throughout the world.

The meaning of the key word, "scattering", makes it clear that the radiation scattered by an object need not travel in the same direction as the primary radiation. The real problem appears when one realizes that matter is made by many – a great, great many – atoms and electrons, and that collective oscillatory phenomena generate interference. This is at the same time a curse and a blessing: a curse, because the problem of finding the amplitudes of the scattered radiation becomes very complicated; a blessing, because a study of the interference pattern is what allows the reconstruction of the reciprocal positions of the scattering bodies, that is, the structure of the diffracting medium at an atomic level.

The above ideas may now seem commonplace even in sophomore chemistry, but they certainly were not obvious in the early decades of the twentieth century, when ideas about electrons and electromagnetic waves were still far from clear. As usual, it takes a person "who knows, and knows what he or she knows", that is, has an understanding not only of the nature of a particular idea or phenomenon, but also of its consequences. Max von Laue knew what atoms and X-rays were, and speculated that atoms in the periodic symmetric arrays in a crystal should be the ideal grating to provide a rich and significant interference pattern, the necessary condition being that the wavelength of the primary radiation be of the same order of magnitude as the spacings of the diffracting arrays. The first diffractograms from crystalline matter were recorded on ordinary photographic film. When they came out as an alternation of narrow black regions over much lighter background, rather than as fuzzy mezzotints, the scientific community had hit one of the greatest jackpots in its history.

Once the bright idea had been proved right, it became a matter of working out the mathematical details for going from a set of black spots on a film to the positions of atoms in the crystal. Relating the X-ray diffraction patterns to crystal cell geometry and to crystal symmetry requires much geometrical manipulation, while the backwards synthesis of the electron distribution that is at the origin of the diffracted beams requires a large amount of dull numerical computation. The large number of diffraction features that could be recorded for crystals of even modest molecular complexity was again a curse and a blessing: a blessing because it meant that the problem was over-determined, the number of observables comfortably exceeding the number of variables; a curse, because each and every feature had to be taken into account in the calculation. Besides, the entity that describes the effects of diffraction is a wave and therefore by necessity it carries both an amplitude and phase information, but only the first can be directly retrieved from the photographic plates. The geometrical aspects of X-ray diffraction by crystals could be developed almost by pencil and paper, but the solution of the phase problem and the synthesis of the electron density, in those times of zero electronic computing power, were almost in excess of human possibilities. But in spite of all the difficulties that stood in the way, it was immediately clear that the new technique would bring about a revolution in physics and chemistry. And as is always the case when curiosity and motivation drive human enterprises, no effort was spared. By the early 1950s the detailed crystal structures of a good number of inorganic and organic compounds were already known. In the 1960s the first really efficient electronic computers became available, and that was a giant leap forward. The main task of crystallography shifted from the classification of macroscopic crystals to the study of the inner structure of matter [2].

5.4 The structure factor

5.4.1 *Scattering by one or two charge points*

Consider a charge e accelerated by the electric field of an electromagnetic radiation. Using standard electromagnetic theory, the ratio of the scattered intensity to the primary intensity, $I/I°$, is a function of distance and of the angle α between the scattering direction and the direction of acceleration of the diffracting charge. Electromagnetic radiation is an oscillating electric field, and the representation of an oscillatory phenomenon needs an amplitude and a phase. For any diffracting charge unit at a given location k the diffracted radiation is thus written as

$$f = f_k \exp(i\phi_k) \tag{5.11}$$

where f_k is the amplitude, i is the imaginary unit (i = $\sqrt{-1}$) and ϕ_k is the phase of the wave at point k. The f_ks are intrinsic scattering amplitudes, and depend on the physical nature of the scattering object.

When more than one charge unit is involved, interference sets in and must be taken into account through a phase difference (Fig. 5.9) between the diffracted beams. Consider two charge units, A and B, and an arbitrary scattering direction denoted by angle θ (see Fig. 5.10). The path difference between the wave scattered by A and that scattered by B is

$$\text{path difference} = 2R \sin \theta \tag{5.12}$$

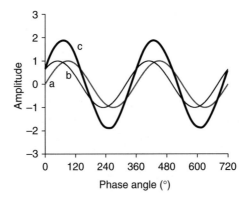

Fig. 5.9. Curve **a** is a sine wave of zero phase, curve **b** is a sine wave of phase 40°. Curve **c** is the sum of the two: there is constructive interference because curve **c** has a peak midway between the peaks of **a** and **b**. Had the phase difference been 180°, the amplitude of the resulting wave would have been zero everywhere.

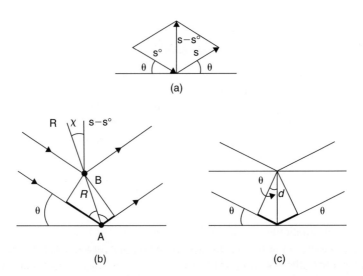

Fig. 5.10. (a) Definition of unit vector s° (incoming radiation), of unit vector s (scattered radiation), and of angle θ; $|s - s^\circ| = 2\sin\theta$. (b) Scattering by two points, A and B at distance R, with scattering angle θ. The path difference, the sum of the two heavy segments, is equal to $2R\sin\theta\cos\chi$, which is also equal to the dot product of vectors $s - s^\circ$ and $\mathbf{R}$, or $|s - s^\circ||\mathbf{R}|\cos\chi = 2R\sin\theta\cos\chi$. Without loss of generality one can take $\chi = 0$. (c) The setting for the discussion of Bragg's law: the horizontal lines are traces of Bragg planes, with the vector $\mathbf{R}$ in (b) coinciding with the interplanar distance d.

THE STRUCTURE FACTOR

The phase difference between any two points separated by vector $\mathbf{r_k}$ is given by 2π times the path difference between the two points, divided by the wavelength λ:

$$\phi_k = (2\pi/\lambda)(\mathbf{s} - \mathbf{s}°)\, \mathbf{r_k} = 2\pi \mathbf{r}^* \cdot \mathbf{r_k} \qquad (5.13)$$

$$\mathbf{r}^* = (\mathbf{s} - \mathbf{s}°)/\lambda \quad |\mathbf{r}^*| = 2\sin\theta/\lambda \qquad (5.14)$$

where $\mathbf{r_k}$ is the position vector of point k, and $\mathbf{s}$ and $\mathbf{s}°$ are unit vectors along the primary and the scattered beam directions, respectively. The diffracted wave resulting from the simultaneous presence of two scattering points is given by the sum of their respective waves:

$$F_{A+B} = f_A \exp(i\,\phi_A) + f_B \exp(i\,\phi_B) = f_A \exp(2\pi\,i\,\mathbf{r}^* \cdot \mathbf{r_A}) + f_B \exp(2\pi\,i\,\mathbf{r}^* \cdot \mathbf{r_B}) \qquad (5.15)$$

This is the key expression for combining the waves scattered by an array of two points and, by obvious extension, of N points. Note that the scattered wave depends on the intrinsic scattering power through the f_ks, on the scattering angle θ through vector $\mathbf{r}^*$, and on the internal structure of the diffracting object through the position vectors of the scattering points, $\mathbf{r_k}$. Exactly what was needed. This expression does not depend in any way on symmetry and/or periodicity; it applies to a set of scatterers in any arrangement.

A further very useful relationship gives the expectation value of the exponential term in a wave, when averaged over all possible orientations in space:

$$<(\exp(2\pi\,i\,\mathbf{r}^* \cdot \mathbf{r})>= \sin(kr)/k\,r \qquad (5.16a)$$

$$k = 4\pi\,\sin\theta/\lambda \qquad (5.16b)$$

5.4.2 The atomic scattering factor

An atom consists of a nucleus plus an electron density distribution, given in the form of a radial function $\rho(x, y, z)$:

$$\rho(x, y, z) = |\psi(x, y, z)|^2 \qquad (5.17)$$

The atomic wavefunction $\psi(x, y, z)$ can be calculated to a very high degree of accuracy by quantum mechanics (Section 3.5). If each infinitesimal unit of a continuous electron density is taken as a scattering unit, the summation of equation 5.15 becomes an integral. In addition, the scattered wave is averaged over the spherical distribution around the nucleus, and application of 5.16 gives

$$f(\theta) = \int 4\pi R^2 \rho(\mathbf{R})(\sin kR)/(kR)\, dR \qquad (5.18)$$

A sum of partial scattered waves amounting to the total scattering power of an object is called the structure factor of that object. Equation 5.18 is the structure factor of an

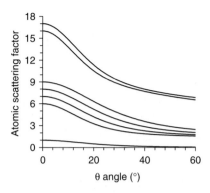

Fig. 5.11. Atomic scattering factors as a function of θ. Each curve starts from the atomic number at $\theta = 0°$.

atom, commonly called the atomic scattering factor. The calculation of these scattering factors has been performed once and forever for all atoms and ions, numerical values are tabulated and have been adapted into the form of polynomials in θ [3]. Some of them are shown in Fig. 5.11. The sharp decay at high θ angles is due to interference among scattering units.

5.4.3 The molecular structure factor

Calling f_k the atomic scattering factor of atom k, an obvious extension of the previous reasoning leads to the following expression for the structure factor of an array of atoms, or a molecule containing m atoms:

$$F(\theta) = \sum_{k=1,m} f_k \exp(2\pi i \, \mathbf{r}^* \cdot \mathbf{r}_k) \tag{5.19}$$

This expression neglects deformations due to chemical bonding (see Section 5.7) The intensity of a diffracted beam is proportional to the square of the module of the structure factor. Since the structure factor is a complex number, the square of the module is obtained by multiplying the structure factor by its complex conjugate:

$$I \propto |F(\theta)^2| = F(\theta)F^*(\theta) = \sum_{k=1,m} f_k \exp(2\pi i \, \mathbf{r}^* \cdot \mathbf{r}_k) \sum_{n=1,m} f_n \exp(-2\pi i \, \mathbf{r}^* \cdot \mathbf{r}_n)$$

$$= \sum_{k,n=1,m} f_k f_n \exp(2\pi i \, \mathbf{r}^* \cdot \mathbf{r}_{kn}) \tag{5.20}$$

where $\mathbf{r}_{kn} = \mathbf{r}_k - \mathbf{r}_n$. This expression depends on distances between all atom pairs, including zero distances when $k = n$. In order to calculate the intensity of a beam diffracted by a lump of matter composed of N molecules in any random orientation, one can proceed as before, i.e. by summing all contributions and averaging through equation 5.16:

$$I(\theta) = <|F(\theta)^2|> = N \sum_{k,n=1,m} f_k f_n (\sin kr_{kn})/(kr_{kn}) \quad (5.21)$$

This is known as the Debye (pronounced "deb-ee-a") equation, and can be used to calculate the diffraction pattern from any specimen, either gaseous or liquid, provided that molecules are randomly oriented. It can be used also to calculate the diffraction pattern of a crystalline specimen in the form of a finely ground powder, where small crystallites are oriented at random with respect to the incoming radiation.

5.4.4 The structure factor for infinite periodic systems

Consider an infinite array of diffracting points, with regular spacings in three directions of length a_1, a_2, and a_3 along x, y, and z respectively (Fig. 5.12). After choosing one point as the origin, the position of any other point in the array can be expressed as

$$\mathbf{R}_{nmp} = n\,\mathbf{a_1} + m\,\mathbf{a_2} + p\,\mathbf{a_3} \quad (5.22)$$

with n, m, and p integers. This space is called real space. Vector $\mathbf{r}^*$, equation 5.14, has the dimensions of the reciprocal of a length, and can be expressed as

$$\mathbf{r}^* = h\,\mathbf{a_1}^* + k\,\mathbf{a_2}^* + l\,\mathbf{a_3}^* \quad (5.23)$$

where $\mathbf{a_1}^*$, $\mathbf{a_2}^*$, and $\mathbf{a_3}^*$ are vectors spanning a space that, for obvious reasons, is called reciprocal space. These unit vectors are conveniently chosen so that the following condition is satisfied:

$$\mathbf{a_k} \cdot \mathbf{a_j}^* = \delta_{kj} \quad (5.24)$$

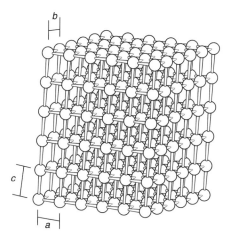

Fig. 5.12. A three-dimensional lattice: the three periodicities a, b, and c and the angles between their directions are arbitrary.

The symbol δ_{kj} means that the product is unit when $k = j$ and zero otherwise. It is easy to show that any distance d in real space has its counterpart in reciprocal space:

$$d = 1/d^* \tag{5.25}$$

With the artifice of reciprocal space, finding the conditions for in-phase scattering by a periodic array of points of scattering factor f becomes very easy. In analogy with equation 5.19:

$$F_{hkl} = \sum_{n,m,p} f \exp(2\pi i\, \mathbf{r}^* \cdot \mathbf{R_{nmp}}) = \sum_{n,m,p} f \exp(2\pi i\,(hn + km + lp)) \tag{5.26}$$

The quantity $(hn + km + lp)$ must be an integer for constructive interference. Since n, m, and p are integers, it follows that also h, k, and l must be integers. Diffracted intensities are non-zero only for integer values of h, k, and l, that is, when the $\mathbf{r}^*$ vector lands on a node of the reciprocal lattice.

In real crystals the translationally independent unit is not a single scattering point, but an ensemble of molecules in the unit cell, each of which is described by the molecular structure factor. When the molecular structure factor, equation 5.19, is combined with the condition 5.26 for constructive interference of a translationally periodical lattice, the final expression for the crystallographic structure factor results:

$$F_{hkl} = F(\theta) = \sum_{j=1,m} f_j \exp(2\pi i\,(h\,x_j + k\,y_j + l\,z_j)) \tag{5.27}$$

In this expression, h, k, and l are three integer numbers, and x_j, y_j, z_j are the fractional coordinates of each of the m atoms in the unit cell.

Equation 5.27 is fundamental in X-ray crystallography. Consider the following derivation:

$$\exp(i\phi) = \cos\phi + i\sin\phi \tag{5.28}$$

$$\begin{aligned}
F_{hkl} &= \sum_{j=1,m} f_j \exp(2\pi i\,(h\,x_j + k\,y_j + l\,z_j)) \\
&= \sum_{j=1,m} f_j \cos(2\pi(h\,x_j + k\,y_j + l\,z_j)) \\
&\quad + i \sum_{j=1,m} f_j \sin(2\pi(h\,x_j + k\,y_j + l\,z_j)) \\
&= A_{hkl} + i\,B_{hkl}
\end{aligned} \tag{5.29}$$

$$|F_{hkl}| = (A_{hkl}^2 + B_{hkl}^2)^{1/2} \tag{5.30a}$$

$$\phi_{hkl} = \arccos(A_{hkl}/|F_{hkl}|) \tag{5.30b}$$

When the coordinates of all atoms within the cell (actually, within the asymmetric unit, Section 5.1) are known, the structure factors can be calculated in modulus, $|F_{hkl}|$, and phase, ϕ_{hkl}; these are called $F_{calc,hkl}$. The single-crystal X-ray diffraction experiment starts by measuring the observed diffracted intensities:

$$I_{obs,hkl} \propto |F_{obs,hkl}|^2 \quad (5.31)$$

A number of corrections allow the retrieval of the observed structure factors, $F_{obs,hkl}$, and the game of X-ray crystallography, also known as the solution of the phase problem, is to find the set of atomic coordinates that give a close match between the $F_{calc,hkl}$ and the $F_{obs,hkl}$. A first guess must be somehow prepared; this problem is now routinely and robustly solved by a technique called direct methods (see Section 5.9). The agreement between observed and calculated structure factors is coded into a discrepancy index traditionally called the R-factor:

$$R = \sum |(|F_{obs,hkl}| - |F_{calc,hkl}|)| / \sum |F_{obs,hkl}| \quad (5.32)$$

Atomic coordinates can be refined by automatic least-squares to lower the R-factor as much as possible. Weights can be assigned to each structure factor according to statistical considerations.

One final point concerns thermal motion. Atoms are never at rest, and thermal motion causes a further decay in the intensity of diffracted beams. The atomic structure factor can be rewritten as

$$f_j = f_j^\circ \exp(-B_j \sin^2 \theta / \lambda^2) \quad (5.33)$$

where f_j° is the scattering factor in the absence of thermal motion, and the exponential term is called the Debye–Waller correction. B is an isotropic temperature factor for atom j. More sophisticated, anisotropic expressions for this correction are normally used: the thermal factor becomes a tensor and is represented by an atomic displacement ellipsoid, requiring six independent parameters for each atom, to be refined by least-squares along with atomic coordinates [4].

5.5 Miller indices and Bragg's law

The genius of sir Lawrence Bragg provided a simple physical model of constructive scattering of X-rays by crystals, an immediate counterpart of the mathematics in term of simple phenomena. The conditions for constructive interference in a crystal will now be rewritten using Bragg's approach.

Let any plane going through the crystal unit cell be labeled by three integers, h, k, and l, defined in the following way (see Fig. 5.13). Consider a cell axis a_k; for $h = n$, the plane intersects the axis at $1/n$ of its length. If $h = 0$, the plane is parallel to the axis. The numbers h, k, l are called Miller indices. There are an infinite number of parallel (hkl) planes, as there are an infinite number of unit cells in the crystal.

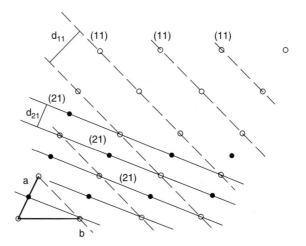

Fig. 5.13. A two-dimensional illustration of Miller indices and Bragg planes: "planes" are in fact lines. Larger open circles denote lattice points, smaller black dots indicate the midpoint of the a periodicity. Planes (11) (heavy dashed lines) pass through the points $(a,0)$ and $(0,b)$, planes (21) (continuous lines) pass through the points $(a/2,0)$ and $(0,b)$. The d_{21} spacing is less than the d_{11} spacing. The (11) and (21) lines are the traces of the (110) and (210) planes in the three-dimensional equivalent.

Let d_{hkl} be the spacing between two members of the family of (hkl) planes. Bragg's idea was to consider the scattering of X-rays as a reflection from these planes. The condition for constructive interference is that the path difference between the waves scattered (reflected) by any two planes of the (hkl) family be an integer number, n, of wavelengths. Returning to Fig. 5.10(c), this condition is written as

$$n\lambda = 2 d'_{hkl} \sin\theta \quad \text{or} \quad \lambda = 2 d_{hkl} \sin\theta \quad (d_{hkl} = d'_{hkl}/n) \qquad (5.34)$$

where the second formulation is more handy, and considers the n-th order reflection from plane (hkl) as the first-order reflection from plane $(nh\ nk\ nl)$. Just a matter of different notation. These planes have no special connotation in terms of chemistry, i.e. they need not contain atoms, nuclei, electrons or whatever; they are just slices through the periodic electron density of the crystal and reflect the translational periodicity of the crystalline edifice.

Bragg's law gives at a glance the number and type of possible "reflections" (in Bragg's law jargon, this term is the equivalent of "constructively scattered wave"); a crystal irradiated by a radiation of wavelength λ will produce all diffracted beams for any triplet of integer values of the Miller indices, up to $(\sin\theta)_{max} = 1$, or $d_{hkl,min} = \lambda/2$. It is also a pictorial view of von Laue's idea, because it shows that in order to get such a diffraction pattern the radiation wavelength must be of the same order of magnitude of the spacing between diffracting planes, which is in turn of the same order of magnitude as the spacing between atoms and molecules in the crystal. Incidentally, this is the reason why if one wants to observe reflections from planes with very high Miller indices, radiation of shorter wavelength must be used.

The link with the reciprocal lattice treatment of the same problem is given by the following expression, which is Bragg's law in vector form (again, recall Figs 5.10 and 5.13):

$$(\mathbf{s} - \mathbf{s}°)/\lambda = \mathbf{d}^*_{hkl} = 1/d_{hkl} \qquad (5.35)$$

Since $\theta_{hkl} = \arcsin(\lambda/(2\,d_{hkl}))$, in order to find the angle θ for any Bragg reflection one needs to find the interplanar spacing for any triplet of Miller indices, either in direct or in reciprocal space. Working out the direction of reciprocal space vectors is easy in orthogonal coordinates but becomes more and more tedious for monoclinic and triclinic crystals [5].

5.6 The electron density in a crystal

It can be shown, at the price of a short course in Fourier analysis, that the structure factors are the coefficients of the Fourier synthesis of the electron density in the cell. We will not give the derivation here, and the reader should be satisfied that the following relationship holds [6]:

$$\rho(x, y, z) = 1/V_{cell} \sum_{h,k,l} F_{hkl} \exp(-2\pi i\,(hx + ky + lz)) \qquad (5.36)$$

The use of this equation obviously requires a knowledge of the phases of the structure factors. Once the phase problem is solved, the moduli of the observed structure factors can be used in a Fourier synthesis with calculated phases, to find the centroids of the electron density peaks, which indicate the positions of atomic nuclei. If the phases are nearly correct, the Fourier map will reveal the position of most if not all of the atoms. The Fourier synthesis is, in brief, a convenient way of transforming the phase information in terms of phase angles into the same information in terms of atomic positions (recall equations 5.28 to 5.30). If the first try does not reveal the position of all atoms, then a Fourier synthesis using as coefficients the differences between observed and calculated structure factors, plus the nearly correct phases, will reveal the position of the missing atoms.

5.7 The atomic prejudice

Throughout the above treatment, it has been more or less implicitly assumed that the atom is the smallest diffracting unit that can be conveniently considered. In the all important formulas, such as equations 5.21 or 5.27, the f_js are assumed to be the atomic scattering factors. This is equivalent to considering a molecule as formed by a superposition of spherical, undeformed free atoms. This model has met with undisputed success over the decades and over the hundreds of thousands of crystal structure determinations so far carried out. Nevertheless, it is an approximation, and, chemically speaking, a rather crude one, because it neglects everything about chemical bonding and all other polarization effects that occur when many atoms join together into a molecule. The proper way of doing things would have been to consider the interference pattern of the scattering not from a collection of atoms, but from the collection of all the infinitesimal elements of the molecular electron density. However, this would have required laborious integrations instead of simple summations, and it turned out that the improvement one could get by using such a more refined model were not so outstanding, as far as X-ray crystal and molecular structure determination were concerned.

So, by long tradition, the theory and practical methodology of crystal structure analysis by X-ray diffraction is based on atoms as fundamental units. As much as X-ray crystallography is the parent of chemical structural theory, this way of thinking has sneaked into the way chemists think about the nature of intermolecular interactions in organic condensed phases. As will be discussed at greater length in the second part of this book, this way of thinking has many pitfalls and intermolecular interactions should be more properly framed in terms of molecular electron densities.

5.8 Structure and X-ray diffraction: Some examples

X-ray diffraction works wonders when applied to crystals, but is also useful in the study of other materials. In order to provide some understanding of how the internal structure of a condensed phase is portrayed in its diffraction pattern, a few computational examples on typical states of matter will now be given.

The succinic anhydride molecule is used as a computational guinea pig. Consider first the unit cell of the succinic anhydride crystal, with four molecules with all atomic coordinates known. Consider then a slab carved out of the succinic anhydride crystal structure (Fig. 5.14(a)) containing 224 unit cells (896 molecules, 9,856 atoms). For the sake of the computational experiment, consider also a box made by 864 succinic anhydride molecules (9,504 atoms) arranged in a more or less regular fashion (Fig. 5.14(b)); this is a semi-crystalline system. The latter box has then been thermalized by a molecular dynamics calculation (see Chapter 9), and frames have been extracted from the simulation after 1 ps (Fig. 5.14(c)), 5 ps, and 46 ps (Fig. 5.14(d)) [7]. The progressive loss of structure is clearly visible, until in Fig. 5.14(d) molecules are completely at random, a true liquid state. The position of all atoms in the crystal unit cell and in each of these computational boxes is known to the computer.

Consider first the true crystalline state. Using the unit cell information, all the structure factors, equation 5.27, can be calculated and a graph can be prepared where each Bragg reflection appears with its intensity at the proper θ location in a θ/intensity plot. A visually conspicuous, overall landscape view of the distribution of Bragg reflections in θ-space is thus obtained. This corresponds to the pattern obtained experimentally by diffraction from finely ground crystalline material, in which crystallites are oriented at

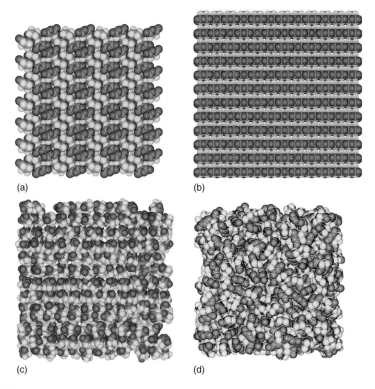

Fig. 5.14. (a) The succinic anhydride crystal. (b) An ordered array of arbitrary structure and density, containing 864 succinic anhydride molecules, built as a starting point for a computer simulation. The pronounced layer spacing is 4.9 to 5.5 Å. (c), (d), the array in (b) after increasing times of molecular dynamics randomization, 1 ps, 44 ps.

random and the three-dimensional information in Miller indices is contracted into one-dimensional θ information only. These powder patterns are often used as extremely sensitive and unequivocal fingerprints in analytical chemistry applications. Taking an experimental powder pattern is a relatively simple matter if the crystalline material is available in finely ground form.

For the simulation, intensities $I_{hkl}(\theta)$, equation 5.31, are calculated, using an average thermal B factor. The complete powder profile is then prepared by placing each Bragg peak at its calculated position $\theta°$, with an exponential spread:

$$I(\theta) = \sum_{h,k,l} I_{hkl}(\theta°) \, \exp(-t(\theta - \theta°)^2) \tag{5.37}$$

where t is an adjustable sharpening coefficient.

The calculation of the powder diffraction pattern can be done also using Debye's equation 5.21, as the position of all atoms and hence all the atom–atom distances

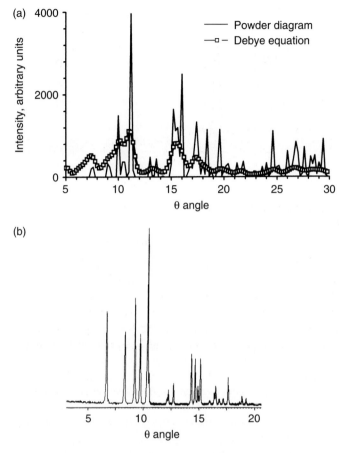

Fig. 5.15. (a) The simulated powder diffraction pattern of the succinic anhydride crystal, using the calculated structure factors or as obtained by using the Debye equation on a box containing 224 crystal unit cells (intensity multiplied by 1.8 for scaling). Copper radiation wavelength. (b) The experimental powder diffraction pattern of succinic anhydride. Copper radiation. (Courtesy of Dr Lucia Carlucci.)

within the crystal box are known. The result should be the same, but the resolution of the resulting pattern will be different: the Debye equation is a brute force summation, which in principle requires an infinite number of terms up to very high interatomic distances; on the contrary, the periodical lattice interference condition 5.26 is incorporated in equation 5.27, and thus the infinite lattice periodicity is implicitly taken into account. Figure 5.15 shows the results of the two calculations, and, for comparison, also the experimental pattern measured on a powdered crystalline sample, which coincides with the one calculated using equation 5.27

STRUCTURE AND X-RAY DIFFRACTION: SOME EXAMPLES

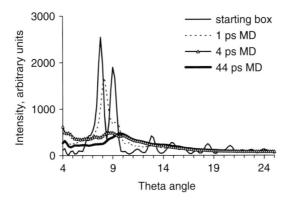

Fig. 5.16. The diffraction patterns calculated by the Debye equation for the boxes in Fig. 5.14(b)–(d).

because the unit cell used in the simulation is the correct crystal cell. The relative intensities of the peaks in experimental patterns may vary, being a sensitive function of many experimental conditions such as diameter of the ground powder particles, quality of the crystalline material, and possible preferential orientation of the powder particles due to crystal morphology. In turn, the simulations are only approximate because the structure factors have been calculated assuming an overall isotropic thermal factor B for all the atoms in the molecule. The pattern calculated by Debye's equation has the same features, but the resolution is much lower, as expected.

Figure 5.16 shows the profiles calculated by Debye's equation on the non-crystalline frames in Fig. 5.14(b)–(d). The starting frame produces two large peaks at $\theta = 8°$ and $\theta = 9°$, roughly corresponding to the interplanar distances of $d = \lambda/(2 \sin \theta) = 4.9$ and 5.5 Å ($\lambda = 1.54$ Å), which are present in its structure. The intensity of these two peaks is extremely high because the alignment of molecules at these d separations is by far the most salient structural feature of the molecular ensemble. After 1 ps, a faint trace of these spacings can still be seen, and the two peaks are still discernible. They disappear as soon as does any trace of regular arrangement of molecules into layers; in the fully randomized, liquid-like frames what is left is just one broad hump that shifts to higher θ as the simulation proceeds in time, because the box is contracting and the spacing between molecules decreases. Such simulated diffraction patterns can help to recognize the amount of crystalline structuring in a given state of a given material.

With time and effort, the radial distribution curves (Section 9.4) for atom–atom contacts in the liquid could be deconvoluted from experimental diffraction patterns taken on liquid samples. The task is not an easy one, as is evident if one just compares the large amount of information contained in the numerous and sharply defined peaks from a crystal with the scarce and ambiguous information in a pattern from a liquid, which usually has just one or two broad humps. The complexity and number of features of the diffraction pattern increases with the amount of structuring in each of the various forms of aggregation.

5.9 Historical portraits: Training of a crystallographer in the 1960s

In those days, Italian Universities did not have a PhD program, and all I could have was a "laurea" which came with the title of "dottore" after five years, one of which to be spent working full time (and that meant full time indeed) on a scientific project in one of the university laboratories under a supervisor. (So I do not have a PhD, and no Italian professor of about my age does, unless his or her family was prepared to spend the money needed to live for four years in Oxford, for example. In fact, in those days we did not even know that such a thing as a PhD existed). The choice of the laboratory and supervisor was quite critical, being full of consequences for one's professional future. When I heard that the Chemistry department in Milano had two well equipped X-ray diffraction laboratories, one for organic and one for organometallic materials, the deal was made: in the summer of 1967 I entered the organic X-ray diffraction group under the supervision of Massimo Simonetta.

The crystallography lab was in the basement, because that was the only place where the massive X-ray generators of the time could be located without danger of crashing through the floor. The place was narrow and badly lit, and smelled bad because it stood just above the drain pipes that came down from the organic chemistry laboratories on the floors above, from where people in those days happily threw nearly anything into open and unchecked washbasins. The equipment included two X-ray generators, one of which Japanese, with an instruction manual that recommended wearing slippers in the generator room to avoid the risk of carrying in harmful dust. The other generator I was told to stay away from, because it was rather old and the X-ray tube was leaking; only people who knew exactly where the leak was were allowed near it

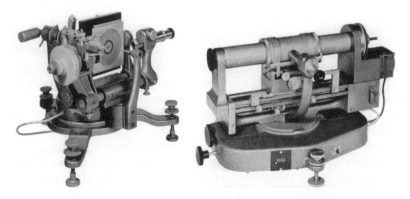

Fig. 5.17. Vintage crystallographic apparatus: a precession (left) and a Weissenberg camera.

(safety rules were schematic, to say the least). We had several machines working with photographic methods, at least two Weissenberg and one precession (Fig. 5.17), and of course there was a darkroom with all the gear for developing photographic film. But, lo and behold, in the middle of the largest room of our den stood the beauty of the day, an IBM 1620 computer. It had hundreds of colored lights, a few sturdy switches, a bulky typewriter, and a card reader–puncher. If one opened its back, one could see the huge arrays of tiny ferrite rings with two sets of crossed wires that were its memory: the rings, magnetized to represent a bit, the wires, one for writing, the other for reading. The whole thing was the size of two large cupboards and its computing power was negligible, but I at once felt that such a machine would be the real game of the trade. And, as it turned out many years later, it – or rather, its progeny – indeed was. In any event, on a table proudly stood a heavy Olivetti mechanical calculator that could do the four arithmetic operations, although it took it about thirty seconds of scratchy noises for the division of two five-digit numbers.

In those days, a professor was a professor, and Simonetta was one. He would occasionally appear, distribute jobs, be kept abreast of developments, and then vanish again. The X-ray working group included Angelo Mugnoli, Carlo Maria Gramaccioli, Carla Mariani, Gianluigi Casalone, Riccardo Destro, and Giuseppe Filippini, roughly in that order of seniority. Everybody wore white lab coats, except Carlo Maria who had spent a year at CalTech with Dick Marsh, and sported Californian nonchalance. After a few weeks I was already imbued with the rudiments of the trade, Miller indices, Bragg's law, reciprocal space, the geometry of an oscillating crystal photograph. I was just beginning to understand something about crystal symmetry. Everything had to be slowly and carefully explained to me by complex drawings on paper, and I remember my first days there as days of pencil sharpeners and rulers. Even getting good writing paper was not trivial; bits of scrap paper were carefully preserved and their backs reused. Getting a photocopy of a scientific paper meant filling in a form, having it approved by an officer, going to the library, and waiting until the personnel there had time to make the copy for you, usually on smudgy paper from which characters disappeared at the touch of a finger. Not that Italy was a particularly poor country; hardware was what it was, but at least things were given their real value and needless waste, the trademark of our plethoric civilization, was carefully avoided.

Every undergraduate had to be associated with a project, meaning a molecule. In those days, one molecule or crystal constituted a project that would last at the very least many months, and at the very worst an eternity, there always being a good chance that the structure determination might not come to a happy ending at all. I was put on 1,1-di-p-tolylethylene (DTE), a name that caused much laughter in my family because it sounded as if they had put me on something Chinese. As I realized only later, the project was a risky one because the molecule was a 16-atom hydrocarbon, no heavy atom, and hence no easy escape through Patterson methods. Choosing the crystal for the X-ray work was also no small matter, because it was a long term investment. Of course I was not involved in this because I had too little experience, but at last, a couple of months after my baptism, a crystal of DTE was put under the X-rays in a Weissenberg. It was quickly seen, to everybody's joy and relief, although I still did not understand why, that the structure was centrosymmetric, in fact $P2_1/c$, the familiar space group of the organic crystallographer. For the benefit of junior readers, I recall that in those days many computer and working routines were designed for one or a few space groups only, and had to be largely rewritten to deal with a space group that had never been treated previously.

The two crystallography groups in Milano shared a rather new data-measuring machine, a semi-automatic diffractometer called PAILRED. The thing was still looked on with some suspicion, because its performance was somewhat shaky. Once the crystal had been laboriously mounted, centered, and oriented, the idea was that the machine would automatically record the intensities of diffraction spots, night and day. In practice, the damn thing had an astonishing ability to find novel ways of breaking down, and it was seldom the case that the collection could be continued for more than a few hours without requiring manual intervention on one of the many things that could go wrong. The most common fault was a jam in the card-punching machine onto which the diffractometer recorded the intensity data; when this happened at night, the result was that the data collection went on and on, while the numbers were punched all together onto the one jammed card, which was retrieved the morning after looking as if it had gone through a threshing machine. To avoid this, blank card packs had to be stoutly bent and stretched to prevent them from sticking together, and I still remember the long times we used to spend in the evenings, flipping the blank cards like poker players before putting them on the input stacker for the coming night's use.

After a few weeks of struggle, the data we had collected were found to be very poor, and so it was decided to go back to film collection. At that point I came in handy, because the data were to be recorded on photographic plates. The procedure started with preparation of the films for exposure, with long and embarrassing sessions in the darkroom, making up thick packets of four or five film plates, folding them into black paper, and coaxing the whole thing into a metallic cylindrical film holder, making sure that the film corner with a small cut would be at the proper corner of the camera as an orientation check. Many times, after sweating for hours, after coming out of the darkroom one found that the mounting was upside down, or that a small scratch in the black paper had let light in and spoiled the whole thing. After that, the films were exposed for days or weeks, and if that small scratch had not been spotted before exposure, it was spotted after, resulting in grim faces around the lab for a couple of days, and much shame for the poor all-thumbs undergraduate.

In the long hours spent beside the humming generator, with the soft noise of the camera rolling on its gears and the clicks of the protection switches that inverted the rotation of the driving motors, I studied computer programming. One could use Fortran on the IBM 1620, but then one could allocate only 7,500 memory positions for program and data all together (today a normal personal computer has some 500,000,000). So I started learning machine language and symbolic programming. That meant addressing, allocating, flagging, operating upon, almost pampering, one number or one memory location at a time. Yet, it was considered a worthy task. Crystallographers were one of the first scientific communities to be absolutely compelled to use electronic computers, and as a consequence, compelled to become computer programmers. The other community was, of course, that of theoretical chemists, with whom we shared the graces of the 1620.

After another couple of months, a set of neat Weissenberg photographic plates for DTE were ready for processing. It was then that the undergraduate crew came in even more handy, because "processing" meant sitting at a desk with one plate and a set of reference spots of decreasing darkness, visually estimating the "total blackness" of each spot on the plate, and marking down with a pencil (pencils, again!) a number that would later be transformed into the diffracted intensity. This had to be done for thousands of such spots. Later on, one would sit at a card-punching machine and punch in the three Miller indices and the intensity: one set for each punched card. After about six months altogether, the data collection was considered complete and the computer processing of the data could start.

The computing center of the Faculty of Sciences also maintained a mainframe computer, which in these days was an IBM 7040. That machine was becoming gradually available to crystallographers, but the problem was that there were no programs good and ready for crystallographic computing. Gramaccioli had been instrumental in setting up the famous program suite called CRYRM (the last two letters standing for Richard Marsh) at CalTech, but unfortunately it was written in machine code, and transfer from the machine it had been written for was a problem: the days of tumultuous computer development had not yet begun, and people still invested their time in writing programs dedicated to only one machine. Our group could count on a mish-mash of base language, symbolic and Fortran programs for the 1620, but that clearly was not the future. So it was decided that it was time to start a project for assembling a series of programs in transferable Fortran code for doing the whole crystal structure analysis business from scratch. People had noticed that I spent hours filling the back of scrap paper sheets with computer code, so I was asked to join. The benefit was that I was forced to learn everything in every possible detail.

In the meantime, the data for DTE were waiting. We decided to go ahead with the existing programs while the new ones were being developed, a wise decision, because the project was not one of weeks nor of months, but possibly of years. We merged the diffractometer and the film data, and then the first thing to do was sorting; the data cards had to be placed in a predetermined order of variation of the Miller indices, as required by all existing programs in order to save memory allocation and computing time. So one morning we took our several thousand cards to a certain Department at the Engineering School, where they had a sorting machine. It was indeed a mind-boggling piece of equipment: it rolled thousands of cards over a set of turning cylinders at an astonishing speed, dropping them into separate bins numbered from 1 to 10 according to the position of the hole in one of the 80 columns. How you worked through it to reach your sorting goal, was your business. Carla Mariani and I came out after a few hours, many false starts and false ends, a few torn cards, and a lot of swearing, but with a beautiful pack of cards ready for input into the data reduction and merging program.

Rather than hitting the return key or clicking an icon or a pulldown menu item, and waiting a few seconds, as is done today, the process involved reserving a full day of 1620 use, loading the program (also on punched cards), hitting the end-of-file button, positioning a few hardware switches, hitting the start button, waiting for the green reader-no-feed light to light up, placing the heap of punched cards over the "in" tray of the card reader, pushing the start button on the card reader, and waiting for the swishing sound of the first card being sucked into the slit and digested. After a due time, one would hear the motor of the card puncher come to life, and the machine would slowly start to drop a new set of cards into the "out" stacker – the cards with the set of merged independent intensities. The typewriter would also come to life, and start typing out pairs of sets of indices and intensities, along with the decision taken (merge, reject, check). Watches were pulled out and the time interval between card inputs would be measured in order to estimate the total time required. If that was beyond regular working hours, it was usually the undergraduate's duty to keep watch. If anything went wrong – a badly positioned card, a jam in the reader or in the puncher, not to speak of power failures – everything had to be started again from scratch, possibly re-punching the maimed cards.

At that point, the intensity data were ready but nobody had the faintest idea of how to get around the phase problem. The Patterson was no help, as expected. Massimo Simonetta had very close ties with Californian structural chemists, having spent a few years in Linus Pauling's surroundings in the 1950s. So Ken Trueblood sent us a copy of a new computer program, originating from David Sayre's group, for the determination of phases by direct methods, a technique of which very few people in Italy knew anything. Having noticed my affinity with computer codes, the group gave it to me to analyze. Actually running the program was something that, as usual, might require weeks of careful planning.

I got hold of a roll of used graph paper from the PAILRED's pen writer (a device whose main purpose was to splash red ink all over the place) and used its blank back side to sketch out a block diagram. After a couple of days I realized I was in the presence of the work of a genius. The way the information was handled and sorted in order to spare every single bit of memory was literally amazing. The programmer (probably David himself) had shrunk the tasks of thousands of instructions into a few hundred. After a week and a few meters of paper roll, I had the full block diagram. What was exceedingly attractive to us was that the program could run on a 7040; my mouth started watering right away.

We got moving. In those days, submitting a job at the computing center meant going to a room where the input card packs were collected, all carefully wrapped up with rubber bands, waiting for the pack to be picked up by operators, and coming back after a few hours (or the morning after) to pick up your original pack, your output punched cards (if any) and a paper output. One walked there holding one's breath: a two-page printout meant that something had gone wrong, anything from a silly mistake in the ordering of the cards to a major program failure; too big an output meant that the program had run into an unexpected loop and had wasted a lot of expensive computing time.

The first thing to do was to compile the Fortran code. This we did, and dutifully got our punched cards with the executable. We then chose a good number of intensity data, and submitted our complete pack, executable first, data next. Being totally inexperienced, we selected all the standard running options. One had to specify an upper time limit, after which the program was killed by the operators; we just set something reasonable and hoped for the best, having no idea whether the task would require an hour or a day. I went to the computer center in the evening with hopes, but the senior people in the lab had grown up with the idea of struggling with a structure for at least six months before anything happened, so they were far more skeptical about this attempt using a so-far untested method.

The morning after, my pack was back on the shelf with a reasonable paper output – not too thin, not too thick: it consisted of about one hundred 30 cm × 40 cm paper sheets, as was customary for computer output in those days. At the end, there was a table with a summary of the sixteen solutions, with a consistency index for each of them. Knowing nothing of the real difficulty of the problem, I was unduly hopeful. I was dying to do a Fourier synthesis and a structure factor calculation using the phases given by the solution with the highest consistency index; but a Fourier synthesis calculation was something of agonizingly slow speed, the 1620 was very busy in these days, the project of an undergraduate did not have a very high priority, and the seniors in the group had a more realistic idea of the chances that the solution might be right – i.e. close to zero, so they were not particularly anxious to go ahead. In particular, they rightfully considered that the best solution was not so "best" after all, its consistency index being just slightly higher than the others (see Table 5.1). My pack of cards with the input for a Fourier synthesis based on 167 strong reflections and the phase signs of the best solution had to wait for many days before making it through the card reader of our computer.

When it did, I was left alone with the output, further confirming the lack of confidence the seniors had in this rather cavalier attempt. The Fourier synthesis was printed on bad paper and with scant attention to scaling. So I decided to try and find

Table 5.1 The top part of the final output of David Sayre's phase-determination program (year 1967). Note that for a centrosymmetric structure phase angles, equation 5.30, can only be 0 or 180° so the phase of a structure factor is just a plus or a minus sign

Set number	Number of cycles	Number of +signs	Number of −signs	Consistency index
1	11	85	112	0.79873
2	9	105	92	0.75657
3	11	93	104	0.74502
4	13	92	105	0.71139
5	10	92	105	0.70068

coordinates for the most prominent peaks and to resort to a different method of visualization. I drew on a large piece of paper (pencils, again!) a grid for the coordinates in the *ac* plane, with the correct monoclinic angle, and then found a set of long knitting needles and a few black paper balls. I stuck a needle through each ball, punched the needle into the base grid at the proper x and z, and measured the y coordinate on each needle for the ball that was to represent an atom. After locating about eight balls, the result was something I still could not interpret at all. At that moment Riccardo Destro walked into the room and said, "Oh, I see you've found a phenyl ring", and pointed it out to me. At that point the interest of our seniors sprang back towards my molecule.

Shortly afterwards, we were sitting in front of the 1620 waiting for the final output of the first structure factor calculation. When the last data card had been pulled in, we knew it would take another minute or so to calculate the R factor. Then the typewriter came to life and – click-click-click – typed "R=15.64", an excellent disagreement factor for the start of the refinement process, meaning that the structural hypothesis was correct and the crystal structure was solved. There was a round of cheers. It was the reward for months and months of tedious work, with the added bonus that a major paper would be ready to be shipped off and accepted by a reputable journal in no time at all. I don't know to what extent the more experienced people in the room realized that such an event was the beginning of a new era in crystallography, one in which the phase problem no longer existed. I certainly didn't then, but I do now, and I consider it a privilege to have witnessed in person such a great turning point in crystallography and in chemistry.

Nowadays, all the trouble we went through is back in the faraway past, and my whole doctoral thesis could be completed by a good technician in a matter of a few hours. The collection of diffracted intensities is now done on CCD plates in one afternoon, and is also a hundred times more accurate and reliable than the pathetic blackness grades handwritten in my ruled pad; data processing including solution of the phase problem would be a matter of milliseconds, and refinement a matter of perhaps a few minutes. So the story told in numbers and formulas in Sections 5.4–5.6 has been recast above in colorful words.

References and Notes to Chapter 5

[1] If **x** is a column positional vector, the orthogonalization operation is written as **x**(*orthogonal*) = **P** **x**(*fractional*). One possible form of the **P** matrix, the orthogonalization matrix, is:

$$t = 1 - \cos^2(\gamma) - \cos^2(\beta) - \cos^2(\alpha) + 2\cos(\gamma)\cos(\beta)\cos(\alpha)$$

$$\mathbf{P} = \begin{matrix} a\sin(\gamma) & 0.0 & \cos(\beta) - \cos(\gamma)\cos(\alpha)c/\sin(\gamma) \\ a\cos(\gamma) & b & c\cos(\alpha) \\ 0.0 & 0.0 & t^{1/2}c/\sin(\gamma) \end{matrix}$$

[2] For an introduction to some basic elements of chemical crystallography see for example Hammond, C. *The Basics of Crystallography and Diffraction*, 2001, Oxford University Press, Oxford. Giacovazzo, C., Monaco, H. L., Viterbo, D., Scordari, F., Gilli, G., Zanotti, G. and Catti, M. *Fundamentals of Crystallography*, 1992, Oxford University Press, Oxford has a more thorough mathematical treatment of many details of X-ray diffraction and of crystallography, and also an introduction to crystal chemistry and crystal structure determination. A must for the organic chemical crystallographer is Dunitz, J. D. *X-ray Analysis and the Structure of Organic Molecules*, 1995, 2nd corrected reprint, Verlag Helvetica Chimica Acta, Basel. A quick but useful and complete introduction to practical X-ray crystal structure determination is in Glusker, J. P.; Trueblood, K. N. *Crystal Structure Analysis, A Primer*, 1985, Oxford University Press, Oxford. See also ref. [5]. Several aspects of X-ray crystallography and related topics are presented in a series of tutorial pamphlets published under the auspices of the International Union of Crystallography. These pamphlets can be downloaded for free from www.iucr.com.

[3] Atomic scattering factors are written as follows ($x = \sin(\theta)/\lambda$);

$$fs(x) = f(1)\exp(-f(2)x^2) + f(3)\exp(-f(4)x^2) + f(5)\exp(-f(6)x^2) \\ + f(7)\exp(-f(8)x^2) + f(9)$$

The value of the nine coefficients $f(1)$ to $f(9)$ for some common atoms in organic molecules are:

Hydrogen: 0.49300, 10.5109, 0.32290, 26.1257, 0.14020, 3.14240, 0.04080, 57.7997, 0.003

Carbon: 2.31000, 20.8439, 1.02000, 10.2075, 1.58860, 0.5687, 0.8650, 51.6512, 0.21600

Nitrogen: 12.2126, 0.00570, 3.13220, 9.89330, 2.01250, 28.9975, 1.16630, 0.5826, −11.529

Oxygen: 3.0485, 13.2771, 2.2868, 5.7011, 1.5463, 0.3239, 0.8670, 32.9089, 0.2510.

[4] Details of the structure determination procedure can be found in books cited at ref. [2].

[5] See Stout, G. H.; Jensen, L. H. *X-ray Structure Determination*, 1968, Macmillan, New York. For an orthorhombic crystal, $1/d_{hkl}^2 = h^2/a_1^2 + k^2/a_2^2 + l^2/a_3^2$. Calling i,j and m any permutation of the cell edges a, b, c or of the cell angles α, β, γ, the expressions are:

$$i^* = jm\,\sin(i)/V$$
$$\cos(i^*) = [\cos(j)\cos(m) - \cos(i)]/\sin(j)\sin(m)$$
$$V = 1/V^*$$
$$= ijm[1 - \cos^2(i) - \cos^2(j) - \cos^2(m) + 2\cos(i)\cos(j)\cos(m)]^{1/2}$$

V is the cell volume. The general expression for the d spacing in a triclinic system is:

$$\sin(\theta)/\lambda = 1/2d_{hkl}$$
$$= (1/2)(h^2 a^{*2} + k^2 b^{*2} + l^2 c^{*2} + 2hk a^* b^* \cos\gamma^* + 2hl a^* c^* \cos\beta^* + 2kl b^* c^* \cos\alpha^*)^{1/2}$$

[6] See one of the textbooks mentioned in ref. [2], for example Giacovazzo *et al.*, pages 169 ff.

[7] The molecular dynamics simulation is a mathematical stick that stirs the system and progressively randomizes the positions and orientations of the molecules within the computational box, under the action of a computational equivalent of thermal energy, molecular momenta. See Chapter 9.

6

Periodic systems: Crystal orbitals and lattice dynamics

Chemists need not enter into a dialogue with physicists with any inferiority feelings at all.
Hoffmann, R. *Solids and Surfaces: A Chemist's View of Bonding in Extended Structures*, 1988, VCH, Weinheim.

6.1 The mathematical description of crystal periodicity

6.1.1 *Equivalent positions and systematic absences in diffraction patterns*

In Sections 5.1 and 5.2 a molecular crystal was defined as an array of molecular objects related by symmetry operations, including pure translation: a system endowed by periodic translational symmetry. The asymmetric unit is the symmetry-independent part of the crystal, and it is repeated within the unit cell into a number of equivalent positions, by as many symmetry operations. The unit cell contents are then repeated in space by pure translation.

Succinic anhydride is a very common chemical, and can be bought in pure crystalline form for a few euros a gram. It can be recrystallized from nearly any solvent (acetone, acetonitrile, chloroform) yielding a nice white material, which under the microscope reveals a rough needle-like morphology. To be sure, some exterior features of the crystals, like the faces and angles between them, reflect the inner symmetric structure of the material, but at first sight there is little direct experimental evidence of symmetry ordering. How can one be satisfied that the molecular arrangement is indeed symmetric at a microscopic level? A first proof comes from the fact that if a piece of succinic anhydride crystal is irradiated by an X-ray beam of the proper wavelength, Bragg diffraction spots appear; and, recalling what was said in Section 5.4, periodic symmetry is the condition for this to happen. But there is more. Consider the following list of observed structure factors for succinic anhydride:

h	k	l	$F_{obs,hkl}$
0	2	0	285
0	3	0	0.0
0	4	0	243
0	5	0	0.0
0	6	0	21
0	7	0	0.0
0	8	0	28

All 0k0 reflections with k odd have zero net intensity. The explanation is easy. The space group of the succinic anhydride crystal is $P2_12_12_1$, and one of the equivalent positions of this space group is $x, y, z; -x, 1/2 + y, 1/2 - z$. Consider now equation 5.27: the summation has terms from all atoms in the unit cell, including atoms in equivalent positions. Therefore, for 0k0 reflections and for any atom j:

$$f_j(\exp[2\pi i k y_j] + \exp[2\pi i k (y_j + 1/2)]) = f_j \exp[2\pi i k y_j](1 + \exp[2\pi i k/2] \quad (6.1)$$

Whenever $k = 2n+1$, with n integer, $\exp[\pi i k] = -1$ and the structure factor vanishes. Thus, the above list of observed structure factors is indeed a direct experimental proof that in the crystal of succinic anhydride any scatterer at x, y, z has an equivalent scatterer at $-x, 1/2 + y, 1/2 - z$. The same applies to $(h00)$ and $(00l)$ reflections because of the other two equivalent positions of the space group. Internal symmetry is revealed by destructive interference of scattered waves from symmetry-related objects. In fact, the analysis of systematic absences is the method normally used for determining the space group from diffraction patterns.

In the same manner, all properties in a perfectly ordered crystal are identical at all equivalent positions within the unit cell and in translation-related unit cells. A typical example, quite relevant to the further developments in this chapter, is the wavefunction and hence the electron density. Vector and tensor properties are identical in modulus and their components transform according to the symmetry transformations. A relevant example is the gradient of the crystal potential, or the restraining force acting at any point (in classical vibrational treatments, at any nucleus). These simple facts, which are all consequences of the periodic translational symmetry of the crystal, suggest that many properties of crystals and many phenomena occurring in crystalline systems can be described by just studying the properties and the phenomena within the asymmetric unit or within the unit cell, with the addition of some mathematical tool that takes into account symmetry and translational periodicity. This chapter considers two theoretical developments along these lines: (1) the crystal orbital approach for the construction of the periodic wavefunction and electron density of a crystal, and for the derivation of its electronic structure, and (2) the description of periodic collective molecular oscillations, in an approach which has appropriately been called lattice dynamics.

6.1.2 *Reciprocal space, wave vector, Brillouin zone*

The quantum theory of the electronic structure of organic molecules has been developed in a purely chemical conceptual environment, through the works of people like Pauling, Pople, and Dewar. Its tools are atomic nuclei and their positions, the Born–Oppenheimer approximation, atomic orbitals, and the LCAO method, which were illustrated in Chapter 3. By contrast, the theory of the electronic structure of solids has been developed in the context of solid-state physics, and in its early stages has been concerned mainly with objects whose chemical identity was as little specified as possible: a typical example being the absolutely abstract – at least for a chemist – jellium model, a collection of nondescript nuclei in a magmatic electron sea [1]. At best, these models and ideas could be applied to metallic solids, simple ionic salts, graphite, or diamond. At the same time, some of the principles and tools of this discipline were given names that did not coincide with those devised by theoretical chemists, structural chemists, or crystallographers. This has led to some misunderstanding and confusion. It is the main purpose of this chapter to give a chemically oriented description of the methods adopted in the quantum chemistry of periodic systems.

Consider a molecular crystal, like, for example, the one whose layers are depicted in Fig. 5.5. Each molecule is a recognizable entity composed of a number of atoms, N_{at}, each one in turn described by a number, $N_{orb,i}$, of atomic orbitals expressed in gaussian expansions as in equations 3.40–3.42. There are Z entire molecules in the unit cell. A reference unit cell (called the Ref-cell) is chosen, which contains a basis set of N_{bs} atomic orbitals, χ_j, with $N_{bs} = Z \sum N_{orb,i}$. The corresponding real and reciprocal space can then be defined (Section 5.4):

$$\mathbf{R} = n\,\mathbf{a}_1 + m\,\mathbf{a}_2 + p\,\mathbf{a}_3 \qquad (6.2) = (5.22)$$

$$(\mathbf{r}^*) = \mathbf{k} = t\,\mathbf{a}_1^* + u\,\mathbf{a}_2^* + v\,\mathbf{a}_3^* \qquad (6.3) = (5.23)$$

$$\mathbf{R}_i \cdot \mathbf{k}_j = 2\pi\,\delta_{ij} \qquad (6.4) = (5.24)$$

For historical reasons that will become clearer later on, the reciprocal lattice vector $\mathbf{r}^*$ is instead called the wave vector, symbol $\mathbf{k}$, and real to reciprocal-space dot-products are normalized to 2π instead of unity. Besides, the boundaries of the independent volume unit of reciprocal space are defined as follows: after choosing a reciprocal lattice point as the origin, vectors are drawn to its nearest-neighbor points, and planes are drawn perpendicular to these vectors and passing through their midpoint. The space enclosed within these planes is called the first Brillouin zone and its content is translationally invariant.

6.1.3 *Bloch functions*

The quantum problem is, as usual, the solution of a steady state Schrödinger equation like 3.4. The translational symmetry peculiar of the crystalline state must somehow be embedded into this equation. The translational invariance of the potential energy at a generic point **r** in real space entails a translational invariance of the Schrödinger problem:

$$V(\mathbf{r}) = V(\mathbf{r} - \mathbf{R}) \tag{6.5}$$

$$\hat{H}\psi(\mathbf{r}) = E\psi(\mathbf{r}) \quad \text{or} \quad \hat{H}\psi(\mathbf{r} - \mathbf{R}) = E\psi(\mathbf{r} - \mathbf{R}) \tag{6.6}$$

Translational symmetry is conceptually similar to any other symmetry, and the eigenfunctions which are proper solutions of equation 6.6 must be symmetry-adapted to translation. These symmetry-adapted functions are called Bloch functions and must have the form [2]:

$$\Phi(\mathbf{r} + \mathbf{R}, \mathbf{k}) = \exp(i\mathbf{k} \cdot \mathbf{R})\Phi(\mathbf{r}, \mathbf{k}) \tag{6.7}$$

where **k** is for the moment just an index that consecutively labels the eigenfunctions. How many of these should one expect? A very large number, ideally an infinite number in a crystal with an infinite number of atoms. **k** is a vector whose components have the dimensions of a reciprocal length, and the dot product in the exponential is a pure number. In a way similar to that of quantum numbers in the wavefunctions for simple systems, as for example in equation 3.13, **k** contains information on the number of nodes in the eigenfunction, and the higher its absolute value, the higher the number of nodes. Other easily proved properties of the **k**-vector are (1) that its unique values span the first Brillouin zone only, (2) that $E(\mathbf{k}) = E(-\mathbf{k})$, and (3) that since $\exp(i\mathbf{k} \cdot \mathbf{R})$ is a wave, **k** expresses its wavelength (hence the name of wave vector).

The next problem is normalization. While this is easy in a finite system (see, for example, equation 3.12), for infinite systems recourse is made to a mathematical expedient called periodic boundary conditions. It consists of writing the mathematical condition for the Bloch function to be identical after any translation of the entire system of N_j crystal cells along the real space direction j. Calling m the integer number of such translations, the condition leads to the following derivation:

$$\Phi(\mathbf{r} + mN_j\mathbf{a}_j, \mathbf{k}) = \Phi(\mathbf{r}, \mathbf{k}) = \exp(i\, mN_j\mathbf{a}_j\, \mathbf{k}_j)\, \Phi(\mathbf{r}, \mathbf{k}) \tag{6.8}$$

$$\exp(imN_j\mathbf{a}_j\mathbf{k}_j) = 1 \rightarrow \mathbf{k}_j = \mathbf{a}_j^* \, n_j/N_j \tag{6.9}$$

with n_j an integer [3]. Therefore, **k** is in all respects a reciprocal lattice vector, although it need not fall on one of the reciprocal lattice points as happened when the conditions for Bragg diffraction (equation 5.26) were derived. For N_j very large, ideally infinite, the **k**-vector spans the first Brillouin zone with as fine a mesh as desired, and eventually merges into a continuous variable. In practice, however, the gradient of the eigenvalues and eigenvectors of the Schrödinger equation is relatively smooth so that the Brillouin zone can be sampled with good confidence by considering a finite and relatively small number of **k**-points.

6.2 The electronic structure of solids

6.2.1 *The crystal orbital approach*

So far, no specification of the actual form of Bloch functions in terms of atomic orbitals has been given. Their definition, equation 6.7, forces them to be composed of a product of a plane wave and a function, $\Omega(\mathbf{r}, \mathbf{k})$, which has the same translational periodicity of the lattice:

$$\Phi(\mathbf{r}, \mathbf{k}) = \exp(i\mathbf{k} \cdot \mathbf{r})\Omega(\mathbf{r}, \mathbf{k}) \qquad (6.10)$$

It can easily be verified [4] that the above equation satisfies the Bloch condition 6.7. With the Schrödinger problem written in **k**-dependent form, the solutions are sought in the form of a linear combination of Bloch functions 6.10:

$$\hat{H}\psi_m(\mathbf{r}, \mathbf{k}) = E\psi_m(\mathbf{r}, \mathbf{k}) \qquad (6.11)$$

$$\psi_m(\mathbf{r}, \mathbf{k}) = \sum_j c_{jm}(\mathbf{k})\Phi(\mathbf{r}, \mathbf{k}) = \sum_j c_{jm}(\mathbf{k})[\exp(i\mathbf{k} \cdot \mathbf{r})\Omega_j(\mathbf{r}, \mathbf{k})] \qquad (6.12)$$

with coefficients $c_{jm}(\mathbf{k})$ to be determined. For all the N_{bs} atomic orbitals $\chi_j(\mathbf{r} - \mathbf{r}_j)$ in the unit cell, each of them centered at position $\mathbf{r}_j$, in a crystal with N unit cells, the combination which has the same translational periodicity as the lattice is

$$\Omega_j(\mathbf{r}, \mathbf{k}) = N^{-1/2} \sum_\mathbf{R} \exp[i\mathbf{k} \cdot (\mathbf{R} - \mathbf{r})]\chi_{j\mathbf{R}}(\mathbf{r} - \mathbf{r}_j) \qquad (6.13)$$

R being any lattice translation, and $\chi_{j\mathbf{R}}$ a translated AO, the above equation is nothing else but the replacement of a single AO in the Ref-cell with a sum over all translationally equivalent AOs in the extended crystal. Equations 6.12 and 6.13 define the crystal orbitals. Substitution of 6.13 into 6.10 yields the final form of the Bloch function in the crystal orbital approach:

$$\Phi(\mathbf{r}, \mathbf{k}) = N^{-1/2} \sum_\mathbf{R} \exp[i\mathbf{k} \cdot \mathbf{R}]\chi_{j\mathbf{R}} \qquad (6.14)$$

The solution of equation 6.11 with the hamiltonian 3.39 and under the conditions 6.12–6.14 is conceptually similar to the solution of equation 3.4 under the LCAO assumption 3.46, although (not unexpectedly) a number of computational complications arise [5]. Averaging over **k**-space and transformations between real and reciprocal space have to be painstakingly carried out. A major difference is that while for the molecular problem one solution of the Fock equations is sufficient, for the crystal problem the periodicized Fock equations are a function of **k** and therefore the $N_{bs} \times N_{bs}$ variational problem must be solved a number of times equal to the number of sampling points within the first Brillouin zone. At the time of writing, these computational difficulties limit the applicability of the crystal orbital method to rather small molecules and unit cells [6].

While a molecular calculation yields one discrete sequence of molecular orbitals, for a crystal $E(\mathbf{k})$ plots are obtained, called energy bands. Like molecular orbitals, bands are partly filled with electrons and partly empty. The energy of the highest occupied state within the Brillouin zone is called the Fermi level, the crystal equivalent of the molecular HOMO energy. The energy difference between the Fermi level and the energy of the lowest unoccupied band is called the band-gap. Metals have half-filled electronic energy bands and a zero band-gap, so they are good conductors because electrons can easily move within the band. Organic crystals usually have wide band-gaps and are therefore semiconductors or electric insulators because electrons have very little or no mobility at all.

The solution of the crystal Schrödinger equation does not distinguish between intra- and intermolecular terms, and, in principle, provides at the same time a description of bonding among atoms in the constituting molecule and of bonding among molecules in the crystal. However, the methodological problems encountered in the description of intermolecular interactions in the LCAO-MO approach carry over to the crystal orbital method. In particular, since Bloch functions are one-electron functions and the treatment is at the Hartree–Fock level, no account of dispersion energy is possible, and the method applies best to crystals where the dispersion contributions play a lesser role, like metallic, covalent, or ionic crystals. The periodic function $\Omega(\mathbf{r}, \mathbf{k})$ can be taken in forms that do not necessarily depend on gaussian atomic orbitals; for example, plane waves can be used instead. The whole method also can be adapted to a density functional theory (DFT, see Section 3.7) implementation, and appropriate choices of the density functional may lead to at least a partial inclusion of electron correlation effects. The introduction of perturbational or configuration interaction methods like MP2 requires a strenuous computational effort, but is currently being considered [7]. Besides, since a crystal is a multimolecular system and the independent basis set is a molecular-based one, the crystal calculation is also prone to basis set superposition errors, and complex procedures have to be introduced to apply the counterpoise corrections [8].

In all the above treatment, no explicit use has been made of either molecular point-group symmetry or of the equivalent positions within the unit cell. These will have to take care of themselves in the calculation. The Bloch function approach accounts for translational symmetry only, and therefore the independent part of the whole calculation is the unit cell contents. A combination like 6.13 must be done for each of the N_{bs} atomic orbitals in the unit cell, and the resulting total energies are referred to the molecular contents of the whole unit cell. Nevertheless, molecular symmetry and equivalent positions within the cell will determine the asymmetric unit of the Brillouin zone, thus reducing the amount of independent $\mathbf{k}$-space to be sampled.

6.2.2 Band structures: Complicated but not difficult

Equation 6.14 is in nearly all respects equivalent to a linear combination of atomic orbitals, except that the coefficients are complex numbers and the whole Bloch function is a complex function. A molecular theoretical chemist is accustomed to looking at

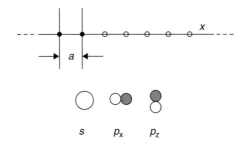

Fig. 6.1. A one-dimensional lattice of s or p atomic orbitals. White lobes denote a positive sign of the wavefunction, shaded lobes denote a negative sign.

MOs by drawing pictures that show the nodal and directional properties of each orbital in terms of the component atomic orbitals; combinations without nodes between nuclei are called bonding, those with one node are called antibonding, and the strength of the bond is proportional to the overlap between the interacting atomic orbitals. Visualizing a Bloch function is a bit more complicated because (1) there is an infinite number of terms, and (2) in general both the real and complex part of the function must be considered. However, a simple interpretation of the Bloch functions and of the band structure can be obtained by considering a one-dimensional system and looking at a few points strategically chosen along the band. This will illustrate the basic principles, although of course band structures of real molecular crystals quickly become very complicated even for relatively small molecules.

Consider a set of lattice points with a spacing a, each point being an "atom" endowed with an s or a p orbital (Fig. 6.1)[9]. There being only one atomic orbital in the unit cell, the one Bloch function formed from this AO is itself an eigenfunction of the hamiltonian operator and hence its eigenvalues are the total energies along the band. Energies are quantized, but bands are represented as continuous lines in the ideal condition of very fine sampling. Any lattice translation is just ma, with m an integer. Neglecting, for simplicity, the normalization constant, the Bloch function 6.14 along this row is:

$$\Phi(k) = \sum_m \exp[ik(ma)]\chi_m \qquad (6.15)$$

In this particular one-dimensional case, k is just a number which may vary (equation 6.9) within the first Brillouin zone between $-a^*/2$ and $+a^*/2$, but since $E(-k) = E(k)$ only the zone between 0 and $a^*/2$ need be considered. Since $aa^* = 2\pi$, one gets:

$$\text{for } k = 0 \quad \Phi(0) = \sum_m \chi_m \qquad (6.16a)$$

$$\text{for } k = a^*/2 \quad \Phi(a^*/2) = \sum_m \exp[i\pi m]\chi_m = \sum_m [-1]^m \chi_m \qquad (6.16b)$$

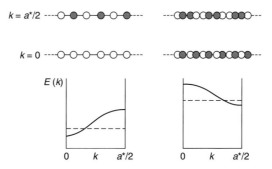

Fig. 6.2. The Bloch functions at special points in k-space for the lattice in Fig. 6.1, and the qualitative shape of the corresponding energy bands, for (left) the s-orbital and (right) the p-orbital. The horizontal lines are the energies of the isolated atomic orbitals. If there is one electron in each orbital, the bands are half-filled and the Fermi level is at half the bandwidth with a zero band gap.

In these two special cases the Bloch function is real and can be readily visualized for the s AO as the fully bonding (all χ_ms with the same sign, no nodes) and the fully antibonding (a node every second χ_m) combinations (Fig. 6.2, left). No wonder that the corresponding energies $E(k)$ are the lowest and the highest possible along the energy band, while all other functions for intermediate values of k will have an intermediate number of nodes and intermediate energies. It is also clear that the case is different for the p orbital, whose symmetry is such that the all-plus combination is the most antibonding one, so that while the s band runs up in energy, the p band runs down in k-space (Fig. 6.2, right). The energy difference between the lowest and the highest level within the band is controlled by overlap, just as the energy splitting between any bonding–antibonding pair of energy levels; the formation of an energy band is the limiting case of the formation of molecular orbitals along a polymer, for an infinite number of monomers. At the limit of $a \to \infty$, the overlap is null and the bonding and antibonding combinations have the same energy, so the bandwidth is zero. The shorter the spacing, and the more the orbitals overlap, the wider is the energy spread and the wider is the band.

Consider now a set of equally spaced biatomic molecules along the same one-dimensional lattice (Fig. 6.3). The complete solution must now be a combination of Bloch functions, but the form of this combination can be guessed by forming the σ and σ^* bonding and antibonding molecular orbitals, and by Bloch-periodicizing these. Consider first the case in which the bond length is equal to one half the periodic translation: this is nothing but a formal doubling of the a periodicity, which means a halving of the length of the Brillouin zone in reciprocal space. The fully antibonding combination is now the $k = 0$, in-phase Bloch function out of the σ^* molecular orbital. At the end of the Brillouin zone (in usual notation, a point denoted by K), the two out-of-phase combinations from the σ and σ^* molecular orbitals are identical and their energies are degenerate. If the bond length becomes smaller than the periodicity,

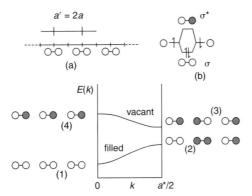

Fig. 6.3. (a) A one-dimensional lattice of diatomic molecules. (b) The formation of the bonding and antibonding molecular orbitals from s-type atomic orbitals. The lower part of the figure shows the four critical forms of the combination (1–4): the formation of the chemical bond lifts the degeneracy of the out-of-phase combinations of σ and σ^* and a gap opens between the so-called "valence" (filled) and "conduction" (vacant) bands. The band diagram is almost equivalent to that of Fig. 6.2, left, "folded back" at half its width (the periodicity is doubled) except for the gap at $k = a^*/2$ (actually, one half of the former $a^*/2$).

as it should be in a real molecular case, this degeneracy is lifted: the Bloch function (2) from the σ MO becomes more bonding, and hence more stable, as the overlap between orbital lobes of the same sign increases, while the combination (3) from the σ^* MO becomes less stable because of increased overlap between lobes of opposite sign. There is now a non-zero band gap, and this one-dimensional "material" has undergone a transition from a "metal" to a molecular "insulator". This is what physicists, beautifully oblivious of chemical bonding, sometimes call a Peierls distortion [10].

Symmetry considerations alone can teach a lot about the nature and properties of crystal orbitals. In a first approximation, each orbital in the unit cell can be considered as giving rise in k-space to a separate energy band, which may be vacant, half-filled or filled, as the original MO is, and whose width and slope can be qualitatively guessed on the basis of simple symmetry and overlap arguments. For example, in a row of π-stacked ethylene molecules the bands arising from s-σ and p-σ molecular orbitals must be very flat because of scarce or null overlap, while the π-π band would develop some width because of π-overlap (Fig. 6.4). Since all the s and the p-π MO are fully occupied, so are also the s and the p-π bands, and the wide gap to the lowest unoccupied band, the p-π^* band, explains the insulating nature of this hypothetical material.

Actual band structures in organic crystals must obviously be much more complicated than these simple examples [11], and yet they are built on the same basic principles, and many of their properties are understandable in terms of the symmetry and properties of the atomic orbitals. In the simple one-dimensional cases considered above, one-dimensional paths through k-space are all there is, but in crystals the band structure is three-dimensional. For visualization, some one-dimensional

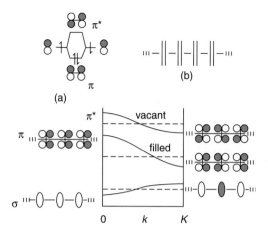

Fig. 6.4. (a) Two π molecular orbitals from *p*-type atomic orbitals. (b) A stack of parallel ethylene molecules, represented by the double bond. The diagram shows the qualitative trends in the band structure; horizontal lines are the energies of the orbitals of an isolated molecule. For clarity the σ manifold is represented by a single $E(k)$ curve. The σ* bands are high above the π* one.

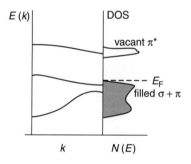

Fig. 6.5. Schematic DOS curves for the electronic bands in Fig. 6.4. The top filled states are mostly π-type orbitals, the lowest vacant states mostly π*-type orbitals.

paths of particular symmetry or significance through the Brillouin zone must then be considered.

6.2.3 *Comparison with experiment; electronic density of states*

For any given electronic energy band, a corresponding density of states (DOS) function can be defined. The function DOS(E) is simply the number of energy levels within any small interval between E and $E + dE$. Since energy levels are equally spaced, the flatter the band, the sharper and the higher the DOS function (see Fig. 6.5). The

DOS function can be considered as an averaging over reciprocal space, and hence no longer depends on the wave vector **k**. The DOS description emerges from reciprocal space into the real world of electronic transitions; for example, for an insulator, the density of occupied states drops to zero at the Fermi level, and the energy gap to the first DOS peak for the unoccupied states corresponds to the energy of the first electronic transition within the material, in the same line of thought and with the same approximations that were introduced for electronic transitions in isolated molecules (Section 3.5.2). This energy gap is measurable by electron spectroscopy experiments. In addition, the electronic DOS can be analyzed by an overlap population procedure quite similar to the one that applies to the molecular case. The contribution from each basis orbital can be evaluated, so that the various parts of the calculated DOS can be assigned to different parts of the molecule [12].

6.3 Lattice dynamics and lattice vibrations

6.3.1 *Periodic vibrations in infinite crystals*

Consider again a three dimensional molecular crystal, concentrating now on atomic nuclei, taking advantage of the Born–Oppenheimer approximation in which the kinetic energies of electrons and nuclei are fully decoupled. Let N_{at} be the number of atoms in the constituting molecule, and Z be the number of molecules in the unit cell. All atoms in the crystal are oscillating about their equilibrium positions under the restraining action of some potential V, which must also obey the translational invariance condition 6.5. Let x_i be the instantaneous displacement of atom i from its equilibrium position in the crystal; consider all the atoms in the Ref-cell and all atoms in the surrounding cells, denoted as before by the real space vector **R**. The displacement force constants (recall equation 2.4) can be written as:

$$D_{ij,\mathbf{R}} = (m_i m_j)^{-1/2} [\partial^2 V / \partial x_{\text{Ref},i} \, \partial x_{\mathbf{R},j}]^\circ \tag{6.17}$$

where the m_is are atomic masses, and the "0" superscript recalls that these derivatives must be calculated at the equilibrium position, where the first derivatives of the potential vanish. The derivatives 6.17 once again have the same lattice periodicity as has the potential, and this immediately suggests the possibility of treating these vibrational entities just as the electronic orbitals had been treated, i.e. to periodicize them through a Bloch theorem expansion:

$$D_{ij}(\mathbf{k}) = \sum_{\mathbf{R}} \exp[i\mathbf{k}\mathbf{R}] \, D_{ij,\mathbf{R}} \tag{6.18}$$

Note the formal equivalence of equations 6.14 and 6.18. The dynamical problem, whose form and solution were described in Chapter 2, must be solved to find the normal coordinates as linear combinations of the basis Bloch functions, and the amplitudes and frequencies of these normal vibrations. These depend on **k**, and therefore the problem must be solved for a number of **k**-points chosen so as to ensure an adequate sampling of the Brillouin zone. A dispersion of vibrational frequencies in **k**-space is obtained, just as the former treatment had given a dispersion of electronic energies in **k**-space [13].

The model potential V is the key to accuracy here. The potential may come from a molecular orbital or crystal orbital calculation, in which case the derivatives must be computed numerically. In another approximation, the potential may consist of a sum of intra- and intermolecular terms in the form of the empirical force fields described in Section 2.2. This is particularly convenient because all the derivatives of equation 6.17 can be computed analytically. In an even coarser approximation, the molecule may be considered as a rigid unit, without allowing for internal deformations. In this case the displacement coordinates are just three coordinates for the center of mass and three coordinates for rotation around the inertial axes. Equation 6.17 is rewritten in terms of these coordinates, the potential is just the intermolecular part and there is no need to define an intramolecular force field, and the problem is reduced from a $3ZN_{at} \times 3ZN_{at}$ one to $6Z \times 6Z$ one [14].

Consider again the one-dimensional lattice of Fig. 6.1, in which each lattice point is occupied by a non-bonding atom (for example, this could be a string of argon atoms). Finding the crystal vibrational normal coordinates is the equivalent of finding the crystal molecular orbitals. The vibrational Bloch functions have the same form as equation 6.15, with atomic displacements instead of atomic orbitals, and for this one-dimensional lattice the Bloch function itself is a normal coordinate. The vibrational

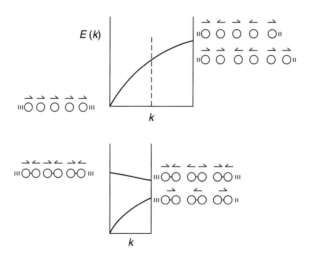

Fig. 6.6. Arrows indicate the direction of instantaneous displacement. Top: the energy band for the lattice vibration of a row of atoms. Bottom: a diatomic molecule: the acoustic branch corresponds to displacements of the rigid diatomic molecule, without bond stretching: at $k = K$ there is a considerable compression of the intermolecular contacts so the band runs "up" in energy. For the optical branch, the intermolecular perturbation is larger at $k = 0$ than at $k = K$ and the band runs "down" in energy. Arrows reverse periodically during the oscillation.

band structure can be explained by considering the displacements, which in this case can only be $+x$ or $-x$ (Fig. 6.6, top). For the in-phase combination of all $+x$ displacements one gets a bare translation of the whole crystal, a vibration of zero frequency. The out-of-phase combination involves a displacement of all atoms from their minimum energy position, either to longer or to shorter intermolecular distance, always with a destabilization, and the corresponding vibration must have a non-zero frequency. So the phonon dispersion band [15] runs up from zero frequency at the origin to some higher value at point K. There are three of them for a real three-dimensional crystal. Obviously, the width of the band is controlled by the second derivatives of equation 6.17, and so ultimately by the strength of the intermolecular potential: at the limit of infinite separation and null potential, the frequency would stay zero throughout the whole band, while for a steep potential and close contact the frequency at point K becomes higher and higher and the band becomes wider and wider. Here the force constants play the same coupling role as that played by overlap in the development electronic bands.

Consider now a one-dimensional crystal made of biatomic molecules, a molecular crystal: note that the following line of reasoning runs strictly parallel to the molecular orbital case illustrated in the preceding sections. The two atoms are connected by a chemical bond with its own proper vibrational stretching frequency. As a guess at the normal coordinates, two combinations of displacements can be formed, $x_A + x_B$ and $x_A - x_B$ (aside from normalization). As before, the lattice parameter is doubled so that the dimension of the Brillouin zone is halved. The Bloch functions at the two extremes of the Brillouin zone are shown in the lower part of Fig. 6.6. For the first combination, the acoustic branch is again obtained; but the displacements at point K (one half of the former Brillouin zone) now involve a collision between unstretched molecules, with a change in intermolecular energy, although at a relatively low price. The corresponding vibration must then have a relatively low frequency. For the asymmetric combination of displacements the bond length is stretched, and the strength of the intermolecular potential determines the amount of spread from the frequency of the intramolecular vibration in the isolated molecule. These modes do not have zero frequency at $\mathbf{k} = 0$, and are called optical modes.

For a comparison of relative bandwidths, consider the corresponding case in which the diatomic molecules are aligned with the same periodicity but perpendicular to the lattice displacement direction (Fig. 6.7). Any up–down motion now involves much less displacement from the intermolecular energy minimum distance, hence the vibration is less constrained than when the molecules are aligned along the lattice periodicity direction, and the band widths are much smaller. If, as is often the case in soft organic crystals bound only by weak dispersion potentials, intramolecular vibrations are much more energetic than intermolecular ones, the coupling can be neglected, off-diagonal force constant matrix elements between inter- and intramolecular vibrations are negligible, and the crystal has very nearly the same vibrational frequencies as the isolated molecule. This is usually true for rigid molecules with stretching and bending degrees of freedom only, but the decoupling may not be legitimate for molecule with torsional or other "soft" intramolecular degrees of freedom.

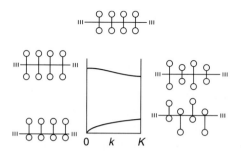

Fig. 6.7. Transverse displacements: compare with Fig. 6.6. These are shear motions with a much smaller intermolecular energy contribution than in the head-on contacts between molecules aligned along the displacement direction, and bands are much narrower.

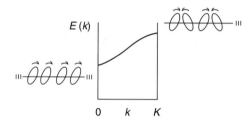

Fig. 6.8. A schematic dispersion curve for orientational oscillations. The band runs up because the $k = K$ mode involves a more critical compression among molecules displaced from equilibrium positions.

Frequencies at $\mathbf{k} = 0$ are never zero when molecular rotations are involved. Consider (Fig. 6.8) a collective molecular oscillation in a one-dimensional array, with in-phase rotations at $k = 0$ involving a relatively small displacement from equilibrium positions, and out-of-phase rotations at $k = K$, with a relatively larger displacement from equilibrium. The corresponding band starts at some non-zero frequency and runs up a little.

What are the atoms really doing during a lattice vibration? First of all, a real lattice vibration must be a complex combination of atomic displacements in the three directions of space, but still its nature and physical meaning do not change from those of the simple one-dimensional example given above. Molecules in crystals are very constrained, so that oscillations are restricted to relatively small displacements from equilibrium. This is the reason why the harmonic assumption can be successfully applied in lattice dynamics. We are now dealing with a dynamic, time-dependent phenomenon, in which atomic displacements are periodic in time (equation 2.19). The illustrations in Fig. 6.2 and 6.6 differ in one crucial point: the picture of atomic orbital combinations is static, while vibrational modes have an additional phase term that depends on time, and describes the periodic oscillation of the nuclei around the equilibrium positions. The Bloch phase term describes the phase of the translational adaptation and thus the overall symmetry of the vibration, the time-dependent phase term describes the periodic oscillation. Thus, the arrows in Figs 6.6–6.7 describe one direction of the oscillation, and revert as the oscillation proceeds. In other words, what one gets is a collective, periodic oscillation of all atoms within the crystal along the crystal normal coordinates. At $\mathbf{k} = 0$ all unit cells oscillate in phase, just as all atoms are displaced in phase in a molecular normal vibration. At $\mathbf{k} \neq 0$ there is a phase shift among unit cells, described by the value of $\mathbf{k}$, and molecular motion is best described as a travelling vibrational wave.

Of these lattice vibrations, the low-frequency ones can be excited thermally and contribute to the heat capacity of the crystal, making it much larger than that of the gas phase (Section 7.3). For example, it can be shown [16] that the contribution from a crystal containing N molecules is:

$$C_\mathrm{V} = 6N k_\mathrm{B} \int g(\nu) e^u / (e^u - 1)^2 \, d\nu \tag{6.19}$$

where $g(\nu)$ is the vibrational density of states, $u = h\nu/(k_\mathrm{B} T)$, and k_B is Boltzmann's constant. This expression incorporates a sum over all branches and over all wavevectors in the Brillouin zone. Aside from technical problems in renormalization to the unit cell contents and in the actual numerical integration, the evaluation of this quantity is fairly straightforward.

6.3.2 Comparison with experiment; measuring lattice-vibration frequencies

Just as gas-phase IR-Raman spectroscopy can measure the frequencies of molecular vibrations, solid-state IR-Raman experiments can measure the vibrational frequencies of the crystalline material for $\mathbf{k} = 0$ [17]. In the limit of a completely decoupled intra- and intermolecular vibrations the lattice sees the molecule as a completely rigid object, and the molecule sees the lattice as a negligible perturbation. There will therefore be a set of so-called "internal" modes, in which atomic positions change without a significant change in the position of the center of mass and in the overall molecular orientation, and whose frequencies are not much different from the frequencies of the gas-phase molecular vibrations. The "external" modes, or genuine lattice vibrations, result from collective displacements and rotations within the regular assembly of rigid objects, which is the crystal. External modes must have much lower frequencies than internal ones, if one considers the relative magnitude of the intra- and intermolecular restraining potentials and hence of the corresponding force constants. The region of pure lattice vibrations for organic crystals (see an example in Fig. 6.9) is the far infrared, 150–50 cm^{-1}, as compared with 500–1,000 cm^{-1} for the weakest bond stretching vibrations. The experimental measurement of these very low vibrational frequencies poses a number of experimental problems, not to speak of the severe problems encountered in assigning a particular observed frequency to a particular vibrational mode. This latter task is usually accomplished by

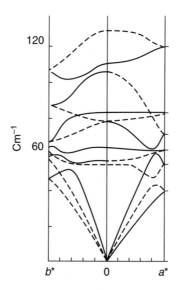

Fig. 6.9. Schematic, representative phonon dispersion curves for the naphthalene crystal. Three acoustic branches and several optical branches can be seen. Note that the shape of the bands is different in the two different paths through the Brillouin zone, to a^* and b^*. This is due to the anisotropy of the crystalline environment. Adapted from data in ref. [13], p. 264.

using appropriately oriented crystals and polarized radiation, in a complex and very critical experiment that requires in the first place the growth and manipulation of suitably large single crystals. All these facts considered, it is not surprising that the available library of accurately determined and symmetry-assigned lattice vibrational frequencies for organic crystals is not large [18]. Entire phonon dispersion curves can be determined by inelastic neutron scattering experiments, in an even more awkward experimental setting [19]. Nevertheless, available data are enough to allow a quantitative comparison between observed lattice modes and their calculated counterparts for a selection of representative organic molecules treated with different intermolecular potentials [20]. The comparison, within the above described limits, is surprisingly favorable even when using relatively unrefined model potentials, but only for molecules without internal degrees of freedom that might couple with lattice modes.

References and Notes to Chapter 6

[1] Jellium is a neutral system consisting of a number of dynamic electrons in a background of uniform positive charge density.
[2] Equation 6.7 is a mathematical expression of the so-called Bloch theorem.
[3] The last derivation is verified because $\exp[im(n_j/N_j)N_j\mathbf{a}_j \cdot \mathbf{a}_j^*] = \exp(imn_j 2\pi) = 1$ when m and n_j are integers.

[4] In fact if $\Omega(\mathbf{r},\mathbf{k}) = \Omega(\mathbf{r}+\mathbf{R},\mathbf{k})$, the periodicity condition, then $\Phi(\mathbf{r}+\mathbf{R},\mathbf{k}) = \exp(\mathrm{i}\mathbf{k}\cdot(\mathbf{r}+\mathbf{R}))\Omega(\mathbf{r}+\mathbf{R},\mathbf{k}) = \exp(\mathrm{i}\mathbf{k}\cdot\mathbf{r})\exp(\mathrm{i}\mathbf{k}\mathbf{R})\Omega(\mathbf{r},\mathbf{k})$; then $\Phi(\mathbf{r}+\mathbf{R},\mathbf{k}) = \exp(\mathrm{i}\mathbf{k}\cdot\mathbf{R})\Phi(\mathbf{r},\mathbf{k})$, q.e.d.

[5] A clear and exhaustive description of the method is in Dovesi, R.; Civalleri, B.; Orlando, R.; Roetti, C.; Saunders, V. R. Ab initio quantum simulations in solid state chemistry, *Revs. Comp. Chem.* 2005, **21**, 1–125. This review contains a discussion of all aspects of periodic calculations, including differences between conducting and non-conducting systems, Brillouin zone sampling, basis set dependence, etc., and a survey of available computer packages. For the CRYSTAL software package for periodic MO calculations in crystals, developed by the authors, see www.crystal.unito.it.

[6] For the first example of fully ab initio periodic SCF study of a molecular crystal see Dovesi, R.; Causà, M.; Orlando, R.; Roetti, C.; Saunders, V.R. Ab initio approach to molecular crystals: a periodic Hartree–Fock study of crystalline urea, *J. Chem. Phys.* 1990, **92**, 7402–7411. The unit cell contains sixteen atoms and the basis set employed is 6-21G**. The band structure, density of states, and deformation electronic densities are obtained. The lattice energy is -117 kJ mol^{-1} without, and 67 kJ mol^{-1} with basis set superposition error correction [8]. The latter value reflects the absence of dispersion contributions in the HF approach (experimental heat of sublimation 88 kJ mol^{-1}). Such a study would now be comfortably feasible on any standard PC, but clearly the computational load of a periodic crystal calculation increases with the increasing number of atoms in the unit cell and size of the atomic orbital basis set. A good critical entry point to the literature is Spackman, M.A.; Mitchell, A.S. Basis set choice and basis set superposition error (BSSE) in periodic Hartree–Fock calculations on molecular crystals, *Phys. Chem. Chem. Phys.* 2001, **3**, 1518–1523.

[7] Pisani, C.; Busso, M.; Capecchi, G.; Casassa, S.; Dovesi, R.; Maschio, L.; Zicovich-Wilson, C.; Schulz, M. Local MP2 electron correlation method for nonconducting crystals, *J. Chem. Phys.* 2005, **122**, 094113.

[8] See ref. [5], p. 50ff., and ref. [6].

[9] The following examples owe much to the clear and amusing treatment of the matter given in Hoffmann, R., *Solids and Surfaces: A Chemist's View of Bonding in Extended Structures,* 1988, VCH, Weinheim, absolutely recommended reading. See also ref. [5], pp. 21ff.

[10] Ref. [9], p. 92ff.

[11] For a two-dimensional example on graphite including the visualization of real and imaginary parts of Bloch functions see ref. [5], pp. 23ff. For a clear example of the dependence of band structure and band width on intermolecular interaction, see Widany, J.; Daminelli, G.; Di Carlo, A.; Lugli, P.; Jungnickel, G.; Elstner, M.; Frauenheim, T. Electronic band structure and intermolecular interaction in substituted thiophene polymorphs, *Phys. Rev.* 2001, **B63**, 233204. One of the three polymorphs has planar stacking of the aromatic rings, with relevant intermolecular interactions; it is calculated to be a semiconductor with a band gap of 1.35 eV, and a band dispersion of about 0.5 eV, very large for organic

compounds. Another polymorph has a much less tightly packed crystal structure, the band gap is above 2 eV, and the band structure is comparatively flat.

[12] The energy levels at the top of the band can usually be probed by ultraviolet photoelectron spectroscopy (UPS). For a brief description of how the molecular population analysis concepts (equation 3.47) apply to periodic wavefunctions, see ref. [9], pp. 32ff.: the procedure requires a summation and averaging of the overlap density sums over the Brillouin zone.

[13] A detailed description of the principles and the mathematical equations of lattice dynamics in the harmonic approximation can be found in a number of standard references. For example, Pertsin, A.J.; Kitaigorodski, A.I. *The Atom–Atom Potential Method. Applications to Organic Molecular Solids*, 1987, Springer-Verlag, Berlin, Chapter 5, includes a detailed derivation of all the necessary lattice energy derivatives and a number of examples of typical molecular crystals; Chapter 6 has the thermodynamic derivations. Another useful reference is Califano, S.; Schettino, V.; Neto, N. *Lattice Dynamics of Molecular Crystals, Lecture Notes in Chemistry*, 1981, Springer-Verlag, Berlin.

[14] The rigid-body approximation in the atom–atom potential energy approximation is accessible even to modest computing resources, and was therefore used in early applications: see e.g. Filippini, G.; Gramaccioli, C.M.; Simonetta, M.; Suffritti, G. B. Lattice-dynamical calculations on some rigid organic molecules, *J. Chem. Phys.* 1973, **59**, 5088–5101. A lattice-dynamical treatment of organic crystals also opens the way to a theoretical estimation of the atomic and molecular thermal displacement parameters, which can also be measured by X-ray diffraction (as described in Chapter 5). For an example of this, where internal vibrational frequencies are also considered, see Gramaccioli, C.M.; Filippini, G. Lattice-dynamical evaluation of temperature factors in non-rigid molecular crystals: a first application to aromatic hydrocarbons, *Acta Cryst.* 1983, **A39**, 784–791. These two references also include cross-references to parallel work by other groups.

[15] Physicists use the imaginative term "phonons" for collective molecular motions in crystals. The name "acoustic" branches comes from the fact that molecular oscillations are related to the transport of mechanical energy and hence to the propagation of sound waves within the material. The name "optical" branches is used when the $\mathbf{k} = 0$ frequencies are not zero and hence can be measured in an optical (i.e. spectroscopic) experiment.

[16] See the derivation in ref. [13], Chapter 6.

[17] Lattice vibration frequencies are directly measurable by IR-Raman spectroscopy only for $\mathbf{k} = 0$ (for simplicity all equations will here be written with scalar k). To see why this is so consider the expression $w = \exp(ikR) = \exp(i\varphi)$ as a wave where φ is the phase angle, R is the current displacement along the wave propagation direction, and $k = 2\pi/\lambda$, where λ is the wavelength. For $R = 0$ the phase angle is zero, for $R = \lambda/2$ the phase angle is π, and so on. Since $\exp(i\varphi) = \cos\varphi + i\sin\varphi$, the amplitude is $ww^* = (\cos^2\varphi + \sin^2\varphi)^{1/2} = 1$. Whenever the wave represents a quantum phenomenon, wavelength is associated with

momentum through the de Broglie relationship $\lambda = h/p$ where h is Planck's constant. In these terms a typical IR radiation with frequency 1,000 cm^{-1} has a wavelength of 0.001 cm and hence k is of the order of 10^3 cm^{-1}. A crystal phonon has $0 < k < a^*$, where a* is a typical reciprocal lattice parameter, or 10^7 cm^{-1} for a typical real cell translation of $10\text{Å} = 10^{-7}$ cm. The wave vector **k** also measures momentum and this is why **k**-space is also called momentum space.

Quantum chemical momentum, like all energy-related quantities, must be conserved. In an ordinary gas phase spectroscopic experiment the incoming photon of appropriate wavelength is absorbed by the molecule and the corresponding normal mode vibration is excited, and the condition that the momentum of the wave be equal to the momentum of the vibration (momentum conservation) is automatically satisfied. A crystal phonon has an additional translational wavelength and the momentum associated with high-k modes is very high, so the incoming IR photon does not have enough momentum to excite those modes. This is why only very low-k (ideally, $k = 0$) phonons can be probed in an ordinary IR experiment.

[18] See ref. [13], Section 5.6.

[19] In order to probe the entire Brillouin zone of **k**-space, inelastic neutron scattering experiments must be carried out, where the excitation of lattice phonons draws whatever momentum is needed from the high momentum of the neutron quanta. In addition, the **k**-vector of the incoming radiation can be properly matched with the phonon **k**-vector in modulus and direction, so that **k**-dependent intensities can be obtained, the various branches can be probed separately, and the entire dispersion curves can be experimentally determined. Such experiments have been carried out only for some particularly simple crystals. For a review see e.g. Gonze, X.; Rignanese, G.-M.; Caracas, R. First-principle studies of the lattice dynamics of crystals and related properties, *Z. Krist.* 2005, **220**, 458–472.

[20] Day, G. M.; Price, S. L.; Leslie, M. Atomistic calculations of phonon frequencies and thermodynamic quantities for crystals of rigid organic molecules, *J. Phys. Chem. B* 2003, **107**, 10919–10933. The molecules considered are hexamethylenetetraamine, naphthalene, pyrazine, imidazole, and glycine.

7

Molecular structure and macroscopic properties: Calorimetry and thermodynamics

The molecular interaction energy is a major factor determining the disposition of the molecules in a lattice. While voluminous data and, often, great detail are available on the geometry of the molecular placements we have in general only a semiquantitative appreciation of the energy functions which have led to the lattice structure and often only crude qualitative interpretations can be offered. To that extent chemical crystallography is still a descriptive science.

Davies, M. Studies of molecular interactions in organic crystals
J. Chem. Educ. 1971, **48**, 591–593.

Measurements of sublimation enthalpies and vaporization enthalpies have been made for over a century ... despite the longevity in interest in these measurements and the significant technological improvements in instrumentation and assessment of purity, examination of recent data still show significant discrepancies in data published by different laboratories and different techniques.

Chickos, J. S. Enthalpies of sublimation after a century
of measurement: a view as seen through the eyes of a collector
Netsu Sokutei 2003, **30**, 116–124.

7.1 Molecules and macroscopic bodies

Spectroscopic and diffraction experiments can give a detailed picture of molecular objects. The dimensions of a water molecule are know with extreme reliability and accuracy. In real life, however, one is never confronted with a single water molecule, but, rather, with bodies composed of a very large number of molecules. We deal with moles, not with molecules.

A mole of water vapor occupies 22.4 liters at room conditions. Each nucleus is described by three positional parameters and three components for its momentum, these coordinates spanning what is briefly called phase space. Molecules are vibrating, and are zipping around and rotating very fast. Assuming that all, or very nearly all, of them are in the ground electronic state, the energies belonging to this molecular ensemble are vibrational potential energies, plus vibrational, rotational, and translational kinetic energies. In addition, there may be a minor contribution from intermolecular potential energies, whenever two molecules happen to come close enough to form a fleeting hydrogen bond. Temperature is the external manifestation of molecular velocities and kinetic energies. Pressure is related to the amount of momentum that crosses a surface within the gas per unit of time. The sum of all the energies in the ensemble is the thermodynamic internal energy of the system. The adjective, "internal", denotes that all these energies are defined in a local reference frame; the temperature or the internal energy of a glass of water are the same at rest on the ground or in flight at 1,000-plus kilometers per hour in a jet plane.

There are several kinds of energy, and molecules are continuously interchanging and redistributing them at a very high rate. A molecule has no special preference, so the available energy will be distributed at random among vibrational, translational, and rotational degrees of freedom. In the absence of external intervention the energy distribution that can be obtained in the greatest number of ways will eventually be the most frequently observed one, because molecules are distinct but not distinguishable. For such large numbers of molecules as are present in macroscopic bodies, this distribution with the highest statistical weight is overwhelmingly more frequent than all other distributions, and is called the equilibrium distribution.

When water condenses into the liquid or crystalline state, one mole occupies only about 0.018 liters, and this is because a cohesive force (Chapter 4) holds the molecules together much more tightly than in the gas phase. The internal energy now has a substantial contribution from the intermolecular potential energy. In addition, intermolecular vibrational degrees of freedom provide a large number of extra pockets into which potential energy can be stored: in the crystalline state, molecules oscillate in collective lattice vibrations (Section 6.3).

Chemical systems do not tend to equilibrium by minimizing their internal energy. Another driving force is provided by spontaneous randomization of the position of molecules in space and of the distribution of energies among available energy levels. This fundamental fact is embodied in the macroscopic property called entropy.

Temperature, pressure, internal energy, and entropy are the basic functions of chemical thermodynamics. They can be explained in principle by considerations based on quantum mechanics, or on structural science, in terms of the properties of the constituting molecules, but they are defined only as averages over large molecular ensembles, not for single molecules. Doing chemistry with quantum levels, electrons, nuclei, bond lengths, and torsion angles is a fascinating intellectual adventure, but, sooner or later, one must come to grips with the real world in terms of ordinary objects, on which a measurement of some macroscopic thermodynamic property is carried out: molecular energetics meets chemical thermodynamics. Energy can be detected by an experimenter only when it is exchanged, and the most easily measurable form of energy exchange is heat, so that experimental thermodynamic evidence comes from measurements done with a calorimeter. The basic experiment carried out with this invaluable instrument looks at the amount of exchanged heat against a change in temperature – a heat capacity. It will be shown in this chapter how heat capacity measurements indeed constitute one of the pillars, not only of thermodynamics, but of much of physical chemistry as we understand it today [1].

7.2 Energy

The total amount of energy available to any given system spreads over all accessible energy levels of the system. All distributions are equally likely, so that at equilibrium the system will choose the distribution that has the largest statistical weight. This brilliant intuition is one of the key concepts in thermodynamics, in chemistry, and perhaps in the whole of modern science; it is due to Ludwig Boltzmann. Molecular energies are expectation values of quantum mechanical operators, and the Boltzmann distribution applied to quantized energy levels leads the way to the theoretical calculation of thermodynamic functions for chemical systems.

7.2.1 The partition function: Molecules

The partition function is the key quantity in the calculation of the Boltzmann equilibrium distribution of all molecular energies, and paves the way to a calculation of the total internal energy of any molecular system at equilibrium. Consider a system made of N molecules, each of which has a set of quantized energy levels ε_i, known by the solution of some quantum chemical secular equation, as shown in Chapter 3. Energies are distributed over the accessible energy levels. With a modicum of mathematical derivation the following basic equations can be obtained:

$$n_i^* = N \exp(-\beta \varepsilon_i) / \sum_i \exp(-\beta \varepsilon_i) \tag{7.1}$$

$$\beta = 1/(k_B T) \tag{7.2}$$

$$q = \sum_i \exp(-\beta \varepsilon_i) \tag{7.3}$$

Equation 7.1 defines the equilibrium Boltzmann distribution, that is, the distribution with the highest statistical weight, and n_i^* is the number of molecules in energy level ε_i in that distribution. k_B is the Boltzmann constant, and equation 7.3 defines the partition function. The physical meaning of the partition function is easy to grasp. The first level is taken as the zero of the energy scale, so the first term in the summation is equal to 1. Other terms are numbers between 0 and 1 and are small if the temperature is low or if the energy of the level is high; briefly, the partition function is a number that tells how many levels in a given set are likely to be populated. At a given temperature, the population of higher levels decreases exponentially as ε_i increases; as temperature rises, the population of higher-energy levels increases and that of lower-lying levels decreases. As a rule of thumb, a quantum level is expected to be significantly populated only if its relative energy is not much higher than $k_B T$. Energy levels are a ladder that molecules climb with the help of an input of energy or a rise in temperature.

Each overall energy level ε_i is in fact a combination of translational (Tr), rotational (Ro), vibrational (Vi), and electronic energy levels. As a first approximation, Tr, Ro, Vi, and electronic energies are considered as uncoupled, so that any overall molecular energy state ε_i is represented just by a sum of various kinds of quantized energies:

$$\varepsilon_i = \varepsilon(Tr)_k + +(Ro)_l + +(Vi)_m + +(El)_n \tag{7.4}$$

$$\exp(-\beta\varepsilon_i) = \exp(-\beta\varepsilon_k)\exp(-\beta\varepsilon_l)\exp(-\beta\varepsilon_m)\exp(-\beta\varepsilon_n) \tag{7.5}$$

The molecular partition function is then obtained by summing over all possible ε_i levels, which is equivalent to separately summing over all possible indices $k, l, m,$ and n in equation 7.5:

$$\sum_i \exp(-\beta\varepsilon_i) = q(\text{Mol}) = q(\text{Tr})q(\text{Ro})q(\text{Vi})q(\text{El}) \tag{7.6}$$

The next step is to obtain an explicit form for the four terms in the product, each of which has the form of equation 7.3. For most organic and organometallic molecules the spacing between the ground and the first excited electronic state is so high that in all reasonable temperature ranges $E(\text{El}) \gg k_B T$ and $q(\text{El}) = 1$. The results of Section 3.1 on the quantum mechanical energy levels for translation, rotation, and vibration now apply. The quantized Tr energy levels for a particle of mass m_P in motion within a linear space of dimension X are given by equation 3.11. Then:

$$\varepsilon_n(X) = (n^2 - 1)h^2/(8m_P X^2) \tag{7.7}$$

$$q(\text{Tr}, X) = \sum_n \exp(-\beta\varepsilon_n) = \int \exp[-\beta n^2 h^2/(8m_P X^2)]\, dn = (2\pi m_P/\beta h^2)^{1/2} X \tag{7.8}$$

Tr levels are so closely spaced that the summation can be replaced by an integral. Besides, n is always so large that $n^2 - 1 = n^2$. When the three components of displacement are considered, one gets

$$q(\text{Tr}) = q(\text{Tr}, X)q(\text{Tr}, Y)q(\text{Tr}, Z) = (2\pi m_P k_B T/h^2)^{3/2} V \tag{7.9}$$

where V is the total volume of the system.

Equation 3.21 is used to calculate the rotational partition function. Each rotational energy level is $2l+1$ times degenerate because of the vectorial quantization of angular momentum, so it must be counted as many times in the summation. On the other hand, whenever the molecular symmetry is such that identical atoms exchange their positions because of a molecular rotation, the summation must be divided by an appropriate symmetry number, σ [2]. Then:

$$q(\text{Ro}) = \sum_l \exp(-\beta\varepsilon_l) = (1/\sigma)\sum_l (2l+1)\exp[-(\beta/2I)h^2[l(l+1)]] \tag{7.10}$$

Ro levels are much more spaced than Tr ones but the replacement of the summation by an integral is still legitimate, at least for molecules with large moments of inertia (Section 1.3.1). Solution of the integral corresponding to the summation 7.10, with due extension to polyatomic molecules with three rotational moments of inertia, I_x, I_y, I_z, yields:

$$q(\text{Ro}) = (\sqrt{\pi})/\sigma (2k_B T/h^2)^{3/2} (I_x I_y I_z)^{1/2} \qquad (7.11)$$

The quantized vibrational energy levels for a single harmonic oscillator are given by equation 3.24. Taking as usual the first level at zero energy, the vibrational partition function is

$$q(Vi) = \sum_v \exp(-\beta v h\nu) = [1 - exp(-\beta h\nu)]^{-1} \qquad (7.12)$$

A non-linear molecule with n atoms has $3n - 6$ independent normal vibration modes with frequencies $\nu_1, \nu_2, \nu_3, \ldots, \nu_{3n-6}$, and each normal mode contributes one addend to vibrational energies and one factor to the corresponding partition function:

$$\varepsilon(Vi) = \varepsilon(\nu_1) + \varepsilon(\nu_2) + \ldots + \varepsilon(\nu_{3n-6}) \qquad (7.13)$$

$$q(Vi) = q(\nu_1)q(\nu_2)\ldots q(\nu_{3n-6}) \qquad (7.14)$$

Equations 7.9, 7.11, and 7.12 show how all the necessary partition functions for a molecule can be calculated. The Tr part requires a knowledge of molecular mass and of the volume in which the system is contained; an easy task. The Ro part requires the moments of inertia; these can be calculated if the molecular structure is known (Section 1.3.1). The Vi part depends on normal mode frequencies, which can be measured or anyway obtained by the procedures outlined in Section 2.1.

7.2.2 The partition function: Macroscopic systems

Consider now a mole of molecules in a real state, be it gaseous, liquid, solid, or any other. The energy levels of such a system comprising a myriad of molecules are very complex; each quantum state would be a mixture of Tr, Ro, and Vi energies of all molecules, plus contributions from intermolecular interactions. These states, whose energy will be designated as E_i, should be found by solving a Schrödinger equation and finding the expectation value of a hamiltonian with some 10^{23} terms – a task no one could ever dream of accomplishing. As usual, ways around these difficulties are found by successive approximations, which in this case will be rather drastic. The first approximation is to neglect intermolecular interactions. This at once rules out the calculation of condensed phase properties and limits the treatment to the ideal gas. The second approximation, coherent with the first one, is to assume that each energy E_i be just a sum of energies of single molecules. Let $\varepsilon_i(k)$ be the contribution of molecule k to the energy level E_i:

$$E_i = \varepsilon_i(1) + \varepsilon_i(2) + \varepsilon_i(3) + \cdots + \varepsilon_i(N) \qquad (7.15)$$

Each term in the above summation is in turn a sum of molecular quantized energies. Summing up to obtain the macroscopic partition function Q is now a matter of rearranging summation indices:

$$Q = \sum_i \exp(-\beta E_i) = \sum_j \exp[-\beta \varepsilon_j(1)] \sum_j \exp[-\beta \varepsilon_j(2)] \ldots \sum_j \exp[-\beta \varepsilon_j(N)] \quad (7.16)$$

Molecules are distinct, but each factor in the above expression is identical and equal to the molecular partition function q(Mol), equation 7.6.

In a crystal molecules preserve their spatial identity, so to speak, so labels $1, 2, 3, \ldots N$ could be imposed upon them and preserved as long as the crystal preserves its rigid spatial construction. Then equation 7.16 would be all right as it is, but, unfortunately, it cannot be used in practice since the energy levels would be wrong because intermolecular terms are missing. In a gas, on the other hand, molecules are distinct but undistinguishable, and they may exchange their position, identity, and energy. Two configurations in which two molecules interchange their energy and position are undistinguishable and should not be counted separately. Equation 7.16 should be toned down by a factor that accounts for this, and combinatorial algebra tells that the appropriate factor is the factorial $N!$. Therefore:

$$Q = (1/N!)q(Mol)^N \qquad \text{ideal gas} \quad (7.17)$$

7.2.3 Internal energy I: From statistics and quantum mechanics

Consider a system with Avogadro's number of molecules, N, at constant volume. Consider now M identical copies of the system, which can freely exchange energy among them. Let m_i be the number of times each energy E_i appears; the frequency of appearance of energy E_i is $P_i = m_i/M$. The most likely distribution of energies at equilibrium is given by a Boltzmann distribution:

$$m_i^* = M \exp(-\beta E_i)/Q \quad (7.18)$$

$$P_i^* = m_i^*/M \quad (7.19)$$

where the m_i^*s have the same meaning as the n_i^*s in equation 7.1 The total energy of the system, or the average internal energy of the system at equilibrium, in a statistical sense, must be equal to the sum of the energy of each state times the probability of finding the system in that energy state:

$$<E> = \sum_i P_i^* E_i = 1/Q \sum_i E_i \exp(-\beta E_i) \quad (7.20)$$

$$<E> = -(1/Q)\sum_i \{d[\exp(-\beta E_i)]/d\beta\} = -1/Q d[\sum_i \exp(-\beta E_i)]/d\beta$$

$$= -1/Q(dQ/d\beta) \quad (7.21)$$

Absolute energies are not known, because among other things the first level of each quantum set was arbitrarily taken as zero energy level. Let U be the symbol for the internal energy, and let $U°$ be the (unknown) difference between U and the absolute energy. Then the internal energy from statistical thermodynamics is:

$$U - U° = -(\partial \ln Q/\partial \beta) \qquad \text{ideal gas, at constant volume} \qquad (7.22)$$

From now on the reference energy $U°$ and the constant volume label will be considered implicit and will no longer appear. Then:

$$U = -\frac{\partial \ln[q(\text{Mol})^N/N!]}{\partial \beta} = -N\partial/\partial\beta[\ln q(\text{Tr}) + \ln q(\text{Ro}) + \ln q(\text{Vi})] \qquad (7.23)$$

The total internal energy is a sum of contributions from translation, rotation and vibration. The translational part of U is given by

$$U(\text{Tr}) = -N\partial[\ln q(\text{Tr})]/\partial\beta = -N[1/q(\text{Tr})]\partial[q(\text{Tr})]/\partial\beta$$
$$= 3/2 N\beta^{-1} = 3/2\,RT \qquad (7.24)$$

where $Nk_B = R$, the ideal gas constant. When a similar calculation is done for the rotational contribution, if many rotational levels are occupied the result is the same, that is $U(\text{Ro}) = 3/2RT$. This is true if the temperature is high, and if the molecule is large and its rotational moments of inertia are very large, so the spacing between rotational quantized levels is small. For example, benzene at room temperature qualifies.

Vi levels are much more spaced than Tr or Ro levels, but if the same assumption as above is made, the result is $U(\text{Vi}) = RT$ for each vibrational mode. Tr and Ro energies are kinetic energies; as velocity has three components, kinetic energies consist of three quadratic terms. Vi energies have a kinetic energy part along the vibration coordinates, as above, plus a potential energy contribution, also in a quadratic dependence on displacement coordinates, for a total of two quadratic terms for each vibrational mode. These facts illustrate and justify the following principle of energy equipartition:

Whenever a great many quantum levels are populated and the energy spectrum merges into a continuum, every quadratic term in molecular energies contributes $1/2RT$ to the internal energy.

The equipartition principle is consistent with the loosening of quantum restrictions at higher temperatures or for heavier molecules – in fact, a high temperature means a population of a great many levels and hence a lesser importance of quantum effects. In summary, the equipartition principle says that if the spacing of quantum levels becomes for any reason very small compared with available energies, each degree of freedom takes up the same amount of energy: very reasonable. If quantum separations were prices, this would explain why at the grocer's store only a few people buy caviar while everyone buys more or less the same amount of bread.

Equipartition holds at room temperature for ordinary organic molecules for Tr and Ro energies. When (as is often the case) equipartition does not apply for the vibrational part of the energy spectrum, the explicit calculation of $U(\text{Vi})$ must be carried out. Let v be the frequency of one molecular normal mode of vibration:

$$u = -hv\beta, \ du = -hv d\beta; q(\text{Vi}) = (1 - e^u)^{-1}$$
$$dq(\text{Vi})/du = e^u(1 - e^u)^{-2}; dq(\text{Vi})/d\beta = -hv \, dq \, (\text{Vi})/du \quad (7.25)$$
$$U(\text{Vi}) = Nhv e^u/(1 - e^u)$$

Of course this is the expression for the vibrational energy relative to the first level, and tends to zero as temperature tends to zero. But the absolute value of the vibrational energy is never zero (recall equation 3.24). For an order of magnitude evaluation, the zero-point energy $1/2hv$ for a $100\,\text{cm}^{-1}$ frequency, a typical value for a lattice vibration, is about $0.5\,\text{kJ mol}^{-1}$.

7.2.4 Internal energy II: From thermal and mechanical experiments

Joule and Clausius knew nothing about molecules, quantum mechanics and computers. Just consider now a macroscopic system like 1 kg of liquid acetone. If no chemical bonds are broken or formed and no energy is exchanged by irradiation of electromagnetic light, the system may only exchange heat, q (thermal energy) or work, w (mechanical energy) with its surroundings. The amounts of energy thus exchanged can be measured fairly easily, as Joule did. For any given process the variation in internal energy of the system, $\Delta U(\text{system})$, must be the sum of the two kinds of exchange:

$$\Delta U(\text{system}) = q + w \quad (7.26)$$

and since energy is conserved, what the system takes the surroundings give, and *vice versa*:

$$\Delta U(\text{surroundings}) = -\Delta U(\text{system}) \quad (7.27)$$

These two equations were in fact proposed long before the existence and the properties of molecules were discovered, solely on the basis of thermal and mechanical experiments on macroscopic systems. Conceptually, the key of the first principle of thermodynamics is not in the balance of equation 7.26, but in the fact that (as recognized by Joule) U is a state function, that is, its value in a given state is always the same, no matter how that state has been reached or what kind of energy is exchanged. This is quite understandable in the light of equation 7.22, which only depends on energy levels of the system as it is, not as it has ever been.

7.3 Heat capacity

There is no way of directly measuring absolute values of internal energies, and there is no way of calculating them either, because $U°$ is not known. But suppose a controlled amount of heat, q, is supplied to the molecular system, and the ensuing change in temperature, ΔT, is measured. The volume is kept constant so there is no exchange of energy in terms of work. These are all easy experiments. Then all of the thermal energy supplied is used to increase the internal energy of the system. C_V is the heat capacity at constant volume:

$$C_V = (q/\Delta T)_V = (\partial U/\partial T)_V = -k_B \beta^2 (\partial U/\partial \beta)_V \qquad (7.28)$$

$$C_V(\text{Tr}) = C_V(\text{Ro}) = 3/2\,R \qquad \text{equipartition} \qquad (7.29)$$

$$C_V(\text{Vi}) = (3n_{\text{atom}} - 6)R \qquad \text{equipartition}$$

or

$$C_V(\text{Vi}) = R \sum_k [u^2 e^u / (e^u - 1)^2]_k \qquad \text{spaced Vi levels} \qquad (7.30)$$

In equation 7.30, index k runs on all normal modes of vibration, and $(3n_{\text{atom}} - 5)R$ holds for linear molecules.

C_V can thus be calculated, at least for an ideal gas, and the whole construction can be put to a stringent scientific test by comparing it with measured heat capacities. No need to say, it turns out that statistical thermodynamic calculations provide extremely accurate evaluations of the heat capacities of diluted gases. In fact, these calculations are so reliable that for small molecules the C_Vs of gases found in thermodynamic repertories are usually calculated from experimental vibration frequencies, rather than measured. On the other hand, heat capacities for condensed phases cannot be calculated, but are much more easily measured than for gases. In this case, calculation and experiment match and complement each other perfectly.

Chemists like to operate at constant pressure, rather than at constant volume, and the use of enthalpy, H, gets rid of the energy exchanges due to expansion (volume work). The expression for the heat capacity at constant pressure is then:

$$H = U + PV; \quad dH = (dq)_P \qquad (7.31)$$

$$C_p = (dH/dT)_P \qquad (7.32)$$

7.4 Entropy

A spontaneous (irreversible) process occurs, without any external intervention, with an increase in dispersion within a system left on its own: dispersion in the location of molecules in space; dispersion of molecular velocities in all possible directions; dispersion of available energies into available quantized energy levels; in other words, "irreversible" means progress towards the statistically most probable configuration across phase space. Irreversible processes are irreversible just because molecules have no compass and no North, and behave in a random fashion. As a consequence, any conclusion or law derived on such premises will be of an entirely probabilistic character: going against prescriptions is not forbidden, it is just very, very unlikely. When

ENTROPY

the equilibrium configuration has been reached, the system oscillates by microscopic displacements in any direction, without a driving force for further evolution; this condition is also called a reversible condition. Chemical equilibrium must be discussed in terms of decrease in potential energy together with an increase in dispersion.

The task of finding a mathematical formulation of the second law of thermodynamics was accomplished by Sadi Carnot and Clausius on the traditional, macroscopic side, and, again, by Boltzmann from a molecular, statistical perspective. What is sought is a state function that quantitatively describes the degree of dispersion in a chemical system. This is *entropy*, and its symbol is S. It must increase in any irreversible process.

7.4.1 Classical entropy

In order to find how entropy is related to other thermodynamic quantities such as heat and temperature, the following facts may be considered.

1. Entropy must be a state function, otherwise it would not be useful.
2. Dispersion increases with increasing available energy, just as there are many more ways of distributing 50 coins than just 5 coins among five children. Hence, it is reasonable to assume that entropy may be directly proportional to heat absorbed by a system.
3. On the other hand, molecules and energies within a body are more and more disperse as temperature increases, because molecular velocities are more distributed and more quantum levels are occupied. The increase in entropy brought about by the intake of an amount of heat dq is larger at lower temperature, so dS is likely to be inversely proportional to the temperature at which heat is exchanged.
4. The entropy of a system changes both in reversible and in irreversible processes, but reasonably entropy changes in the latter should be larger than in the former.
5. A system that does not exchange heat, work, or matter with the surroundings is called an isolated system. Some of the previous definitions have been derived in this hypothesis, which however is not so convenient, since no truly isolated system exists in reality. The only system that, as far as we know, does not admit external intervention is the whole universe, or the ensemble system+surroundings. Formulas for entropy should be such that any spontaneous process produces an increase of the entropy of system+surroundings.

The classical form of the mathematical definition of entropy is along these lines. For a system:

$$dS > dq/T \qquad \text{irreversible processes} \qquad (7.33a)$$
$$dS = dq/T \qquad \text{reversible processes or equilibrium} \qquad (7.33b)$$

For system+surroundings:

$$dS > 0 \qquad \text{irreversible processes} \qquad (7.34a)$$
$$dS = 0 \qquad \text{reversible processes or equilibrium} \qquad (7.34b)$$

Joule's achievement was to realize that U is a state function. Realizing that q is not a state function, but q/T is, was another great accomplishment. dq(system) = $-dq$ (surroundings), and therefore in a reversible condition dS(system) = $-dS$(surroundings), and entropy behaves like energy, being exchanged with a net balance of zero. In irreversible processes the system may raise or lower its entropy, according to the sign of dq, provided that the surroundings change their entropy to produce a net overall increase, equation 7.1c. So there is no conservation of entropy, but an everlasting increase.

7.4.2 *Statistical entropy*

In deriving equation 7.1, it was assumed that a system without external intervention would assume the most probable distribution of energy among energy levels. The partition function is the key to the equilibrium distribution or to the equilibrium amount of dispersion in molecular energies. It is a fairly straightforward matter to link entropy with the partition function.

Let M be the total number of molecules and m_i^* the number of molecules whose energy is ε_i at equilibrium. The statistical weight of the equilibrium distribution is given by

$$W^* = M!/(\Pi m_i^*!) \qquad (7.35)$$

Clearly entropy must increase with increasing W^*, which is also the number of ways in which the best distribution can appear. But if entropy were just directly proportional to W^* itself, the Universe would have gone to an ocean of degraded energy a long time ago; W^* is a huge number and increases astronomically with increasing available energy. In relating entropy to W^*, Boltzmann's intuition led him to find the appropriate mathematical damper, so to speak, and the appropriate proportionality constant. He proposed the natural logarithm for the first, and the constant that took his name for the second:

$$S = [k_B/N]\ln(W^*) \qquad (7.36)$$

N here is the number of particles in the system, or Avogadro's number for one mole.

7.4.3 *The calculation of entropy for chemical systems*

In the Clausius approach to entropy, one gets:

$$(\partial S/\partial P)_T = -(\partial V/\partial T)_P \qquad (7.37)$$

$$S(P_2) - S(P_1) = \int -(\partial V/\partial T)_P \, dP \qquad (7.38)$$

For a real gas or for a condensed phase, $(\partial V/\partial T)_P$ is the thermal expansion coefficient of the material at constant pressure, a measurable quantity, so entropy variations with pressure can be easily calculated. $(\partial V/\partial T)_P$ is very small for condensed phases, and entropy is hardly sensitive to pressure variations in liquids and solids. In any case, the standard entropy $S°$ is defined as the entropy at unit pressure (usually 1 atm, but sometimes also 1 bar). For a reversible process in which only P–V work is exchanged,

$$(dS)_P = (dH/T)_P = (C_p/T)\,dT \tag{7.39}$$

$$S(T_2) - S(T_1) = \int (C_p/T)\,dT \qquad \text{at constant } P \tag{7.40}$$

Equations 7.37–7.40 allow the calculation of entropy changes for a pure substance in any state of aggregation [3].

The calculation of entropies on the basis of the Boltzmann definition, equation 7.36, is only a matter of algebra. The final result is:

$$S = -k_B \sum [(m_i^*/M)\ln(m_i^*/M)] = -k_B \sum P_i^* \ln P_i^* \tag{7.41}$$

With m_i^* as in equation 7.18, Q as in equation 7.17, and $q(\text{Mol})$ as in equation 7.6, the entropy of an ideal gas can be calculated from a knowledge of molecular parameters only.

The statistical definition of entropy gives

$$S = 0 \text{ at } 0\,\text{K} \tag{7.42}$$

At zero temperature, all molecules have the same ground-state energy, and are fixed in an ordered crystal lattice, so there is only one way in which the configuration can appear, and $W^* = 1$. All ordered crystals have zero entropy at zero temperature, irrespective of crystal symmetry. This is sometimes called the third law of thermodynamics.

7.5 Free energy and chemical equilibrium

7.5.1 *Chemical potential*

So far, only pure compounds have been considered, that is, systems composed of only one molecular species in one given state of aggregation. With entropy in Clausius form and expansion work only, equation 7.26 becomes:

$$dU = T\,dS - P\,dV \tag{7.43}$$

More generally, a system may undergo a change in chemical composition through a chemical reaction or a phase change. Molecules may change their structures, so there may be changes in quantum levels, in bond energies, in intermolecular interactions, in nearly everything. From a macroscopic point of view, one can just write a derivative called the chemical potential of component i with respect to the number of moles of the i-th component, n_i:

$$\mu_i = (\partial U/\partial n_i)_{S,V,n_j} \tag{7.44}$$

The chemical potential is the rate of change in internal energy for a change in the number of moles of component i, when entropy, volume, and number of moles of all other components are constant. Taking this into account, the total differential of equation 7.43 becomes

$$dU = T\,dS - P\,dV + \sum_i \mu_i dn_i \qquad (7.45)$$

7.5.2 Free energy

The Gibbs free energy, symbol G, is defined by the following equations:

$$G = H - TS = G(T, P, n_i) \qquad (7.46)$$

$$dG = V\,dP - S\,dT + \sum_i \mu_i\,dn_i \qquad (7.47)$$

$$\mu_i = (\partial G/\partial n_i)_{T,P,nj} = (\partial U/\partial n_i)_{S,V,nj} \qquad (7.48)$$

Equations 7.33–7.34, the conditions for spontaneous evolution, when written in terms of G at constant T and P, lead to the following expression:

$$\left[\sum_i \mu_i\,dn_i\right]_{T,P} \leq 0 \qquad (7.49)$$

This is the fundamental equation of chemical equilibrium. It applies to any chemical system (not just isolated systems) at constant temperature and pressure, that is, the normal working conditions of chemistry. To understand how it works, consider the simple example of a transformation of a pure substance between states A and B, with $dn_A = -dn_B$; equation 7.49 says that $(\mu_A - \mu_B)\,dn_A \leq 0$, so if $\mu_A > \mu_B$ then n_A must decrease – hence μ as chemical potential. At equilibrium, $\mu_A = \mu_B$ so that the sign of dn_A is irrelevant; the system has no driving force to evolve either way.

The Helmholtz free energy, $A = U - TS$, is an equivalent form, more useful in conditions of constant temperature and volume and hence of lesser use in practical thermodynamics. Its application will be discussed in Section 9.7.

7.5.3 Chemical potentials in practice

Free energy cannot be measured directly, and a practical way of estimating the derivatives in equation 7.48 must be found. The most useful and successful form of the chemical potential for any substance, label i, in any system is the following:

$$\mu_i = \mu°_i + RT \ln a_i \qquad (7.50)$$

In this expression $\mu°_i$ is called the standard chemical potential, and a_i is called the activity of component i in any multi-component system.

Where did this particular functional form come from? To start with the simplest example, consider as usual a pure ideal gas. For n moles of the gas, then $G = n\,G_m$, where G_m is the molar free energy. Then:

$$\mu = (\partial G/\partial n)_{T,P} = G_m$$
$$\mu = \mu° + RT \ln(P/P°) \quad \text{(from } (\partial G/\partial P)_T = -V, \text{ if } PV = RT) \quad (7.51)$$
$$a = P/P°$$

The standard chemical potential is then the molar free energy of the gas at unit pressure. The above equation subdivides the chemical potential of a gas into a temperature-dependent part, $\mu°_i$, and a concentration-dependent part, $\ln(a_i)$ – pressure is equivalent to concentration for a gas.

The functional form of equation 7.51 is so convenient that it is also preserved in practice for condensed phases, even if a formal derivation is no longer possible in general. Consider an ideal liquid solution in which x_i is the mole fraction of component i. The form of the chemical potential is then

$$\mu_i(\text{liq}) = \mu^*_i(\text{liq}) + RT \ln(x_i) \quad (a_i = x_i) \quad (7.52)$$

$\mu^*(\text{liq})$ is now the standard chemical potential, or the molar free energy, of the pure liquid, whose mole fraction x_i is unit by definition. The asterisk (*) is used here instead of the symbol (°) to remind one that the standard is now not one of unit pressure, but one of unit concentration. For a pure crystalline solid the standard chemical potential $\mu^*(\text{cr})$ is defined in the same way; there is no concentration-dependent part in the chemical potentials of pure liquids and solids, which formally have a unit activity. The condition for liquid–gas equilibrium is:

$$\mu^*(\text{liq}) = \mu(\text{g}) = \mu° + RT \ln[(\text{svp})/P°] \quad (7.53)$$

where the partial pressure of the vapor in equilibrium over its liquid is called the saturated vapor pressure (svp) at that temperature. A similar expression holds for the gas–solid sublimation equilibrium.

Finally, for any chemical system with components in any state, liquid, solid, or solution:

$$\mu_i = \mu^*_i + RT \ln(a_i) \quad [a_i = \gamma_i C_i / C^*_i] \quad (7.54)$$

where C_i is the concentration (molarity or molality, for example) of each component in the mixture, and C^*_i is a reference concentration. γ_i are dimensionless activity coefficients. The price to be paid for saving the functional form of the chemical potential is that activity coefficients must be specified for any temperature, pressure, and chemical composition. These coefficients incorporate all the non-ideality of the system.

For an equilibrium involving several molecular species, each with its own chemical potential within the system, equation 7.49 can be recast in the more familiar form:

$$\Delta G^* = \Delta H - T\Delta S^* = -RT \ln K \quad (7.55)$$

where ΔG^* is the difference of standard chemical potentials, and K is the equilibrium constant in terms of the concentration-dependent parts of the chemical potentials. For example, for a chemical reaction:

$$C(cr) + 1/2 O_2(g) = CO(g) \qquad \text{formation reaction for carbon monoxide}$$

$$\mu(C, cr) = \mu^*(C, cr)$$

$$\Delta G°(r) = \Delta G°(f, CO, g) = \mu°(CO, g) - 1/2\mu°(O_2, g) - \mu^*(C, cr) = -RT \ln K$$

$$K = [p(CO)/p(O_2)^{1/2}]_{eq}$$

For liquid–gas equilibrium, equation 7.55 becomes:

$$\Delta G°(\text{evap}) = \mu°(g) - \mu^*(\text{liq}) = -RT \ln(\text{svp})$$

$$K = p(g)_{eq} = \text{saturated vapor pressure, svp}$$

For liquids and solids the chemical potential has no concentration-dependent part, so the solid–liquid equilibrium condition at constant pressure is written as

$$\mu^*(\text{liq}) = \mu^*(\text{cr}) \quad \text{or} \quad \Delta G^* = 0 \tag{7.56}$$

The equilibrium melting temperature $T(\text{melt})$ is the temperature at which the above equation holds:

$$T(\text{melt}) = \Delta H(\text{melt})/\Delta S(\text{melt}) \tag{7.57}$$

7.6 Thermodynamic measurements

7.6.1 *Heat capacity*

There is no doubt that the very fundamental quantity for the description of the thermodynamic behavior of pure substances is the heat capacity at constant pressure, C_p. The molar heat capacity of gaseous substances increases steadily with molecular size, as more and more vibrational degrees of freedom become available to store energy. The heat capacities of liquids and solids are much larger than gas phase ones, as

Table 7.1 Heat capacities of benzene derivatives at 298 K (J K^{-1} mol^{-1})

	C_p(cr)	C_p(liq)	C_p(g)		C_p(cr)	C_p(liq)	C_p(g)
benzene	118	136	82	—NH$_2$		192	108
—CH$_3$		157	104	—COOH	147	204	
—C$_2$H$_5$		185	128	—OH	127	198	104
—OCH$_3$		199		—CONH$_2$	156		
—F		146	94				
—Cl		150	98	dinitrobenzenes	186–192		
—Br		154	98	dicyano isomers	161		
—CN		165	109				
—NO$_2$		186		dicarboxylic acids	199–202		

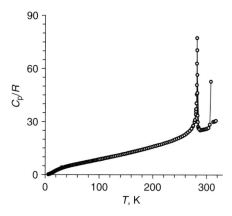

Fig. 7.1. A plot of the heat capacity of solid 1-bromoadamantane showing three anomalies corresponding to solid–solid phase transitions: at 31.0 K (hardly noticeable), at 282, and 310 K. The melting point of the material is 391.8 K. Prepared using data from Bazyleva et al., J. Chem. Thermod. 2005, **37**, 643.

energy in condensed phases can be stored in intermolecular vibrations. The C_ps of condensed phases can rather easily be measured by differential scanning calorimetry (DSC). Table 7.1 shows some examples.

The heat capacity is the key to the evaluation of enthalpy and entropy changes:

$$H(T_2) = H(T_1) + \int C_p \, dT \tag{7.58}$$

$$S(T_2) = S(T_1) + \int (C_p/T) \, dT \tag{7.59}$$

The heat capacity trace as a function of temperature (a thermogram) reveals phase transitions, as C_p shows abrupt changes or as heat is absorbed at constant temperature and C_p rises sharply, leaving an unmistakable mark of the thermal event (see a typical example in Fig. 7.1).

Equations 7.28–7.30 explain trends in heat capacity (recall that $C_p = C_v + R$ for an ideal gas). The C_v of a gas has a fixed contribution of $3R$ from translational and rotational equipartition, plus vibrational contributions: for a 100 cm^{-1} vibration the heat capacity contribution is 8.2 J K^{-1} mole^{-1} at room T; for a 2000 cm^{-1} vibration, the same contribution is about 0.05 J K^{-1} mole^{-1}. Figure 7.2 shows the steady increase in C_p with increasing molecular size. Heat capacities tend to zero at very low temperature, and increase with temperature because more and more vibrational energy levels become accessible and available for storage of the incoming heat. The lower the implied frequencies, that is, the less the material is internally constrained, the higher the heat capacity. The structural sensitivity is very low, as Fig. 7.2 also shows that alkanes, alcohols and acids have roughly the same C_p for the same number of non-hydrogen atoms. Table 7.1 shows a similar indifference to the nature of the

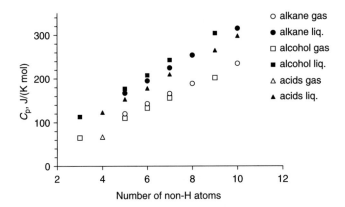

Fig. 7.2. The heat capacities of linear-chain alkanes, alcohols, and acids.

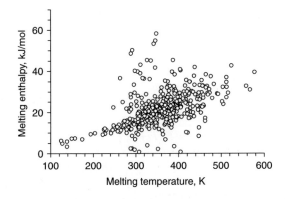

Fig. 7.3. A scatter-plot of molar enthalpies of melting for organic crystals (data from ref. [5]). The least-squares line is $y = 0.0606x$.

substituents. Methods have been proposed for an estimation of heat capacities from group additivity [4].

7.6.2 Melting enthalpies

The enthalpy of melting is readily measurable by direct scanning calorimetry at the melting temperature, and the melting entropy can be obtained by equation 7.57. Data for enthalpies of melting [5] are usually very consistent because the experiment is a comparatively easy and reproducible one. Figure 7.3 shows a scatter-plot of experimental molar enthalpies of melting, whose slope reveals that an average melting entropy for organic compounds is around $60 \, \text{J K}^{-1} \, \text{mol}^{-1}$.

The melting enthalpy is related to the strength of crystal forces, and correlates with molecular structure in a number of simple, if rough ways. For example, linear-chain

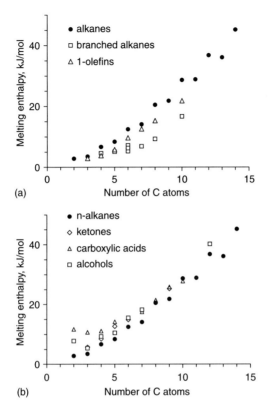

Fig. 7.4. (a) The melting enthalpies of linear alkanes compared with those of the corresponding 1-olefins and branched alkanes as a function of the number of carbon atoms in the molecule. Data from ref. [5]. (b) The same comparison for linear alkanes and the corresponding ketones, carboxylic acids, and alcohols. Data from ref. [5].

alkanes allow reasonably close-packed crystalline structures where saturated chains are aligned and parallel. For these dispersion-dominated crystals, any deviation from the favorable cylindrical shape of the molecule, brought about for example by branching or by the introduction of kinks along double bonds, is a disturbance to close packing and decreases the melting enthalpy, as seen in Fig. 7.4(a). Figure 7.4(b) shows the significant increase in melting enthalpy that results from the introduction of polar groups (ketones) and even more of hydrogen bonding.

Things are not as simple for melting points, whose correlation with molecular structure is quite elusive [6]. Consider, for example, Table 7.2. Melting points increase with increasing melting enthalpy but also with decreasing melting entropy: disordered crystals, which are already "similar" to the liquid state, melt with a low entropy change and thus at a relatively high melting temperature. The extreme case is benzene, where molecules are almost free to rotate in the crystal because of the planar shape (Section 1.4); the presence of any substituent hinders this rotation because the

Table 7.2 Effect of substitution on the melting properties of benzene derivatives

	T(melt)	ΔH(melt), kJ mol^{-1}	ΔS(melt), J K^{-1} mol^{-1}
benzene	279	8.95	32
toluene	178	6.61	37
ethylbenzene	178	9.16	52
propylbenzene	174	9.29	54
—Cl	228	9.54	42
—F	231	10.4	45
—NH$_2$	276	10.9	40
—OH	314	11.5	37
—NO$_2$	279	12.1	44
—COOH	396	18.0	46
naphthalene	354	19.0	54
anthracene	489	28.9	59
phenanthrene	372	18.7	50

molecular shape is no longer circular (compare, for example, benzene with toluene or chlorobenzene) and raises the melting entropy. Otherwise, Table 7.2 shows the already noted effects of branching and hydrogen bonding. The comparison between anthracene and phenanthrene shows the dramatic effect of molecular shape on crystal packing (see further discussion in Section 13.3).

7.6.3 Sublimation enthalpies

Things are more complicated for equilibria involving gas phases, whose experimental handling sometimes poses severe problems. Using equation 7.53, together with the dependence of $\Delta G°$ from temperature (the Gibbs–Helmholtz equation), the following expressions are obtained for chemical equilibria:

$$\ln(K) = -\Delta H/R(1/T) + \text{constant} \qquad \text{van't Hoff equation} \qquad (7.60a)$$

$$\ln(pvs) = -\Delta H_{\text{subl}}/R(1/T) + \text{constant} \qquad (7.60b)$$

Equation 7.60b is the van't Hoff equation written for a crystal–gas phase equilibrium (an identical one holds for the liquid–gas equilibrium), and coincides with the Clausius–Clapeyron equation: (svp) measurements as a function of temperature permit the evaluation of the enthalpies of sublimation or of evaporation. The saturated vapour pressures of liquids are usually of the order of fractions of an atmosphere, so that their measurement is not so difficult.

Equation 7.60 is derived on the assumption of an ideal gas phase, which is certainly true for the low equilibrium pressures over solids, and of constant enthalpy of sublimation, which holds only in a restricted temperature interval. Many crystals have very small saturated vapor pressures, so measurements are carried out at high temperature, while for reference and calibration of intermolecular potentials, room temperature values are more convenient. The dependence of sublimation heats from temperature is:

$$d(\Delta H_{subl})/dT = C_p(gas) - C_p(cr) = \Delta C_p(subl) \tag{7.61}$$

Since it is usually rather difficult to have temperature-dependent values for the heat capacities of both phases, approximate relationships have been proposed [7], either assuming a constant value $\Delta C_p = -60$ to -40 J K^{-1} mol^{-1}, or empirical forms like:

$$-\Delta C_p = 0.75 + 0.15 C_p(cr, 298K) \quad J K^{-1} mol^{-1} \tag{7.62}$$

The heat of sublimation is connected with intermolecular forces in the crystal and with the lattice energy. This last quantity can fairly easily be calculated for crystals even by simple models (Section 8.7), so that a comparison between calculated lattice energies and measured heats of sublimation is a common way of checking the performance of intermolecular potential energy schemes. Unfortunately, heat of sublimation measurements for organic compounds are also hampered by many experimental problems and are often scarcely reliable [8]. Three main methods for their determination are available:

1. saturated vapour pressure measurements and Clausius–Clapeyron equation. The (svp) of a solid can be measured by somehow measuring the mass of the vapour in equilibrium with the solid, for example by condensing the vapor after effusion through a small hole. One has to deal with very low pressures (small fractions of a pascal) and/or very small amounts of matter, or, alternatively, one must work at very high temperatures in order to have significant vapor pressures;
2. direct calorimetry: the sample is sublimated and the heat flow is directly measured. There are problems in ensuring equilibrium conditions between the solid and the gas phase;
3. indirect methods: (a) combustion of the solid to determine its heat of formation, together with the evaluation of the heat of formation of the gas; (b) addition of heats of evaporation and heats of melting, which must be at the same temperature, while these two quantities are usually determined in widely different temperature ranges so that a correct assessment of the temperature dependence of both is needed.

The measurement of enthalpies of sublimation is still an art rather than a consolidated technique. There are no commercially available standard apparatuses, so that over forty different experimental apparatuses have been proposed, all being home-made to a large extent, a combination of calorimeters, pumps, vacuum chambers, cold fingers, each being a potential source of experimental variance. It is not surprising that there are very few sublimation energy standards, especially at high temperatures. Besides, results may be sensitive to sample purity, and to undetected crystal polymorphism or crystal phase transitions. In a survey of available determinations of sublimation enthalpies [8], a plot of multiple measurements against their averages resulted in a standard deviation of 5–6 kJ mol^{-1}. Critical scans of the thermochemical literature on

Table 7.3 Sublimation enthalpy of anthracene through the years [10]

ΔH_{subl}, kJ mol^{-1}	Temperature, K	Method	Year
102.6	313–363	gas saturation	1986
94.3	353–399		1983
91.8 ± 0.9	283–323		1983
104.5 ± 1.5	337–361	mass effusion	1980
98.8 ± 0.4	363–448	head-space analysis	1977
97.2	328–372	mass effusion	1976
101 ± 0.5	353–432	mass effusion	1973
99.7	393	calorimetry	1973
96 ± 6	283–323	Langmuir evaporation	1973
84.1	290–358	mass effusion	1971
98.3 ± 2.1	342–359		1964
90 ± 0.13	327–346	torsion effusion	1960
103.4 ± 2.9	303–373		1958
102.09	338.7–353.4	vapor pressure	1953
97.3 ± 1.2	378–398	Rodebush gage	1949
93.3 ± 4.2	353		1938

Table 7.4 A few examples of sublimation enthalpy ranges (kJ mol^{-1}) from different experimental determinations [10]

Compound	ΔH_{subl} range, kJ mol^{-1}
biphenyl	69–84
pyrene	96–101
phenanthrene	84–96
naphthoquinone	72–91
maleic anhydride	69–85
coumarin	83–86
benzoquinone	63–69

heats of sublimation [9,10] leave an unequivocal impression of overall inconsistency (see Tables 7.3 and 7.4). Besides, one has to make do with what is available rather than choosing what is needed, as heats of sublimation are usually determined for particular classes of compounds (explosives, pesticides or poisonous compounds, food ingredients) and seldom if ever as a part of a research project on intermolecular forces [11].

Figure 7.5 shows a plot of experimental [10] molar sublimation enthalpies, showing a weak correlation with molecular weight. Points above average are for crystals of very polar compounds or for hydrogen-bonded crystals, points below average pertain to non-polar compounds or molecules that, for several reasons, pack less efficiently in the crystal. A typical value for the sublimation enthalpy of a medium-size organic

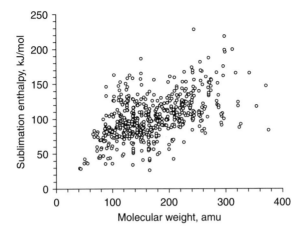

Fig. 7.5. A scatterplot of experimental sublimation enthalpies as a function of molecular weight. Data from ref. [10].

compound is thus about 150 kJ mol^{-1}, to be compared with the heat of sublimation of the simplest ionic solid, sodium chloride, 781 kJ mol^{-1}, or of metals, 400 kJ g atom^{-1} for iron and 853 g atom^{-1} for tungsten. Correlations between heats of sublimation, lattice energies, and molecular constitution will be presented in Chapter 8.

7.7 Derivatives

Many more useful relationships can be obtained by algebraic manipulation of equation 7.43. For a pure, one-component system one gets, for example:

$$T = (\partial U/\partial S)_V \tag{7.63}$$

revealing that temperature is the rate of change of free energy with entropy in a system at constant volume – a rather mysterious conclusion. On the other hand, or when equation 7.43 is combined with the definitions of enthalpy, equation 7.31, and of free energy, equation 7.46, many new relationships are obtained, some of which offer practical ways of calculating thermodynamic properties. For example:

$$(\partial S/\partial P)_T = -(\partial V/\partial T)_P \tag{7.64}$$

$$(\partial G/\partial T)_P = -S \tag{7.65}$$

$$(\partial G/\partial P)_T = V \tag{7.66}$$

Equation 7.64 is obtained, like many others, by taking advantage of the fact that free energy is a state function and therefore second derivatives are insensitive to the order of derivation, i.e. $(\partial^2 G/\partial x\,\partial y) = (\partial^2 G/\partial y\,\partial x)$. Equations 7.65 and 7.66 are useful in explaining free energy changes as a function of temperature and pressure, and hence in describing phase equilibria.

References and Notes to Chapter 7

[1] Chemical thermodynamics is a compulsory part of undergraduate training in nearly all college curricula in the physical and chemical sciences. The principles of chemical thermodynamics are exposed in countless textbooks. This chapter gives a compendium of indispensable concepts for the purpose of studying the physical chemistry of condensed phases. One of the aims is to give a parallel discussion of classical and statistical methods in molecular thermodynamics.

[2] Explaining symmetry numbers is not easy. Just consider the following examples: asymmetric molecules have $\sigma = 1$; sample rotational symmetry numbers are 3 for NH_3, 2 for H_2O, 4 for ethylene, 6 for ethane, 12 for benzene and methane. Molecules with the same symmetry have the same symmetry number, so methane and carbon tetrachloride both have $\sigma = 12$.

[3] This illustrates the use of the concept of reversibility. In a real process heat is not exchanged reversibly, but the analytical function $Cp(T)$ can be determined from calorimetric experiments and it is a continuous function that can be subjected to integration. Integration is a sum of an infinite number of infinitesimal stages, so reversible equations apply. Whatever the actual path of any given transformation, the entropy difference between initial and final state, T_1 and T_2, is always equal to that given by integration of equation 7.40, since entropy is a state function.

[4] Chickos, J. S.; Hesse, D. G.; Liebman, J. F. A group additivity approach for the estimation of heat capacities of organic liquids and solids at 298 K, *Struct. Chem.* 1993, **4**, 261–269.

[5] Acree, W. E., Jr., Thermodynamic properties of organic compounds: enthalpy of fusion and melting point temperature compilation, *Thermoch. Acta* 1991, **189**, 37–56; Acree, W. E., Jr.,Thermodynamic properties of organic compounds. 4. First update of enthalpy of fusion and melting point temperature compilation, *Thermoch. Acta* 1993, **219**, 97–104; Chickos, J. S.; Hesse, D. G.; Liebman, J. F. Estimating enthalpies and entropies of fusion of hydrocarbons, *J. Org. Chem.* 1990, **55**, 3833–3840.

[6] Melting points depend in a complicated way on a number of non-additive structural and energetic features of molecules and of their crystals. See Katritzky, A.; Maran, U.; Karelson, M.; Lobanov, V. S. Prediction of melting points for the substituted benzenes: a QSPR approach, *J. Chem. Inf. Comput. Sci.* 1997, **37**, 913–919, and references therein. Use of a large number of molecular descriptors gives just moderately satisfactory correlations.

[7] Chickos, J. S. A protocol for correcting experimental fusion enthalpies to 298.15 K and its application in indirect measurements of sublimation enthalpy at 298.15 K, *Thermoch. Acta* 1998, **313**, 19–26.

[8] Chickos, J. S. Enthalpies of sublimation after a century of measurement: a view as seen through the eyes of a collector, *Netsu Sokutei* 2003, **30**, 116–124 (see the Website of the author).

[9] The following citations recall the historical development of repertoires and of experimental techniques: Bondi, A. Heat of sublimation of molecular crystals. A catalog of molecular structure increments, *J. Chem. Eng. Data* 1963, **8**, 371–381; Jones, A. H. Sublimation pressure data for organic compounds, *J. Chem. Eng. Data* 1960, **5**, 196–200; Chickos, J. S.; Annunziata, R.; Ladon, L. H.; Hyman, A. S..; Liebman, J. F. Estimating heats of sublimation of hydrocarbons. A semiempirical approach, *J. Org. Chem.* 1986, **51**, 4311–4314; Ouvrard, C.; Mitchell, J. B. O., Can we predict lattice energy from molecular structure? *Acta Cryst.* 2003, **B59**, 676–685; Knauth, P.; Sabbah, R. Thermochemistry of organic compounds. A review on experimental methods and present-day research activities, *Bull. Soc. Chim. Fr.* 1990, **127**, 329–346.

[10] Chickos, J. S.; Acree, W. E., Jr. Enthalpies of sublimation of organic and organometallic compounds 1920–2001, *J. Phys. Chem. Ref. Data* 2002, **31**, 537–698; H. Y. Afeefy, J. F. Liebman and S. E. Stein, *Neutral Thermochemical Data*; and J. S. Chickos, Heat of Sublimation Data, in: *NIST Chemistry WebBook, NIST Standard Reference Database Number 69*, Edited by Linstrom, P. J.; Mallard, W. G. March 2003, National Institute of Standards and Technology, Gaithersburg MD, 20899 (http://webbook.nist.gov).

[11] But see de Wit, H. G. M.; van Miltenburg, J. C.; De Kruif, C. G. Thermodynamic properties of molecular organic crystals containing nitrogen, oxygen and sulphur. 1. Vapor pressures and enthalpies of sublimation, *J. Chem. Thermod.* 1983, **15**, 651–663.

8

Correlation studies in organic solids

As for solids ... in order that among their molecules or particles be made such a tight binding as is cause of their compactness, it is necessary that they may be located in certain given ways, which are appropriate for cohesion to exert its energy.

Translated from G.Brugnatelli, *Trattato delle Cose Naturali e dei loro Ordini Conservatori*, 1837 Pavia, Tipografia Tizzoni, Vol. I, p. 34.

The attractive forces acting between the atoms will cause the portions of space which they respectively appropriate ... to be in contact with one another, at the maximum number of points; and as a result ... the molecules themselves will also pack closely together ...

Barlow, W.; Pope, W.J. *J. Chem. Soc.* 1906, 1675–1744.

8.1 The Cambridge Structural Database (CSD) of organic crystals

A complete single-crystal X-ray structure determination (Section 5.5) provides the cell dimensions, the space group, a list of positional coordinates for the atomic nuclei in the asymmetric unit, and the atomic displacement parameters (ADPs). The experiment is nowadays feasible with a modest investment of equipment money and human time, provided that a suitable crystal is available – often, the bottleneck for the productivity of an X-ray diffraction laboratory is not machine time, but the preparation of good crystal samples. The result is a high resolution picture of the molecular structure and a detailed description of the crystal packing. Through the ADPs, X-ray diffraction also provides some hints at the intra- and intermolecular dynamics.

Of course these pictures are not always perfect and not always unique. The crystal may be disordered or twinned or may have many molecules in the asymmetric unit, in which cases the assignment of space group symmetry may be dubious, with pseudo-symmetries and modulations. Frequently, organic compounds exhibit crystal polymorphism, whereby the same molecule adopts a different conformation and a different packing in different crystals, sometimes at the same temperature and pressure (Section 14.2). Otherwise, crystals can be formed via the interaction of an organic molecule with its solvent, which is carried along into the solvate crystal, or by co-crystallization of different molecules into binary or ternary crystalline compounds.

In the late 1960s the determination of one crystal structure was still a considerable feat, and crystallographers were satisfied with the discussion of the properties of the one compound at hand; imagine the expectations of structural chemists having their first direct look at the bond lengths in the naphthalene molecule after long theoretical debate on delocalization and aromaticity. Very soon, however, molecular and crystal structural data began accumulating very fast, and the idea arose of storing all the structural information in one place, to perform statistical studies on trends in molecular properties over classes of compounds. For organic crystals, this project started in Cambridge with volunteers punching atomic coordinates on cards, and has reached electronic perfection today, as crystallographic data are sent directly to the Cambridge Crystallographic Data Centre (CCDC) [1] through the Internet, by agreement with the editorial offices of all chemical journals, or by spontaneous submission by practicing crystallographers. The data are machine-checked to try to reduce mistakes and inconsistencies.

The availability of such a vast amount of structural data in electronically manageable form has been and still is a tremendous bonus to the community of structural chemists, and to the chemical community in general. The establishment of crystallographic databases has changed the science of molecular structure from a collection of rough qualitative guesses into a coherent and quantitative discipline. In recent times, systematic studies of crystal packing have started to throw some light on the mysterious ways of molecular aggregation. The scientific advantages derived from the use of structural databanks are so evident as to hardly need further explanation or confirmation. Often forgotten, in the enthusiasm for new discoveries, are however the few caveats one should not forget when knocking at the door of the great structural Cassandra.

Databases are socially biased. They collect whatever has been done, rather than deciding what should be done. Organic compounds have been and are synthesized, and crystal structures have been and are determined, for a variety of reasons, anything from pure academic pride in the demonstration of synthetic ability to a specific requirement of characterization for industrial production. Routine crystallographers, on their part, are more inclined towards obvious crystal structure determinations than towards messy ones, so the true occurrence of disorder, multiple molecules in the asymmetric unit, and other undesirable phenomena is probably higher than their percent occurrence in the database. Polymorphs are certainly under-represented. Statistical studies on crystallographic databases are conducted neither on purely random nor on appropriately adapted data sets. Fortunately, the variety of the chemistry represented in the CSD is so large that this is seldom a limiting factor, at least in studies of intramolecular geometry.

Structural data are affected by uncertainties, which are often but not always properly represented in the R-factor (equation 5.32) and in standard deviations on nuclear coordinates. R-factors can to some extent be manipulated by inclusion or exclusion of certain structure factors, sometimes for very good reasons, or by juggling with weights in least-squares refinements. Selecting only crystal structures with an R-factor of below, say, 5% may mean excluding the work of careful authors who have given proper weighting to their results, while accepting the work of less critical authors who did not hesitate to clip a few corners. The reliability of structure data, and of consequent generalizations on bond lengths and molecular structure, should be judged by standard deviations of atomic parameters. However, these critical numbers are also to some extent malleable, and since 90% of the structural work is carried out in black box automatic diffractometers, the reported accuracy of bond lengths depends to some extent on how generous the manufacturer has been with the built-in software.

Structural databases contain mistakes: perfection is not of this world. Checking procedures concern the internal consistency of the data, but subtle errors can pass through the sieve. A typical example, relevant to the calculation of packing factors, is wrong settings, such as space groups indicated as $P2_1/c$ but being in fact $P2_1/a$ or $P2_1/n$ (Section 5.2). Also, checks cannot confirm the physics of the crystal constitution, and a typical example of undetectable errors is crystal structures with an abnormally low density, because some included solvent has been missed. These mistakes are sometimes large, obvious, and thus not dangerous, but sometimes they are small and insidious. In statistical studies over thousands of crystal structures it is positively impossible to check every single one in detail, and faulty data must seep in, so that a certain amount of noise in correlation studies is unavoidable. In this respect, always consider the simpler explanation first: the large majority of outliers in any distribution are the result of wrong data, rather than representative of new physical or chemical facts.

8.2 Structure correlation

The term "correlation" denotes an attempt at bringing together, comparing, and analyzing by statistical methods a large amount of structural data, with the aim of discovering new chemical laws. Structure correlation studies were first carried out on intramolecular parameters, a classic work being that of Bürgi and Dunitz on molecular deformations at a carbonyl center as a function of the intermolecular distance to an incoming nucleophile [2]. In further applications, the term "structure correlation" was extended to include all studies in which molecular or crystal properties are analyzed in a systematic way over the database, to reveal correlations between structural or energetic properties of crystallized molecules [3].

This book is concerned with molecular aggregation, where the problem is one of finding correlations between molecular structure and the structure and energy of the condensed system. Three broad areas can be distinguished in these studies.

1. Bonding categorization: a large number of crystal structures are examined to find recurrent geometrical patterns, usually in terms of coordinates such as lengths of interatomic vectors and angles between them, as is done for intramolecular bonding. Sometimes, macro-coordinates are used, such as for example the distance and reciprocal orientation between recognizable groups, like aliphatic chains or aromatic rings.

2. Space filling: crystal structures are analyzed to find how molecules use the available space, using macro-coordinates such as molecular volumes, positions of molecular centers of mass, packing coefficients and crystal density.
3. The systematic study of crystal energies: while the first two approaches proceed by qualitative statements based on geometry alone, this last approach requires a mathematical model for the intermolecular potential.

8.3 Retrieval of molecular and crystal structures from the CSD

A consideration of the whole bulk of the CSD is both impractical and unnecessary. At the same time, the Cif file format in which the CSD entries are stored requires extensive manipulation of alphabetic data and is impractical for use with computer programs written in the most widely used languages for scientific calculation. For the examples shown in this chapter, some reduced databases have therefore been prepared from the CSD, containing molecular data in numerical form. A full description of the retrieval procedure is given in the Supplementary material (see OPiX manual, Retcif-Coor procedures).

The positions of hydrogen atoms are crucial in all studies of crystal energies, but they are very poorly determined by X-ray diffraction experiments. Sometimes, especially for older determinations, the positions of the hydrogen atoms were not given by the authors and hence cannot be found in the CSD. In such cases the retrieval procedure must assign hydrogen atoms to carbon, oxygen, and nitrogen atoms, on the basis of standard bond lengths in the fragment considered. A check can be performed between the number of assigned hydrogens and the number of hydrogens in the formula of the compound, but this is not a 100% foolproof procedure, and mistakes are frequent. Even when H-atom positions are determined, X-rays see electrons, not nuclei, and nuclear positions are assigned at the centroid of electron density peaks. Because of the low scattering factor of the hydrogen atom (Section 5.4), the peak of the electron density seen by X-rays is located in between the actual position of the hydrogen nucleus and the actual peak of the bonding electron density, so that if H-atom positions are freely refined, C—H distances are too short.

Renormalization is imperative: many crystal structure solution packages assign H-atom positions by some geometrical constraint based on the position of the carbon atoms to which they are attached. The above mentioned Retcif-Coor procedure uses a protocol [4] that has withstood the test of two decades, using a C—H distance of 1.08 Å and an O—H or N—H distance of 1.00 Å. While for hydrogens bound to carbon the standardization is always possible by simple geometrical rules, unsolvable problems arise with hydrogen attached to oxygen or nitrogen, when, as is often the case, their position has not been accurately determined by X-ray diffraction. In alcohols, the hydrogen atom can be anywhere on a cone whose side is the O—H distance; in primary amides, the degree of pyramidalization at the N atom is variable; over the carboxylic acid hydrogen bond, the hydrogen atom position is often poorly determined due to ambiguities between *cis*- and *trans*-arrangement of the O—H bond, although the latter is less frequent, or to disorder by jumps in the space between the two oxygen atoms. In a non-negligible number of cases the position of these hydrogens is assigned by X-ray experimentalists on the basis of the approximate location of residual peaks in difference Fourier syntheses, and is quite unreliable. There is very little that can be done in such cases.

When molecules in a crystal have point-group symmetry that coincides with the crystal symmetry, the appropriate subgroup must be used in the computational simulation of the crystal structure: for example, the crystal structure of a centrosymmetric molecule in space group $P2_1/c$ must be described by space group $P2_1$ or Pc. This is not always obvious in the data stored in the CSD, because CSD entries always carry the full space group symmetry. The retrieval routines must find and discard the redundant symmetry operations, and this is not always a trivial task and is always a potential source of error. Trouble usually appears in the form of a very low calculated density (wrong number of space group symmetry operations), of unreasonably short intermolecular distances, of repulsive/destabilizing lattice energies. All structures affected by these anomalies must be discarded.

Some reduced databases have been prepared out of the CSD for use in intermolecular studies, with renormalized hydrogen positions, and including only crystal structures that pass all the above-described tests. Other general sieves were: heaviest element Cl; number of C atoms < 40; total number of atoms < 50; no disorder, no errors, no powder data, X-ray diffraction only, 3D coordinates determined, only organics. Special sieves were used for the different sets, as follows.

1. The Z(1) database: only one chemical unit in a crystal (no salts, solvates, molecular complexes, etc.); $Z' = 1$ (one molecule in the asymmetric unit); 35,429 hits, of which 27,535 passed all the tests.
2. The Z(1/2) database: $Z' < 1$, or less than one full molecule in the asymmetric unit (mostly, $Z' = 1/2$); 4,585 hits, 3,468 passing the tests.
3. The Z(2) database: $Z' > 1$, two or more identical molecules in the asymmetric unit; 1,501 hits, 825 passing the further tests.

Other reduced databases were prepared with the following specifications.

4. The SMALL database: heaviest element F; number of carbon atoms between 6 and 10; total number of atoms < 20; as above for the rest; 1,026 hits, 775 passing the tests.
5. The Molcomp database: more than one chemical unit in the crystal, but no salts or hydrates; 1,599 hits, 701 passing the further tests.

The above shows that 20–50% of the crystal structures stored in the database do not lend themselves to automatic renormalization of hydrogen atom positions and to immediate and unchecked use in crystal packing analysis studies. True, the checks

8.4 The SubHeat database

The simplest test of the performance of any computational scheme for the calculation of intermolecular potentials is a comparison of the total calculated lattice energy with the experimental heat of sublimation of the crystal. Indeed this is the prime reference test in all parameterizations of the intermolecular potential, as discussed in Sections 4.9 and 7.6. Unfortunately, the simultaneous availability of the crystal structure and the experimental heat of sublimation occurs infrequently. For completeness, for comparison of parameter schemes, and for future reference in parameter optimizations, we have collected a large number of – in fact, nearly all – such occurrences in a database called the SubHeat Database (deposited as supplementary material for this book, see subheat.nam and subheat.oeh). It contains 36 hydrocarbons, 25 oxohydrocarbons (non-H-bonding CHO compounds), 23 azahydrocarbons (non-H-bonding CHN compounds), 36 carboxylic acids, 28 amines, amides, imides, or otherwise N—H···N hydrogen-bonding compounds, 11 alcohols, 11 chlorinated compounds (CHCl), 4 fluorinated compounds (CHF), 9 sulfur compounds (CHS), 8 nitro compounds, and 10 miscellaneous crystals (polyfunctional compounds), for a total of 201 organic crystals. The choice and collection have been conducted using the criteria described in the following.

1. Crystal structures. These have been extracted from the Cambridge Database, choosing only entries unaffected by disorder or other errors, and, in case of multiple determinations, selecting the most recent one, which is usually also the one with the lowest R-factor. Room-temperature determinations have been preferred because any simulation potential should preferably be applicable at room-T conditions; low-temperature determinations have been selected when room-temperature ones were unacceptable for other reasons. When neutron diffraction data were available, these were selected for the positions of the hydrogen atoms.
2. Sublimation enthalpies. The NIST database [5] has been scanned for each compound. When several ΔH_{subl} data were present, which was usually the case for the compounds considered (Section 7.6), the choice was made on the basis of: (2a) internal consistency: when similar values had been obtained by different workers, the average value was assumed; (2b) chemical consistency: when data were available for a series of chemically similar compounds (e.g. the carboxylic acids, the chlorobenzenes) care was taken to select values which consistently fitted within the series; (2c) reliability: from personal judgement of the reliability of the research groups, preference being given to data coming from laboratories with a consistent and well documented experience in the business; (2d)

temperature: room-temperature values were always preferred. In a few cases, a correction was applied to data determined at very high or very low temperature according to methods discussed in Section 7.6, using an average ΔCp of -30 J K^{-1} mol^{-1}.

A major problem in the comparison between lattice energy and sublimation heat is that the calculated quantity cannot easily take into account any energy contribution due to a difference between the molecular conformation in the solid state and in the gas phase. Since the molecule must be at its minimum conformational energy in the gas phase, any conformational change on going from gas phase to the crystal must be destabilizing and the lattice energy calculated assuming a rigid molecule as found in the crystal must always be larger than the observed sublimation heat.

8.5 The geometrical categorization of intermolecular bonding

Correlation studies on intramolecular bonding parameters were a great success in structural chemistry [6]. In a natural extension, statistical studies of the geometrical parameters of intermolecular bonding began to appear in an attempt to rationalize the complex phenomena of molecular recognition and condensation. Not unexpectedly, success in this latter task was much less immediate.

The categorization of the hydrogen bonding patterns began as early as 1951 [7] and was later pursued more systematically with pioneer studies by Leiserowitz [8] (when the CSD was not available, and crystal data had to be painstakingly collected by hand). In a later study [9], which considers also minor components in two- or three-center bifurcated bonds, one reads: "The hydrogen-bonded regions merge with regions of non-bonding arrangements, as α(O—H$\cdots$O) decreases below 90° and d(H$\cdots$O) increases above 3 Å, and no clear separation between these regions can be observed." There lies the seed of doubt on relationships between intermolecular distances and intermolecular bonding. Things become even more problematic for weaker bonds. The paradigm of the geometrical reasoning can be expressed as follows [10]: "Hydrogen atoms that are covalently bonded to carbon have a statistically significant tendency to form short intermolecular contacts to oxygen atoms rather than to carbon or hydrogen atoms...the crystal structures also contain several short, intermolecular C—H$\cdots$N and C—H$\cdots$Cl contacts. It is concluded that the C—H$\cdots$O, C—H$\cdots$N and C—H$\cdots$Cl interactions are more likely to be attractive than repulsive and can reasonably be described as hydrogen bonds." A later contribution on the same topic [11] more cautiously uses quantitative energy considerations and refrains from using the term "hydrogen bond". With modern computing techniques available, the strength–length relationship is severely challenged by MO calculations on O—H$\cdots$O bonded dimers: does it really hold? The answer seems to be "not always" [12].

The above ideas have spurred a continuing flow of statistical studies on all kinds of intermolecular contacts, with a vast literature [13]. The quest for statistical correlations between structural and energetic properties of molecules and their crystals has been undertaken over the years in the author's group [14]. In spite of a number of firm and interesting conclusions, definite molecule-to-crystal predictive power has always remained questionable. The problems connected with the definition of intermolecular bonding are treated in modern perspective in Chapter 12.

8.6 Space analysis of molecular packing modes

8.6.1 *Empty space versus filled space*

The simplest way of studying the molecular arrangement in the compact structure proper of crystalline solids is to analyze volume occupation. The intrinsic molecular volume, V_M, can be evaluated from molecular geometry and approximate atomic radii (equations 1.2 and 1.3). The experimental measure of volume occupation in crystals and liquids is of course the molar volume, V_{mol}, an easily measurable quantity. The density d_x, the molecular mass M_M, V_{mol} and V_M are related as follows:

$$V_{mol} = M_M/d_x = N_{AV} V_M/C_{pack} \tag{8.1}$$

where N_{AV} is Avogadro's number and C_{pack} is a packing (space occupation) coefficient, which can be defined if one assumes that molecular envelopes may not overlap, and that the space that is not occupied by the molecular envelopes is "empty" space. This model has a considerable value in the description and, to some extent, in the prediction of the reciprocal arrangement of molecules in condensed phases. Discussing the compactness of materials in terms of molecular volumes and space occupation factors has the advantage that packing effects can be measured by an unique unit, molecular volume, without recourse to density, a quantity that depends on the number of protons and neutrons concentrated in atomic nuclei, which are irrelevant from the point of view of cohesive effects.

The packing coefficient in a crystal structure can be obtained as:

$$C_{pack} = ZV_m/V_{cell} \tag{8.2}$$

where V_{cell} is the volume of the crystal unit cell and Z is the number of molecules per cell. A more accurate method uses a point-by-point integration procedure: the unit cell is subdivided into a very large number of small volumes, $V°$, whose centers are either inside or outside the molecular envelopes. The number of inside and outside volumes is counted, and the packing coefficient is then:

$$C_{pack} = N_{in}/(N_{in} + N_{out})V_{cell} = N_{in}V°/V_{cell} \tag{8.3}$$

The average value of C_{pack} for organic crystals is around 0.72, or about the same as for a collection of rigid spheres in closest contact. C_{pack} is around 0.5 for liquids far from their freezing point, as seen in Fig. 8.1. Since the ratio of the packing coefficients is the same as the inverse of the ratio of molar volumes, from the above mentioned average values one can conclude that at room conditions a portion of organic matter

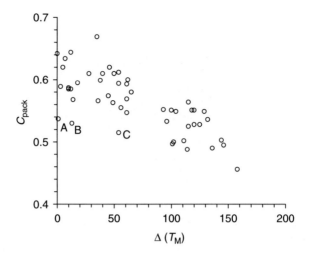

Fig. 8.1. The relationship between space occupation factor in liquids and $\Delta(T_M)$, the distance from the freezing temperature. Sample of 51 of the most common organic liquids, hydrogen bonding or not. C_{pack} is smaller for rotationally disordered crystals of globular molecules: A, *tert*-butanol; B, cyclohexane; C, CF_4.

in the liquid state occupies about 20% more space than the same portion of matter in the crystalline state. Figure 8.1 also shows that near freezing temperature the packing coefficient of organic liquids is close to 0.6. In these conditions the difference between liquid and crystal is considerably reduced, because the crystal is at its highest possible expansion, while the liquid is at its highest possible contraction.

8.6.2 Close packing in crystals

Crystal structures with overlapping molecular envelopes and large voids are not favourable under the conditions of a rather weak and more or less isotropic intermolecular potential, as most of the potentials in organic crystals turn out to be. In fact, compression is equivalent to repulsion (Section 4.6), and no strong attraction between other partners prevents the compressed part of the structure from relaxing to a new structure in which this repulsion is relieved; compare this, for example, with sodium chloride, where repulsion between neighbor ions of the same electrical charge is more than compensated by attraction between ions of opposite sign.

Stable structures with empty spaces are possible only for crystals where some interatomic forces are stronger and directional, that is non-isotropic. One example are zeolites, where covalent bonding forces hold together the molecular scaffold and more than compensate for the very weak attraction across the void zones [15]. Another obvious example is water ice, with its strong hydrogen bonding network that creates

SPACE ANALYSIS OF MOLECULAR PACKING MODES

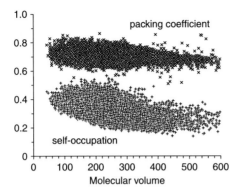

Fig. 8.2. Scatterplots of the crystal packing coefficient (equation 8.2) and the self-occupation coefficient (equation 1.11) for a very large number of organic molecules. Both distributions show a slight decline on increasing molecular volume.

and sustains a cage structure in which the density is less than in the liquid, a unique instance among non-ionic and non-covalent crystal structures.

The statement that a stable crystal will be a crystal with maximum space occupation is the essence of the concept of close packing, a very old principle (see the epigraphs to this chapter), which has been consistently formulated by Kitaigorodski [16] and was a remarkable insight in times when very limited structural data were available. This principle is at least partially quantified by the constancy of the packing coefficients in crystals, and the rule that observed crystals must have a packing coefficient of between 0.6 and 0.8 has been found to suffer no serious breakdown in the hundreds of thousands of crystal structures so far determined, as seen in Fig. 8.2. The data in Table 8.1 allow more insight into the concept and relevance of this packing index. The average packing coefficient decreases on going from flat aromatic hydrocarbons to alkyl-substituted aromatics, where the substituents perturb the regular shape of the packing molecule, or even more on going to *gem*-phenyl substituted compounds, where aromatic rings attached to the same atom impart a highly twisted shape to the molecule. The comparison among oxohydrocarbons of different sizes reveals that the larger the molecule, the less likely it is that it will have a regular shape.

The trend in average packing coefficient follows the trend in average self-occupation coefficient ($C_{\text{self,occ}}$, equation 1.11), as expected for a pure shape effect. However, the strength of attractive potentials also plays a role: in aliphatic hydrocarbons the packing coefficient is very low in spite of the self-occupation coefficients being very high, because the packing forces in aliphatic hydrocarbons are much weaker. The effect is however small: crystals of hydrogen-bonded compounds have packing coefficients that are only marginally less than those for non-hydrogen bonded crystals, in a clear indication that the usual medium-weak $X-H \cdots Y$ (X, Y=O, N) organic hydrogen bond either does not need or is not able to force the appearance

Table 8.1 Average values and root mean square deviation of the average for the packing coefficients in organic crystals of different classes of compounds

Compound class	Average C_{pack} and rms deviation	Number of crystals	Average $C_{self,occ}$
general organic, $Z' = 1$	0.710(30)	27535	0.313
general organic, $N_{atoms} < 20$	0.738(32)	770	0.394
general organic, $Z' = 1/2$	0.720(41)	3401	0.365
general organic, $Z' = 2$	0.715(36)	813	0.254
molecular complexes, solvates	0.729(35)	704	0.312
condensed aromatic hydrocarbons	0.731(28)	122	0.394
aromatic hydrocarbons with aliphatic substituents	0.699(25)	77	0.337
gem-phenyl substituted aromatic hydrocarbons	0.699(18)	24	0.291
aliphatic hydrocarbons	0.672(25)	17	0.424
small oxohydrocarbons	0.712(29)	217	0.338
medium-size oxohydrocarbons	0.705(25)	223	0.312
large oxohydrocarbons	0.698(22)	77	0.266
azahydrocarbons	0.713(29)	742	0.335
chlorohydrocarbons	0.698(28)	181	0.317
amides	0.707(32)	428	0.331
carboxylic acids	0.707(30)	326	0.337
alcohols	0.697(32)	316	0.337

of open crystal structures with voids among the constituting molecules. Hydrogen bonding does not predominate so much over diffuse dispersive interactions.

Another confirmation of the close packing principle can be found in the rather common occurrence [17] of the formation of clathrates, hydrates, and solvates: whenever the shape of the crystallizing molecule is too complex to allow efficient self-recognition with total space occupation, highly mobile solvent molecules slip in between and are incorporated in the crystal structure as space fillers. Incidentally, no numerical indicator of the ability or even the tendency of a given molecule to form solvate crystals has ever been found, a confirmation of the difficulties encountered in the definition of numerical descriptors of molecular shape.

The recognition of the fact that organic crystals are close-packed was a great advancement in the understanding of crystal packing and in the classification of crystal structures. As predicted by Kitaigorodski, organic molecules crystallize only in space groups that contain inversion, screw, or glide operations, because these are the operations that allow close packing. In an even clearer demonstration, space groups that contain non close-packing symmetry operations, like the rotation axes and mirror planes, are populated mainly by molecules in which these operations are point-group operations, that is, internal to the molecule and not preventing compact crystal packing. Thus, Table 8.2 [18] clearly shows that in organic crystal chemistry only a handful

Table 8.2 Frequencies of space groups in organic and inorganic crystal structures

Space group	% distribution, organic compounds	% distribution, inorganic compounds	% organic with point-group non-close packing operation
$P2_1/c$	38.4	8.2	
$P\bar{1}$	20.1	4.4	
$P2_12_12_1$	10.6	1.1	
$C2/c$	7.5	3.8	
$P2_1$	5.8	0.7	
$Pbca$	4.3	1.3	
$Pnma$	1.4	8.3	
$Pna2_1$	1.1	1.1	
$Pbcn$	0.9	0.8	
$C2/m$	0.45	—	99
$P2_1/m$	0.60	—	94
$R\bar{3}$-	0.45	—	73

of space groups need to be actively considered, all the others being just mathematical consequences of group theory, almost irrelevant to chemistry. Not surprisingly, these concepts do not apply to inorganic crystals where stronger covalent forces operate and close packing becomes a less stringent requirement [19].

When close-packing ideas were formulated, there was some hope that these concepts might also help in the prediction and control of crystal structures, but this expectation was not fulfilled. The complexity of crystal packing is too vast to be understood entirely by such simple and wide generalizations. Close packing is a necessary but not sufficient condition for the prediction of crystal structure, because all observed crystal structures are close packed, but not all close packed crystal structures are observed. These points will be taken up again in Chapter 14.

8.7 The calculation of intermolecular energies in crystals

8.7.1 *Lattice energies: Some basic concepts*

The interaction energy between two molecules at infinite separation is zero. When they approach one another, the system is stabilized as the potential energy decreases. Stabilizing interaction energies are negative (less than zero) and destabilizing interaction energies are positive (greater than zero). Attractive intermolecular forces pull molecules together, repulsive intermolecular forces push them apart. Thus, the terms "stabilizing" and "destabilizing" refer to energies, the terms "attractive" and "repulsive" refer to forces. The total energy of a crystal is always stabilizing, and at equilibrium the overall force on any molecule within the crystal should be zero, but some separate intermolecular interactions can be destabilizing and/or repulsive. In an extreme example, as mentioned before, the lattice energy of sodium chloride is largely stabilizing, but the energy between two sodium ions in the crystal is destabilizing and repulsive. This happens also in many organic crystals, especially those with large electrostatic energy contributions – one example is the glycine zwitterion [20].

Consider a molecule, described by a set of invariable internal coordinates; the molecule is rigid – either it has no conformational flexibility, or it is frozen in one conformation. Neglect for the moment internal vibrations and rigid-body translation and libration. Let the gas phase consist of N molecules (with N very large) at infinite distance from each other, so that their intermolecular interaction energy is by definition zero:

$$U(\text{inter, gas}) = 0 \tag{8.4}$$

Consider the same N molecules at rest in a perfectly ordered infinite crystalline array with one molecule in the asymmetric unit. All molecules are equal, and surface or truncation effects are neglected, so the following discussion refers to a bulk crystal. Assuming for the moment that the intermolecular potential is pairwise additive, e.g. in the atom–atom potential approximation, the packing potential energy, PPE, or the interaction energy of any reference molecule m with the surrounding molecules n, is a sum of molecule–molecule terms, each of which is in turn a sum of atom–atom terms:

$$PPE(m) = \sum_n U(m,n) \quad (n \neq m) \quad \text{J/molecule} \tag{8.5}$$

$$U_{kl} = A \exp(-BR_{kl}) - C R_{kl}^{-6} \quad \text{J/atom pair} \tag{8.6}$$

$$U(m,n) = \sum_{k,l} U_{kl} \quad \text{J/molecule pair} \tag{8.7}$$

Atoms k belong to molecule m, atoms l belong to molecule n, R_{kl} is the interatomic distance. Writing now $PPE(m)$ explicitly for all molecules, from equation 8.5:

$$PPE(1) = U(1,2) + U(1,3) + U(1,4) + \cdots + U(1,N) \tag{8.8a}$$
$$PPE(2) = U(2,1) + U(2,3) + U(2,4) + \cdots + U(2,N) \tag{8.8b}$$
$$\cdots \cdots \cdots$$
$$PPE(N) = U(N,1) + U(N,2) + U(N,3) + \cdots + U(N,N-1) \tag{8.8c}$$

$$PPE(1) = PPE(2) = \cdots = PPE(N) \tag{8.9}$$

Obviously $U(m,n) = U(n,m)$, but some of the $U(i,j)$ must also be equal to some $U(k,l)$ due to crystal symmetry, so that equation 8.9 holds, and, as long as the crystal is infinite and perfect, the calculation of PPE is independent of the choice of the reference molecule.

For a mole of molecules in the crystal, that is, when N equals Avogadro's number:

$$PPE = PPE(1) + PPE(2) + PPE(3) + \cdots + PPE(N) = N_{AV}PPE(m) \quad \text{J/mole} \tag{8.10}$$

The above equations imply that PPE can be calculated by picking any molecule m within the crystal as the reference molecule. In practice, the empirical coefficients, e.g. A and C in equation 8.6, may be given in appropriate units that already include the mole factor, so that U_{kl}, $U(m,n)$ and PPE directly result in J/mole from the calculation carried out on a single molecule.

$PPE(m)$ is a negative number for a stable crystal. $-PPE(m)$ is the energy required to transfer any molecule, at rest within the crystal, into the gas phase at rest, or the energy required to create a hole within the crystal lattice without any distortion in the surroundings of the hole:

$$\Delta U(\text{hole}) = U(\text{gas}) - PPE(m) = -PPE(m) \quad \text{J/molecule} \tag{8.11}$$

PPE, in J/mole, as in equation 8.10, is then the energy required for the creation of a mole of holes within the crystal, each at infinite distance from the others; a rather impractical concept. Consider instead the real process of condensing one mole of gas-phase molecules into one mole of molecules in the crystal. The total intermolecular potential energy (PE) involved in this process is the sum of the interactions of all molecules to all other molecules:

$$PE = \sum_{m,n} U(m,n) \quad (m < n) \quad \text{J/mole} \tag{8.12}$$

PE is a negative number, and $-PE$ is the energy required to take Avogadro's number molecules, at rest and in contact with each other in a perfectly ordered crystal, into the gas phase, at rest. The reason why the condition $m < n$ appears is that the interaction between molecules m and n must not be counted twice, once under the m-to-n form, and once under the n-to-m form. By comparing equations 8.10 and 8.12, one sees that in 8.10 each $U(m,n)$ is counted twice. Therefore:

$$PE = 1/2 \, PPE \quad \text{J/mole} \tag{8.13}$$

Therefore, in a static approximation that neglects kinetic energies, the sublimation energy is:

$$\Delta U(\text{subl}) = U(\text{pot,gas}) - U(\text{pot,crystal}) = U(\text{pot,gas}) - PE = -PE \quad \text{J/mole} \tag{8.14}$$

When intermolecular interaction energies are not pairwise additive, the total potential packing energy is not the sum of terms over separate molecule–molecule interactions. For example, the polarization energy at molecule m is the result of the action of an electric field due to the simultaneous influence of all surrounding molecules. Consider a molecule A surrounded by two neighbors B and C. The electric field at molecule A,

ε_A, is the sum of the fields due to molecules B and C, and the polarization energy is (compare also with the discussion around equation 4.33):

$$E_{POL} = -1/2\,\alpha_A\,\varepsilon_A{}^2 = -1/2\,\alpha_A(\varepsilon_B{}^2 + \varepsilon_C{}^2 + 2\varepsilon_B\,\varepsilon_C) \qquad (8.15)$$

The fact that the electric field is a sum of terms from many atoms, and the presence of the last term in this expression, prevent the polarization energy from being additive over molecular pairs. In addition, the field induced by molecule m at molecule n is different from the field induced by molecule n at molecule m and the two polarization energy contributions are different, so there is no double counting. The halving factor described above applies to all pair additive terms, but not to non-pairwise additive ones:

$$PE = 1/2\,PPE(\text{pairwise additive}) + PPE(\text{non-pairwise additive}) \qquad (8.16)$$

Molecules, however, are not rigid; chemists like to use enthalpies rather than internal energies; and kinetic energies should also be taken into account. The enthalpy difference between a mole of molecules in the gas phase and a mole of molecules in the crystal becomes:

$$\Delta H(\text{subl}) = H(\text{gas}) - H(\text{crystal}) = \Delta H(\text{condensation}) + \Delta H(\text{conformation}) \qquad (8.17)$$

where the first term refers to the condensation of molecules in the conformation found in the crystal, and the second term to the difference between the conformation of the molecule in the gas phase and in the crystal. Then:

$$\Delta H(\text{condensation}) = U(\text{pot, gas}) - U(\text{pot,crystal})$$
$$+ P\Delta V(\text{condensation}) + U(\text{kin,gas}) - U(\text{kin,crystal}) + \Delta E^\circ \qquad (8.18)$$

where the last term is the difference in vibrational zero-point energies between gas and crystal. To a very good approximation $U(\text{pot,gas}) = 0$. The difference in the kinetic part of internal energy is the equipartition value, or $3RT - 6RT = -3RT$, and since the volume of the condensed phase is negligible, $P\,\Delta V(\text{condensation}) = P\,V(\text{gas}) = RT$. When $\Delta H(\text{conformation})$ and ΔE° are negligible:

$$\Delta H(\text{subl}) = -U(\text{pot,crystal}) - 2RT = -PE - 2RT \qquad (8.19)$$

The $2RT$ and ΔE° terms are of the order of a few kJ mol^{-1}, and are often negligible with respect to total energies, or otherwise are of the same order of magnitude as the expected uncertainties on experimental sublimation heats. The use of these approximate corrections does not increase the accuracy of the comparison.

The following steps are involved in the practical calculation of the lattice energy of an organic crystal with atom-atom pairwise additive potentials: (1) provide cell parameters, space-group symmetry and atomic coordinates for a reference molecule

THE CALCULATION OF INTERMOLECULAR ENERGIES IN CRYSTALS   211

(labeled as molecule 1) in the crystal; this is easily accomplished using the results of X-ray diffraction experiments; (2) build a cluster of molecules 2 ... N using cell dimensions and space group symmetry and compute the lattice summations:

$$PE = 1/2 \sum_{k=2,N} E(1,k) \qquad (8.20)$$

This summation is pairwise additive and can be broken down into contributions from separate atoms within the reference molecule. These atom–atom energies have no physical meaning, but in close-packed structures, the contribution of each atom to the total lattice energy is more or less constant [21], because the interaction that produces the lattice energy is evenly distributed over the whole molecule. On the other hand, the molecule–molecule interaction energies, $E(1,k)$, have a physical meaning because a molecule is a recognizable chemical entity. These energies can be informative on packing modes, and an examination of molecule–molecule contributions discloses the most cohesive pairs and hence the structure-driving interactions (Section 14.2.3).

The enthalpy difference between two polymorphic crystal structures α and β is:

$$\Delta H(\alpha \text{ to } \beta) = \Delta U(\text{subl}, \alpha) - \Delta U(\text{subl}, \beta) = -PE(\alpha) + PE(\beta) \qquad (8.21)$$

The above equation is also the basis for the simplest approach to crystal structure prediction, that is, the generation of a large number of crystal structures and the calculation of their relative energies, assuming that the structure with the lowest packing energy will be the most probable one (Section 14.4).

Consider a crystal of a pure substance with two molecules, A_1 and A_2, in the asymmetric unit. The molecules are identical but become distinguishable in the solid state because of the different crystal environment. The condensation process can formally be broken down into (1) the formation of a mole of $A_1 A_2$ dimers, cohesive energy $E(A_1 A_2)$ and (2) the condensation of dimers into the crystal, with a packing potential energy $PPE(A_1 A_2)$, which is obtained by lattice summations in the same way as the PPE of a single molecule. If there are q molecules in the asymmetric unit (a q-mer), the same reasoning applies to the formation of $q(q-1)/2$ dimers and to the condensation of the group of q molecules into the crystal. Then:

$$PPE = PPE(A_1 A_2 \ldots A_q) + E(A_1 A_2) + E(A_1 A_3) \cdots + E(A_{q-1} A_q)$$
$$\text{J/mole of } q\text{-mer} \qquad (8.22)$$

$$PE = 1/2 PPE(A_1 A_2 \ldots A_q) + E(A_1 A_2) + E(A_1 A_3) \cdots + E(A_{q-1} A_q)$$
$$\text{J/mole of } q\text{-mer} \qquad (8.23)$$

For comparison with the heat of sublimation, PE must be given per molecule:

$$\Delta U(\text{subl}) = -1/q \, PE \quad \text{J/mole of monomer} \qquad (8.24)$$

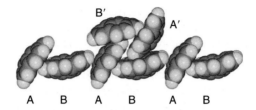

A   B   A   B   A   B

Fig. 8.3. The crystal structure of corannulene, $P2_1/c, Z = 8$. Molecules A and B are symmetry-independent, A′ and B′ are the corresponding molecules related by inversion through a center of symmetry. A and B′ are an equivalent choice for the two molecules in the asymmetric unit.

Equations 8.22–8.23 apply also for an A ... B molecular host–guest complex, where $\Delta U(\text{subl})$ is the energy difference between the crystal and the separate monomers in the gas phase:

$$\Delta U(\text{subl}) = -PE = -[1/2 PPE(\text{AB}) + E(\text{AB})] \quad \text{J/mole of A ... B dimer} \tag{8.25}$$

Note that the definition of the dimer in the crystal is not unique: the location of molecules A and B in the crystal can be chosen at will from among symmetry-related positions, and there is an infinite number of combinations of $PPE(\text{AB}) + E(\text{AB})$ that will add up to the same PE. If $E(\text{AB})$ is to have a chemical significance of its own, the pair should be chosen by supplementary conditions, for example through some specific chemical bonding between the two partners. A clarifying example is given in Fig. 8.3.

8.7.2 Convergence problems in lattice sums

The summation 8.20 should in principle be carried out on an infinite array, while in practice N is a finite number, so the calculation of lattice energies may run into truncation or convergence problems. In the atom–atom approximation the calculation of PE is extremely fast, so one can easily afford N-values up to 1,000–2,000. The summation over short-range repulsive terms has no convergence problem; summations on R^{-6} terms converge readily when N-values of the above order of magnitude are used. For obvious reasons, the summations over long-range, coulomb-type R^{-1} contributions converge much more slowly.

Part of the convergence difficulties are due to the fact that coulombic energies involve a summation of very large terms of opposite sign: the interaction energy between two electrons 10 Å apart is still 138.9 kJ/mol! When summations are carried out using a cutoff on interatomic distances (that is, neglecting all terms with distances larger than the cutoff), large oscillations in the coulombic term appear as a function of the imposed cutoff, because compensation over large terms with different signs is not granted. If the summations are carried out including a large number of entire molecular units, that is electrically neutral units, this compensation is more effective and the coulomb summation converges readily without the need of special acceleration techniques. In fact, the introduction of a cutoff in crystal energy calculations is a relic of times past, when the calculation of lattice sums was a heavy computational task. Nowadays, using a cutoff is quite pointless, because modern computers can handle summations over millions of distances in a fraction of a second.

The problem of convergence of coulombic sums can be acute in crystals with polar space groups, when molecular dipoles are parallel or nearly parallel to the polar axis. Indeed, in such cases even the physical meaning of a coulombic energy is questionable. The problem can be approximately solved by the estimation of an additional term for the coulombic energy, derived by an integration over a uniform distribution of dipolar unit cells [22]:

$$U_{conv} = U_{sums} - 1/(4\pi\varepsilon°)2\pi p^2/(3V_{cell})$$
$$= U_{sums} - 2909.9\, p^2/(ZV_{cell}) \qquad (8.26)$$

where U_{sums} is the energy obtained by lattice sums with a generous cutoff, p is the cell dipole in electrons × Å, V_{cell} is the cell volume in Å^3, and Z is the number of molecules in the unit cell.

8.7.3 Sublimation entropies and vapor pressures of crystals

What are lattice energies useful for? In principle, a larger lattice energy means a more stable material. In practice, mechanical stability often depends more on the directional properties of the cohesive energy, such as planes of easy cleavage. Besides, crystals made of organic molecules are mechanically weak anyway, so that minor differences in total cohesive energy may be quite irrelevant.

The lattice energy is a measure of the cohesion of the crystal, and hence can be related to the attachment energy, or the energy required for a molecule to break away from the lattice, and hence, by very approximate formulations, to the equilibrium solubility of the crystal [23]. The dissolution rate is, on the other hand, scarcely related to the lattice energy, being dominated by kinetic diffusive factors, as in the Noyes–Nernst dissolution model. Attachment energies are also relevant to the calculation of crystal growth rates and the simulation of crystal morphology (see Chapter 13).

The saturated vapor pressure of a crystal can be determined from the standard free energy of sublimation and is related to volatility. The sublimation entropy is:

$$\Delta S(\text{subl}) = S_g - S_c = (S_{int,g} + S_{rot,g} + S_{trasl,g}) - (S_{int,c} + S_{ext,c}) \qquad (8.27)$$

where the suffixes g, c refer to gas and crystal, respectively, and the suffixes int, ext, refer to the intra- and intermolecular vibrational contributions, respectively (the coupling of internal and external modes is neglected). The various contributions to S_g can be calculated as described in Section 7.4. $S_{ext,c}$ can be evaluated from lattice-dynamical calculations as described in Section 6.3. If the molecular conformation is about the same in the gas and in the crystal, $S_{int,g} \approx S_{int,c}$. Hence, the vapor pressure of the solid, P, can be obtained by writing the equilibrium condition (section 7.5.3, where $P°$ is the unit pressure):

$$\mu^*(c) = \mu°(v) + RT \ln(P/P°) \tag{8.28}$$

$$\mu^*(c) - \mu°(v) = -(\Delta G^*) = -(\Delta H - T\Delta S^*) = RT \ln P \tag{8.29}$$

For a medium size organic molecule the sublimation enthalpy is of the order of 100 kJ mol^{-1}, while the sublimation entropies of organic crystals are of the order of 150–200 J K^{-1} mol^{-1}. The corresponding vapour pressure is thus of the order of $P = \exp[-(100000-55000)/RT] \approx 10^{-7}$ atm. Vapor pressure variations are mainly due to changes in sublimation enthalpy, since the entropy term is fairly constant; given the logarithmic dependence, a small variation in free energy corresponds to order of magnitude changes in vapor pressure. Actual values around room temperature are, for example, 0.3 atm for benzene, 10^{-4} to 10^{-5} atm for p-dichlorobenzene and naphthalene (mothballs!), and something like 10^{-11} atm for large condensed aromatic hydrocarbons – fortunately, as these compounds are highly carcinogenic and a low vapor pressure means less poison can be inhaled.

8.8 General-purpose force fields for organic crystals

In correlation studies over large numbers of crystal structures, the calculation of the lattice energy must be relatively fast, and it is very difficult, if not altogether impossible, to calculate accurate point charges or distributed dipoles for thousands of molecules. It is also impossible to analyze all the molecular structures to find the locations of specific charge sites such as lone pairs, bond dipoles, etc.; the formulation must be a strictly atom–atom one, with all interaction centers located at easily recognizable atomic nuclear positions, and should either require no separate coulombic terms, or use a ready recipe for the approximate evaluation of atomic point charge parameters without *ab initio* molecular orbital calculations.

These requirements were met with the introduction of the UNI force field [24], which is in the atom–atom 6-exp form (Table 8.3). For each pair of atomic species (recall equations 4.39 and 4.41) the distance parameter $R°$ was optimized after an analysis of the distribution of the corresponding distances in organic crystals, while the well depth, ε, was optimized by adjustment over about 200 experimental heats of sublimation. The 6-exp functional form in principle allows for three disposable parameters for each atomic pair, but in practice it proved impossible to optimize separately the steepness parameter λ, which was kept constant for all interactions. The parameters were mostly optimized separately for each pair of interacting atomic species, without any averaging for cross interactions, giving more flexibility to the model. The force field works for both non-hydrogen bonding and hydrogen bonding crystals. No point-charge parameters and no separate coulombic terms are needed: this approximation may make the model less accurate in the description of subtle,

GENERAL-PURPOSE FORCE FIELDS FOR ORGANIC CRYSTALS

Table 8.3 UNI Force field potential parameters [24]: A, B, C for $A\exp(-BR) - CR^{-6}$ in kJ mol^{-1}. For cross $i-j$ interactions, when explicit parameters are not given, the combination rules are assumed, $A_{ij} = (A_iA_j)^{1/2}$, $C_{ij} = (C_iC_j)^{1/2}$, $B_{ij} = (B_i + B_j)/2$. Distances in Å. R°, ε, λ: see equation 4.41. Codes: alcO = alcohol O; c = oO carbonyl O; ocoO oxygen in carboxyl groups; nhrN secondary amide N; nh2N primary amide N; sulS C=S sulfur; nitO nitro group O; corresponding symbols are used for H atoms. H(HB) means H bound to N or O

		A	B	C	R°	ε	λ
H	H	24158.4	4.010	109.20	3.36	−0.042	13.5
H	C	120792.1	4.100	472.79	3.29	−0.205	13.5
H	N	228279.0	4.520	502.08	2.98	−0.394	13.5
H	O	295432.3	4.820	439.32	2.80	−0.505	13.5
H	F	64257.8	4.110	248.36	3.29	−0.110	13.5
H	S	268571.0	4.030	1167.34	3.35	−0.458	13.5
H	CL	292963.7	4.090	1167.34	3.30	−0.501	13.5
C	C	226145.2	3.470	2418.35	3.89	−0.387	13.5
C	N	491494.5	3.860	2790.73	3.49	−0.851	13.5
C	O	393086.8	3.740	2681.94	3.61	−0.674	13.5
C	F	196600.9	3.840	1168.75	3.50	−0.350	13.5
C	S	529108.6	3.410	6292.74	3.96	−0.909	13.5
C	CL	390660.1	3.520	3861.83	3.83	−0.678	13.5
N	N	365263.2	3.650	2891.14	3.70	−0.629	13.5
N	O	268571.0	3.860	1522.98	3.49	−0.464	13.5
N	F	249858.9	3.930	1277.90	3.43	−0.435	13.5
N	S	630306.9	3.590	5576.76	3.75	−1.108	13.5
N	CL	462637.0	3.590	4093.13	3.75	−0.813	13.5
N	alcH	23867340.0	7.780	1577.37	1.80	−26.610	14.0
N	ac H	23867340.0	7.780	1577.37	1.80	−26.610	14.0
N	nhrH	30190070.0	7.780	1991.58	1.80	−33.551	14.0
N	nh2H	7547602.0	7.370	690.36	1.90	−8.417	14.0
O	O	195309.1	3.740	1334.70	3.61	−0.336	13.5
O	F	182706.1	3.980	868.27	3.39	−0.320	13.5
O	S	460909.4	3.630	3790.70	3.71	−0.801	13.5
O	CL	338297.3	3.630	2782.36	3.71	−0.588	13.5
F	F	170916.4	4.220	564.84	3.20	−0.293	13.5
SI	SI	972667.1	3.210	16539.35	4.00	−1.458	12.8
S	S	1087673.0	3.520	10757.06	3.83	−1.889	13.5
S	CL	798337.4	3.520	7895.28	3.83	−1.387	13.5
CL	CL	585969.2	3.520	5794.84	3.83	−1.018	13.5
BR	BR	1132190.0	3.280	15179.55	4.00	−1.436	13.1
I	I	1560214.0	3.030	35032.63	4.00	−0.049	12.1
alcO	H(HB)	18868790.0	7.780	1246.83	1.80	−21.031	14.0
c=oO	alcH	22325910.0	8.270	1033.45	1.69	−25.348	14.0
c=oO	ac H	26416400.0	8.750	857.72	1.60	−29.158	14.0
c=oO	nhrH	15095080.0	7.780	995.79	1.80	−16.776	14.0
c=oO	nh2H	15095080.0	7.780	995.79	1.80	−16.776	14.0
ocoO	H(HB)	18868790.0	7.780	1246.83	1.80	−21.031	14.0
nhrN	alcH	23867340.0	7.780	1577.37	1.80	−26.610	14.0
nhrN	ac H	23867340.0	7.780	1577.37	1.80	−26.610	14.0
nh2N	alcH	11933730.0	7.580	928.85	1.84	−13.477	14.0
nh2N	ac H	11933730.0	7.580	928.85	1.84	−13.477	14.0
sulS	H(HB)	7324511.0	5.630	4317.89	2.40	−12.674	13.5
nitO	H(HB)	5501960.0	7.780	362.75	1.80	−6.109	14.0

Table 8.4 Williams potential parameters (ref. [44], Chapter 4). A, B, C for $E = A\exp(-BR) - CR^{-6}$ in kJ mol^{-1}. Distances in Å. A coulombic term must be added with separately determined charge parameters

		A	B	C	$R°$	ε	λ
H	H	11971.0	3.740	136.40	3.31	−0.053	12.4
H	C	66529.6	3.670	576.88	3.61	−0.143	13.2
H	N	55199.3	3.760	433.61	3.49	−0.130	13.1
H	O	52479.5	3.850	391.48	3.36	−0.146	12.9
H	F	65986.0	3.950	339.30	3.39	−0.123	13.4
H	CL	106665.8	3.625	1040.60	3.62	−0.249	13.1
C	C	369743.0	3.600	2439.80	3.90	−0.398	14.0
C	N	306774.1	3.690	1833.85	3.78	−0.360	13.9
C	O	291658.3	3.780	1655.71	3.64	−0.403	13.7
C	F	366721.7	3.880	1434.99	3.65	−0.348	14.2
C	CL	592802.5	3.555	4401.03	3.92	−0.687	13.9
N	N	254529.0	3.780	1378.40	3.66	−0.324	13.8
N	O	241987.5	3.870	1244.50	3.53	−0.361	13.7
N	F	304267.3	3.970	1078.60	3.55	−0.308	14.1
N	CL	491845.5	3.645	3308.00	3.80	−0.624	13.8
O	O	230064.0	3.960	1123.60	3.40	−0.400	13.5
O	F	289275.0	4.060	973.82	3.43	−0.339	13.9
O	CL	467610.7	3.735	2986.64	3.65	−0.702	13.7
F	F	363725.0	4.160	844.00	3.46	−0.288	14.4
F	CL	587958.4	3.835	2588.50	3.67	−0.606	14.1
CL	CL	950430.0	3.510	7938.80	3.95	−1.185	13.9

directional electrostatic effects, but it is extremely advantageous in its simplifying power, making the force field immediately applicable. The model is adequate for the reproduction of heats of sublimation and of stable crystal structures near equilibrium conditions.

Another model that can be applied to large-scale computations is the Williams force field (ref. [44], Chapter 4; see Table 8.4) with atomic point charge parameters calculated by the rescaled EHT approach (equation 4.12). This formulation (which we call here WqEHT) does not apply to hydrogen bonded crystals. The more refined forms of the Williams force field cannot be used, because of the difficulty of automatically allocating site charges far from atomic nuclei.

Using the SubHeat database, Fig. 8.4 shows the agreement between ΔH_{subl} and calculated lattice energy for the UNI and the WqEHT force fields, the latter having been used only for the non-hydrogen-bonded subset of the database. The agreement is acceptable, if one considers all the approximations in the force field and all the uncertainties in the experimental values; in particular, note that the SubHeat database contains a very large number of crystals that were not in the training sets of the two force fields.

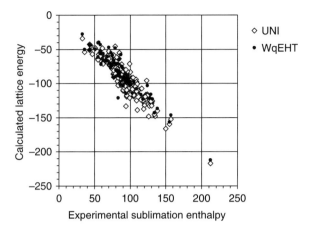

Fig. 8.4. A plot of the calculated lattice energy versus the experimental heat of sublimation for the 201 molecules in the SubHeat database. WqEHT force field: only for non-hydrogen bonded structures.

8.9 Accuracy and reproducibility

The lattice energy of an organic compound whose crystal structure is known in reasonable detail can be calculated very simply in the approximate atom–atom model, and the negative of this energy is a not such a bad approximation to the sublimation enthalpy. The nominal accuracy of the calculation, given the number of significant digits in the numerical representation inside modern computers, is very high, and one does not really even need double precision for atom–atom lattice summations. The practical accuracy and reproducibility of the calculation also depends, however, on a number of other factors, which are seldom if ever considered.

A rather unexpected conclusion emerges from the results of the preceding section: when high transferability of the force field is desired, it is extremely difficult if not impossible to attain a better agreement than that shown in Fig. 8.4 [25], also because errors due to the approximations in the force field are of the same order of magnitude as the uncertainties in experimental values. More generally, it may be said that the comparison between calculated lattice energies and experimental sublimation enthalpies, one of the cornerstones of force field optimization so far, is little more than a rough guideline, and can hardly be taken as stringent proof of accuracy. In particular, the overall accuracy is not significantly improved by correcting for a $2RT$ factor or for zero-point energies, by performing accurate calculations of conformational energy differences, or by forcing the optimization of the parameters to reproduce the observed thermochemical data, as is sometimes done, to very low thresholds like fractions of a kJ.

Since the positions of hydrogen atoms are not well determined by X-ray analysis, they must be recalculated by some kind of geometrical canon, as described in

Section 8.3. Hydrogen atoms are on the rim of all organic molecules, so that they are in close contact with neighbors and small variations in their position does significantly affect the calculated lattice energies. A test calculation has been run with the UNI force field over a sample of 836 crystal structures of the SMALL database, using C—H distances of 1.00, 1.08 (the UNI choice), and 1.10 Å. The lattice energies have been compared by drawing least-squares lines through the whole sample, with the following results:

$$E(1.10\,\text{Å}) = 0.9936\,E(1.08\,\text{Å})$$

$$E(1.00\,\text{Å}) = 1.1092\,E(1.08\,\text{Å})$$

Thus, on a typical lattice energy of 150 kJ mol^{-1}, a change of just 0.02 Å in C—H distance results in an average lattice energy difference of about 1 kJ mol^{-1}, while stabilizations or destabilizations of up to 2–3 kJ mol^{-1} for single cases are frequent. A foreshortening to 1.0 Å brings about a heavy change of up to 10 kJ mol^{-1} on average. Another source of sometimes unresolvable uncertainty is the fact that the position of the hydrogen atoms in methyl groups is never accurately determined by X-ray diffraction, and renormalization is problematic because of torsional libration. Even a small (10–20°) difference in CCCH torsion angles may cause a difference of 5–10 kJ mol^{-1} in the calculated lattice energy.

In carboxylic acids, extensive thermal motion or disorder over the —COOH group quite often causes the carbon–oxygen distances to vary over ranges of 1.30–1.25 and 1.25–1.20 Å for the C=O and C—O bonds, respectively. Test calculations on benzoic acid using either the observed distances of 1.275 and 1.264 Å or renormalized distances of 1.30 and 1.22 Å show oscillations of 1–2 kJ mol^{-1} in the packing energy.

X-ray determinations are nowadays routinely carried out at low temperature, because the resulting determination is more accurate. The effect of the temperature of the X-ray determination on calculated lattice energies has been studied by carrying out UNI calculations on variable-T crystal structures. The results are shown in Table 8.5: a 100 K change in the temperature of the crystal structure determination brings about a variation in lattice energy from 0.1 kJ mol^{-1}, for the crystal with the smallest expansion coefficient, naphthalene, to about 1 kJ mol^{-1} for dinitrobenzene.

X-ray crystal structure determinations do not all have the same level of accuracy. In a typical example, Fig. 8.5 shows the combined effect of temperature and of a change in R-factor for the calculation of the lattice energy of anthraquinone. The crystal structures with higher R-factors, which are less accurate, yield a less stabilizing lattice energy, as expected. A better crystal structure determination lowers the lattice energy by 1–1.5 kJ mol^{-1}. In a further test, the atomic coordinates of a sample of crystal structures were changed by random addition or subtraction of one standard deviation, with lattice energy changes of the order of 1 kJ mol^{-1}.

This discussion highlights some of the many lurking variables that affect the calculation of lattice energies. Calculations can give results that differ by up to 2–5 kJ mol^{-1} for a number of often-hidden reasons; while this is demonstrated here for atom–atom

Table 8.5 Effect of the temperature of the X-ray structure determination on calculated lattice energies. UNI force field

Crystal	$1/V(dV/dT)$, (K^{-1})[a]	dE/dT[b] $(J\,K^{-1}\,mol^{-1})$	Approx $\Delta\,(100)$[c] $(kJ\,mol^{-1})$
anthracene	1.2×10^{-4}	4	0.4
naphthalene	1.2×10^{-4}	0.8	0.1
anthraquinone	1.3×10^{-4}	2	0.2
dinitrobenzene	2.3×10^{-4}	9	0.9
benzoic acid	2.4×10^{-4}	6	0.6

[a] Thermal expansion coefficient.
[b] Lattice energy variation with T; note that this quantity need not compare with the actual heat capacity of the crystal, see Section 7.3.
[c] Expected lattice energy change for a 100 K change in temperature.

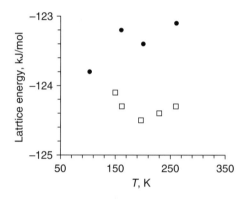

Fig. 8.5. Anthraquinone crystal. Black dots: lattice energies calculated using the crystal structures of Lonsdale *et al.*, Acta Cryst. 1966, **20**, 1, *R*-factors around 0.10. White squares: the same calculated using the crystal structures of Fu and Brock, *Acta Cryst.*, 1998, **B54**, 308, *R*-factors around 0.05.

calculations, it is more than likely that the same holds to be even more true for calculations involving more elaborate potential energy schemes [26]. This is not a serious obstacle for the comparison with experimental heats of sublimation, where such an accuracy level is quite tolerable, but these hidden factors should be kept in mind when comparing calculations made by different methods, or when discussing small energy changes among different crystal structures. For a conservative estimate, the effective reproducibility, and hence discriminating power, of any kind of lattice energy calculations based on theoretical potential models and experimental crystal structures cannot exceed 3–5 kJ mol^{-1}. This is, for example, dangerously close to the energy differences among polymorph crystal structures.

8.10 Correlation between molecular and crystal properties: Fact or fiction?

The Cambridge Structural Database is a very large collection of structural data concerning molecules and the crystals that grow from them. The joint availability of molecular and crystal data naturally leads to the search for correlations. The supreme goal would be the ability to predict the molecular packing just from a consideration of molecular structure, but let it be said at once that this goal has not been reached and is not even in sight. The following sections are devoted to a description of the methods that can be used in data mining through the Cambridge Structural Database.

8.10.1 Bivariate analysis

Molecules and crystals are described by a large number of quantitative indicators. Establishing a correlation between any two indicators is called bivariate analysis, and can be expressed either in a graph or in a linear regression; the first problem is obviously to find out which pairs of properties do indeed correlate.

Figure 8.6 shows that the volume (in $Å^3$) that an organic molecule occupies in a crystal is roughly equal to twice its number of electrons. Clearly, the correlation is only rough because the same number of electrons can be achieved, for example, by a small number of electron-rich chlorine atoms or by a large number of electron-poor hydrogen atoms, with a largely different space occupation. Of course, molecular volume correlates very well with cell volume with a slope of 0.7, the packing coefficient (see Table 8.1).

Another very obvious crystal property is the lattice energy, which can be calculated by potential energy schemes or measured as a sublimation enthalpy. Figure 8.7

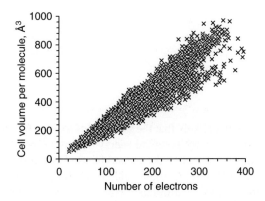

Fig. 8.6. Scatter-plot of crystal cell volume per molecule against number of electrons in the molecule. Least-squares lines are $y = 2.353x$ for general crystals, $y = 2.352x$ for non-hydrogen bonded crystals, $y = 2.317x$ for hydrogen-bonded crystals, and $y = 2.274x$ for crystals with half a molecule in the asymmetric unit. The statistical significance of the small differences in the slopes is questionable.

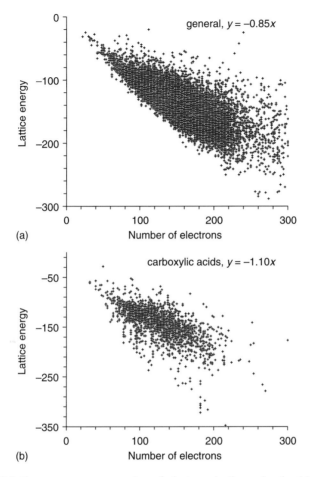

Fig. 8.7. UNI lattice energy versus number of electrons in the molecule: (a) 15,454 non-hydrogen bonded crystals; (b) 1792 carboxylic acid crystals.

shows broad correlations between the lattice energy calculated with UNI atom–atom parameters and molecular size, represented there by the number of electrons in the molecule. The sublimation enthalpy of an organic compound is roughly equal to the number of electrons in the molecule ± 25 kJ mol^{-1}. Not unexpected is the distinction between non-hydrogen bonded and hydrogen-bonded crystals, as demonstrated by the different slopes. Sublimation enthalpies in kJ g^{-1} (Fig. 8.8) are larger for smaller molecules and level out to something around 0.4 kJ g^{-1} at higher molecular sizes. Other mild correlations appear, like the one (Fig. 8.9) found between molecular polarizability, as obtained by a sum of average atomic contributions (ref. [21], Chapter 4), and the lattice energy for non-hydrogen bonded crystals, where presumably the main cohesion energy comes from dispersion, and hence polarizability (Section 4.5).

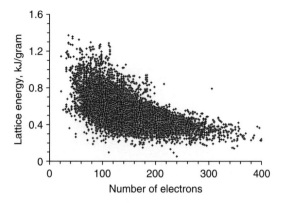

Fig. 8.8. UNI lattice energies in kJ g^{-1} for the Z(1) database (Section 8.3).

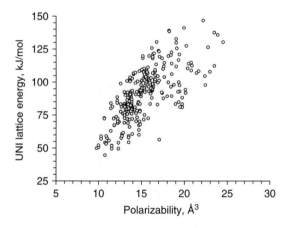

Fig. 8.9. UNI lattice energies for 308 crystal structures (from the SMALL database, no hydrogen bonds) as a function of the overall molecular polarizability computed by sums of approximate atomic increments.

All bivariate correlations between molecular and crystal properties are only broad and approximate, demonstrating that crystal packing is a complex phenomenon. More often that not, one finds that correlations expected on the basis of simple reasoning do not in fact hold. For example, Fig. 8.10 shows that there is no relationship between central molecular dipole moments and point-charge coulombic lattice energies, not even when both quantities are calculated by the same method (the rescaled EHT method). This is a further confirmation that central dipoles are a very poor indicator of molecular electrostatic properties [27].

CORRELATION BETWEEN MOLECULAR AND CRYSTAL PROPERTIES 223

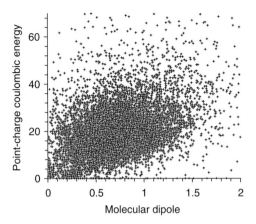

Fig. 8.10. The point-charge lattice coulombic energy (kJ mol^{-1}) for 10,951 crystal structures (from the Z(1) database, 10 to 15 carbon atoms) as a function of the molecular dipole moment in electron $\times$Å (for molecules whose dipole moment is not zero by symmetry). Atomic point charges calculated by the rescaled EHT approach (see text).

8.10.2 *Principal component analysis*

Let the landscape of a given phenomenon be spanned by n samplers, each of which possesses m properties: to be specific, in crystal packing analysis one has n crystal structures each of which is characterized by m molecular or crystal properties. For example, molecular size can be represented by the number of electrons, but also by molecular weight, molecular volume, molecular surface. Common sense tells that these indicators must be highly correlated, and that it would be desirable to have only one dimension in this pseudo-configurational space to describe molecular size, because the number of bivariate correlations increases exponentially with increasing m. A simple and intuitive way of doing this is provided by principal component analysis [28]. The data are arranged in a rectangular matrix **M** with n rows (molecules) and m columns (properties). To obviate the intrinsic difference in numerical size among indicators, each matrix element is renormalized:

$$M_{kp} = (M'_{kp} - P_k)/(\sigma_k) \tag{8.30}$$

where P_k and σ_k are the average of property k and its standard deviation. The $m \times m$ correlation matrix (a square matrix) is then

$$\mathbf{C} = 1/(m-1)\mathbf{M}^T\mathbf{M} \tag{8.31}$$

The eigenvectors and eigenvalues of **C** are then calculated. Calling **V** the eigenvector matrix and **D** the diagonal matrix of the eigenvalues, the composition of the principal components is given as a matrix, **F**, of linear combination coefficients that mix the

Table 8.6 Principal components analysis: the first four principal components for the case described in the text. 27,461 crystals in the Z(1) database

% variance	Molecular weight	Number of electrons	Molecular volume	Molecular surface	ΔH_{subl}	$C_{self,occ}$
78.9 ("size")	0.97	0.98	0.96	0.97	0.73	−0.65
10.7 ("shape")	0.10	0.09	0.07	0.05	0.24	0.75
8.5 ("energy")	0.12	0.14	0.17	0.14	−0.63	0.14
1.6	0.18	0.12	−0.17	−0.13	−0.02	−0.01

single properties into the principal components:

$$\mathbf{F} = \mathbf{M}\,\mathbf{V}\,\mathbf{D}^{1/2} \tag{8.32}$$

Eigenvalues that are significantly different from zero identify the relevant components; the percent contribution of each principal component k to the effective variance of the sample can be determined as follows:

$$m = \sum D_{kk}^2 \tag{8.33}$$

$$k\text{-th component, \%effective variance} = D_{kk}^2/m \tag{8.34}$$

The interpretation of these principal components in terms of combinations of the single properties is then left to the investigator's judgement. As an example, consider Table 8.6, which shows the principal components obtained out of the combination of molecular weight, number of electrons, molecular volume, surface, estimated sublimation enthalpy, and self-occupation coefficient $C_{self,occ}$, for a total variance of six. The four size indicators are always covariant (i.e. appear with the same sign and with similar loadings), and thus count for only one variance unit. The first factor represents 78.9% of the total variance as a covariant combination of all the four size parameters with the sublimation enthalpy, because larger size means larger lattice energy, while $C_{self,occ}$ appears as a contravariant, because a higher size corresponds to lower coefficient (see Fig. 8.2). A second component, 11% of the total variance, has the shape indicator as almost the sole component, but also shows that, to some minor extent, a high $C_{self,occ}$ allows for a higher lattice energy. The third non-zero eigenvalue has hardly 9% of the total variance and is dominated by the sublimation heat. The analysis broadly reveals that one has here six indicators but only three separate effects (non-zero eigenvalues), which can be called "size", "shape", and "energy". The principal component analysis does away at a glance with redundant coordinates in the variational space.

Both in bivariate and multivariate analysis, correlations between molecular and crystal properties are poor, unsure, and scarcely robust. The fundamental reason is that crystal packing is a matter of finely adjusted detail, and global indicators will never reach the subtleness required for a real predictive power.

8.11 Acceptable crystal structures

The canons outlined by the correlations presented in the previous sections of this chapter lack any detailed predictive power, but can be used at least to outline the boundaries within which a crystal structure may be considered acceptable. The two necessary, if not sufficient, requisites for an acceptable organic crystal structure are:

$$0.65 < C_{\text{pack}} < 0.80 \qquad \text{HB and non-HB crystals}$$
$$|[E_{\text{lattice}} + (0.9\,N_{\text{electrons}})]| < 25\,\text{kJ mol}^{-1} \qquad \text{non-HB crystals} \qquad (8.35)$$
$$|[E_{\text{lattice}} + (1.1\,N_{\text{electrons}})]| < 25\,\text{kJ mol}^{-1} \qquad \text{HB crystals}$$

"Acceptable" means here that a crystal structure, either coming from a problematic X-ray diffraction experiment (uncertainties in atomic positions, or cell parameters, or space group; possibly, overlooked disorder; or missing solvate molecules), or from a tentative computational prediction, must fall within the boundaries prescribed by equations 8.35, otherwise it can be confidently discarded as unlikely or altogether impossible.

8.12 Historical portraits: Lattice energies and the phase problem in the old days

In Chapter 5, a description was given of the hard life of X-ray crystallographers in the late 1960s. At more or less the same time, our crystallography group in Milano started playing around with computer programs to calculate lattice energies for organic crystals. The very naive idea was to generate acceptable – in the sense of equations 8.35 – crystal structures and to provide a quick and inexpensive tool for solving the phase problem, since our success with DTE using direct methods was still considered an early (and rare) bird, and, moreover, there was no sign of anything of the same sort that could be used for non-centrosymmetric crystals. Our would-be crystal structure solvers were to use the information on cell parameters and space group, which was easily obtainable even with primitive photographic methods, to try and fit the molecule into the crystal cell by checking that space occupation was even and that there were no offending atom–atom contacts that would result in unacceptable packing energies. We were blissfully ignorant of how ambitious that project could be, witness the fact that some forty years later the problem still cannot be solved that way (it has been solved instead by the astonishingly fast development of direct methods).

Computer resources were outrageously meager, and no one knew for sure if and how the same non-bonded potentials used in molecular mechanics for isolated molecules would work for non-bonded interactions in crystals, or if and how some new potential parameters could be found (again, "got a force constant?"). The way to the solution of that problem, with the calibration of an intermolecular force field, was being set in those years by Don Williams.

We were also experimenting with computer programs that would automatically fit a proposed molecular model to the Patterson function. On one occasion, we had received some beautiful crystals from a very well known organic chemist, a future Nobelist, but the crystal structure was non-centrosymmetric and we could not solve it. Since a visit to our lab by the famous person was soon expected, I worked day and night with my programs and Carlo Maria Gramaccioli's Patterson interpretation routine to try and make the ends meet. One day I came up with a "solution" and calculated a Fourier synthesis with my coordinates, and sure enough there I found the molecule in the crystal. I ran to Carlo Maria with proud expectations, but he looked at my result, shook his head and said, "Have you done an R-factor calculation?" "No, I said, but I don't need one, I can see the molecule in the Fourier, so it must be right!" He shook his head again and said, "Oh, you see, the phases of the structure factors in a non-centrosymmetric structure can be anything, so whatever molecular model you put in will phase the observed intensities so that you get back your model in the density map, that's all." In fact my structure was wrong, and it was quite a take-home lesson. It later turned out, when Angelo Mugnoli eventually solved the structure, that the famous person had sent us the wrong vial and the compound did not have the formula that we had been given, but had a stray sulfur atom in it. I can't remember exactly how Angelo went around solving the structure, but I remember him working with long lists of Miller indices, patiently working by hand the symbolic additions required to expand a trial solution. In those days, direct methods were carried out by hand and it took weeks or months to achieve what is nowadays done in a few seconds.

References and Notes to Chapter 8

[1] The assembling of crystallographic data started with printed repertoires, like the Strukturberichte, and the early printed version of the Cambridge collection. Nowadays, the Cambridge Crystallographic Data Center (www.ccdc.cam.ac.uk) produces and distributes the Cambridge Structural Database (CSD), which contains, in computer-readable form, unit cell dimensions, space groups, atomic coordinates, bibliographic data and other chemical and crystallographic information for organic and organometallic crystals of known structure, plus software for retrieval, visualization, and statistical analysis of the results. Started officially on January 1, 1965, when David Watson joined Olga Kennard's research group at the Department of Chemistry in Cambridge, by mid-2005 the CSD contains more than 335,000 entries and now expands at the rate of about 30,000 structures a year. See Allen, F. H.; Bellard, S.; Brice, M. D.; Cartwright, C. A.; Doubleday, A.; Higgs, H.; Hummelink, T.; Hummelink-Peters, B. J.; Kennard, O.; Motherwell, W. D. S.; Rodgers, J. R.; Watson, D. G. *Acta Cryst.* 1979, **B35**, 2331–2339; Allen, F. H.; Kennard,O.; Taylor, R. *Acc. Chem. Res.* 1983, **16**, 146–153; Allen, F. H.; Kennard, O. *Chemical Design Automation News* 1993, **8**, 1–31. About 30% of the CSD entries contain questionable material; van der Streek, J. *Acta Cryst.*, 2006, **B62**, 567–579.

[2] Bürgi, H. B.; Dunitz, J. D.; Shefter, E. Chemical reaction paths. IV. Aspects of O···C=O interactions in crystals, *Acta Cryst.* 1974, **B30**, 1517–1527, and references therein.

[3] Burgi, H.-B.; Dunitz, J. D. (Eds), *Structure Correlation*, 1994 VCH, Weinheim; Ferretti, V.; Gilli, P.; Bertolasi, V.; Gilli, G. Structure correlation methods in chemical crystallography, *Crystallography Reviews* 1996, **5**, 3–104.

[4] See the Appendix in Gavezzotti, A. Statistical analysis of some structural properties of solid hydrocarbons, *J. Am. Chem. Soc.* 1989, **111**, 1835–1843.

[5] Afeefy, H. Y.; Liebman, J. F.; Stein, S. E. *Neutral Thermochemical Data*; and J. S. Chickos, *Heat of Sublimation Data*, in: *NIST Chemistry WebBook, NIST Standard Reference Database Number 69*, edited by Linstrom, P. J.; Mallard, W. G. March 2003, National Institute of Standards and Technology, Gaithersburg MD, 20899 (http://webbook.nist.gov).

[6] This work was carried out mostly by groups led by F. Allen (see literature quotations in ref. [3]), now director of the Cambridge Crystallographic Data Center.

[7] For an early compilation of H-bonding distances see Brown, C. J. The crystal structure of *p*-aminophenol, *Acta Cryst.* 1951, **4**, 100–103. For some modern examples, see for example Taylor, R.; Kennard, O. Hydrogen-bond geometry in organic crystals, *Acc. Chem. Res.* 1984, **17**, 320–326; Murray-Rust, P.; Shimoni, L.; Glusker, J. P.; Bock, C. W. Energies and geometries of isographic hydrogen-bonded networks. The $R_2^2(8)$ graph set, *J. Phys. Chem.* 1996, **100**, 2957–2967.

[8] Leiserowitz, L. Molecular packing modes. Carboxylic acids, *Acta Cryst.* 1976, **B32**, 775–802.

[9] Steiner, Th.; Saenger, W. Geometric analysis of non-ionic O–H···O hydrogen bonds and non-bonding arrangements in neutron diffraction studies of carbohydrates, *Acta Cryst.* 1992, **B48**, 819–827.

[10] Taylor, R.; Kennard, O. Crystallographic evidence for the existence of C–H···O, C–H···N and C–H···Cl hydrogen bonds, *J. Am. Chem. Soc.* 1982, **104**, 5063–5070.

[11] Berkovitch-Yellin, Z.; Leiserowitz, L. The role played by C–H···O and C–H···N interactions in determining molecular packing and conformation, *Acta Cryst.* 1984, **B40**, 159–165.

[12] D'Oria, E.; Novoa, J. J. The strength–length relationship at the light of *ab initio* computations: does it really hold? *CrystEngComm* 2004, **6**, 367–376.

[13] Desiraju, G. R. *Crystal Engineering, the Design of Organic Solids*, 1989, Elsevier, Amsterdam.

[14] Gavezzotti, A. Statistical analysis of some structural properties of solid hydrocarbons, *J. Am. Chem. Soc.* 1989, **111**, 1835–1843; Desiraju, G. R.; Gavezzotti, A. Crystal structure of polynuclear aromatic hydrocarbons. Classification, rationalization and prediction from molecular structure, *Acta Cryst.* 1989, **B45**, 473–482; Gavezzotti A. On the preferred mutual orientation of aromatic groups in organic condensed media, *Chem. Phys. Lett.* 1989, **161**, 67–72; Gavezzotti, A. Packing analysis of organic crystals containing C=O or C—N groups,

J. Phys. Chem. 1990, **94**, 4319–4325; Gavezzotti, A. Molecular packing and other structural properties of crystalline oxohydrocarbons, J. Phys. Chem. 1991, **95**, 8948–8955; Filippini, G.; Gavezzotti A. A quantitative analysis of the relative importance of symmetry operators in organic molecular crystals, Acta Cryst. 1992, **B48**, 230–234; Gavezzotti, A.; Filippini, G. Molecular packing of crystalline azahydrocarbons, Acta Cryst. 1992, **B48**, 537–545; Gavezzotti, A.; Filippini, G. The crystal packing of chlorine- and sulfur-containing compunds, Acta Chimica Hungarica – Models in Chemistry 1993, **130**, 205–220.

[15] For an impressive series of synthetic results of highly porous materials, using metal-containing subunits, see Yaghi, O.; O'Keeffe,M.; Ockwig, N. W.; Chae, H. K.; Eddaoudi, M.; Kim, J. Reticular synthesis and the design of new materials, Nature 2003, **423**, 705–714. Even for organic materials, crystals of inclusion compounds are known to form metastable phases in which the host scaffold remains intact after release of the guest; see Nassimbeni, L. R. Acc. Chem. Res. 2003, **36**, 631–637.

[16] Kitaigorodski, A. I. *Organicheskaya Kristallokhimiya*, 1961, translated from Russian, Consultants Bureau, New York; Kitaigorodski, A. I. *Molecular Crystals and Molecules*, 1973, Academic Press, New York.

[17] Out of 46,460 analyzed CSD entries, about 20%, were found to contain a solvent. Water takes the lion's share (61%), followed at a distance by methylene chloride (6%), benzene, and methanol. While 95% of the solvate crystals contain only fifteen different solvate molecular species, a total of 147 different cocrystallized solvents were detected. See van der Sluis, P.; Kroon, J., Solvents and X-ray crystallography, J. Cryst. Growth 1989, **97**, 645–656. There is reason to believe that the statistics have not changed substantially since the time of that survey.

[18] Brock, C. P.; Dunitz, J. D. Towards a grammar of crystal packing, Chem. Mater. 1994, **6**, 1118–1127.

[19] Baur, W. H.; Kassner, D. The perils of Cc: comparing the frequencies of falsely assigned space groups with their general population, Acta Cryst. 1992, **B48**, 356–369.

[20] Destro, R.; Roversi, P.; Barzaghi, M.; Marsh, R. E. Experimental charge density of α-glycine at 23 K, J. Phys. Chem. 2000, **A104**, 1047–1054.

[21] Gavezzotti, A. Calculations on packing energies, packing efficiencies and rotational freedom for molecular crystals. Nouv. J. Chim. 1982, **6**, 443–450.

[22] van Eijck, B. P.; Kroon, J. Coulomb energy of polar crystals, J. Phys. Chem. B 1997, **101**, 1096–1100. This paper also has a discussion of convergence acceleration techniques.

[23] The Noyes–Nernst solubility model includes the formation of a very thin saturated layer close to the crystal surface, followed by diffusion into the bulk. The concentration of the saturated layer depends to some extent on the attachment energy, but need not be identical to the bulk saturation concentration (see Section 11.3). Relationships between sublimation enthalpy and solubility rest on some very approximate assumptions.

[24] Filippini, G.; Gavezzotti, A. Empirical intermolecular potentials for organic crystals: the 6-exp approximation revisited, *Acta Cryst.* 1993, **B49**, 868–880; Gavezzotti, A.; Filippini, G. Geometry of the intermolecular X–H···Y (X, Y=N, O) hydrogen bond, and the calibration of empirical hydrogen-bond parameters, *J. Phys. Chem.* 1994, **98**, 4831–4837.

[25] For example, the COMPASS force field (ref. [27], Chapter 2) claims an "excellent agreement" over 24 crystal structures with deviations of up to 19 kJ mol^{-1} and an average deviation of 5.5 kJ mol^{-1}.

[26] When more accurate methods are used, with inclusion of full electrostatic effects, the differences in calculated lattice energy for small differences in molecular geometry are much higher, in the 5–10 kJ mol^{-1} range (Section 14.4).

[27] Whitesell, J. K.; Davis, R. E.; Saunders, L. L.; Wilson, R. J.; Feagins, J. P. Influence of molecular dipole interactions on solid-state organization, *J. Am. Chem. Soc.* 1991, **113**, 3267–3270. A quotation: "attempts to design molecular arrays based primarily on considerations of overall molecular dipole moments have, statistically speaking, a small chance of success."

[28] Murray-Rust, P.; Motherwell, S. Computer retrieval and analysis of molecular geometry. I. General principles and methods, *Acta Cryst.* 1978, **B34**, 2518–2526. A rough application of principal components, or factor analysis, is a shirt that has just one size parameter (S, M, L, XL), instead of a specification of waistline, chest width, arm, or leg lengths, etc., in the assumption that these body parameters are correlated.

9

The liquid state

The basic principle of Monte Carlo methods in molecular simulations is to generate a set of configurations, which is representative for an ensemble, by randomly changing the position of some atoms and accepting the changed configurations based on a criterion that ensures the sampling of a Boltzmann distribution of configurations. In molecular dynamics a set of configurations is obtained by taking instantaneous configurations along a trajectory of the interaction sites calculated by numerically solving the Newton equations of motion.

Mordasini Denti, T.Z.; Beutler, T.C.; van Gunsteren, W.F.; Diederich, F. J. Phys. Chem. 1996, 100, 4256–4260.

9.1 Proper liquids

An ideal gas has by definition no intermolecular structure. Also, real gases at ordinary pressure conditions have little to do with intermolecular interactions. In the gaseous state, molecules are to a good approximation isolated entities traveling in space at high speed with sparse and near elastic collisions. At the other extreme, a perfect crystal has a periodic and symmetric intermolecular structure, as shown in Section 5.1. The structure is dictated by intermolecular forces, and molecules can only perform small oscillations around their equilibrium positions. As discussed in Chapter 13, in between these two extremes matter has many more ways of aggregation; the present chapter deals with proper liquids, defined here as bodies whose molecules are in permanent but dynamic contact, with extensive freedom of conformational rearrangement and of rotational and translational diffusion. This relatively unrestricted molecular motion has a macroscopic counterpart in viscous flow, a typical property of liquids. Molecular diffusion in liquids occurs approximately on the timescale of nanoseconds (10^{-8} to 10^{-10} s), to be compared with the timescale of molecular or lattice vibrations, 10^{-13} to 10^{-12} s.

9.2 Molecular dynamics (MD)

For the crystalline state, a simulation using a static model in which all atoms and molecules are fixed at their average positions is not too far from reality. Such a model

MOLECULAR DYNAMICS (MD)

is quite unsatisfactory for the gaseous and liquid states, whose modeling requires a dynamic approach.

A molecular dynamics (MD) simulation [1] is a mathematical representation of what happens when a chemical system evolves in time under the action of its own internal potentials and forces, with due account taken of kinetic energies, temperature, and pressure. So described, MD simulation seems as close as possible to a representation of the perfect chemical truth. Although a host of practical limitations reduce its scope, MD is an extremely useful tool in chemical analysis, and its applications enjoy a rapidly increasing popularity with theoretical chemists. MD explores phase space, that is, it surveys the landscape of atomic positions and momenta under the action of driving forces, and is therefore a convenient tool for the search of energy minima, orders of magnitude more efficient than just energy minimization algorithms (Section 2.2). It is in fact very successfully applied to conformational searches in large biomolecules like proteins and DNA in aqueous conditions, as well as to the determination of the best recognition modes between macromolecules and biologically active small molecules or between host and guest molecules in supramolecular chemistry. We are interested here in its application to the simulation of the evolution of molecular ensembles representing the structure and properties of condensed phases of organic molecules.

9.2.1 *Equations of motion*

There is no experiment that allows us to see a moving picture of molecules in a gas or in a liquid. In a thought experiment that greatly exceeds the present possibilities of instrumental analytical chemistry, one can imagine blowing up a single isolated molecule of benzene to the dimensions of a macroscopic object, and being able to actually see what nuclei and electrons are doing over a period of 10 seconds at room temperature. One would see a doughnut-shaped object moving at the speed of a jet plane and tumbling about its inertial axes with a spectacular angular speed. Bond distances and bond angles would be seen to perform about ten thousand billion nearly harmonic vibrations in a second. Nothing significant would happen to the electron density since electronic excitation is impossible by the modest thermal action of a room-temperature environment. This visualization may be made using a computer; all that is needed to run a simulation of the motion of a benzene molecule is the potentials and an equation of motion. Model potentials, together with methods of obtaining force constants for bond stretching, bond angle bending, and for other kinds of molecular deformation, from a mixture of experiment and calculation, have been described in Chapter 2. For small displacements from equilibrium, the vibrational potential energy has the harmonic form:

$$E(\text{pot, stretch or bend}) = \sum_m 1/2 \, k_m (R_m - R^\circ_m)^2 + \sum_n 1/2 k_n (\theta_n - \theta^\circ_n)^2 \quad (9.1)$$

where the sums run over all bond distances R_m and bond angles θ_n in the molecule. For benzene, one would need C—C and C—H stretching force constants, plus C—C—C and C—C—H bending force constants, which are readily available from spectroscopic determinations. $R°$ and $\theta°$ equilibrium values are easy to guess, since the shape of the benzene molecule is well known from diffraction and spectroscopic experiments. Other potential energy terms may correspond to torsional modes along the ring, C–C–C–C, H–C–C–C, and H–C–C–H oscillations, described by the proper and improper dihedral functional forms, equations 2.6 and 2.11:

$$E(\text{pot, torsion}) = \sum_k f(\tau_k) + \sum_m 1/2\, K(\tau_m - \tau_m°)^2 \qquad (9.2)$$

Other potential energy contributions may come from non-bonded interactions among atoms in the molecule, described by some function of interatomic distance, as in equation 2.7, possibly supplemented by coulombic terms over some kind of distribution of charge within the molecule (for example, localized point charges on atoms, as in equation 2.8):

$$E(\text{pot,non-bonded}) = \sum_{i,j} f(R_{ij}) + \sum_{i,j} 1/(4\pi\varepsilon°) q_i q_j (R_{ij})^{-1} \qquad (9.3)$$

The sum of equations 9.1, 9.2, and 9.3 is the intramolecular potential field, E(pot, intra). The molecular force field has a classical formulation, and quantum effects are neglected, implicitly assuming an energy equipartition condition (Section 7.2). The neglect of electronic energy changes prevents the simulation of chemical reactions, or of any other chemical process that requires significant electronic rearrangements, like for example tautomery.

When potentials are known, obtaining forces is a simple matter: let $x_{i,k}$ be the k-th cartesian component of the position vector of atom i of mass m_i; the corresponding component of the force at atom i is given by

$$F_{i,k} = -\partial[E(\text{pot})]/\partial x_{i,k} \qquad (9.4)$$

and the calculation is easy, given the simple functional forms of vibrational and non-bonded potentials. Time evolution is derived by solving the equations of motion of classical mechanics:

$$F_{i,k} = m_i a_{i,k} = m_i (\mathrm{d}v_{i,k}/\mathrm{d}t) = m_i (\mathrm{d}^2 x_{i,k}/\mathrm{d}t^2) \qquad (9.5)$$

Given a reasonable starting configuration $(x_{i,k})°$, integration of the above differential equation gives the trajectories of all atoms in the molecule. Of course the integration cannot be carried out analytically, but numerical integration techniques are cheaply and readily available [2].

9.2.2 Temperature

A single isolated molecule of mass m would just travel forever on a linear trajectory with constant velocity v and kinetic energy $E(\text{kin}) = \int mv\,\mathrm{d}v = 1/2\, mv^2$. When two molecules are close to one another, they interact with a potential such as shown in Fig. 4.4, which will here be denoted simply by E(pot, inter). This potential may have one of the many analytical forms discussed in Section 4.9, but for application in MD simulations, the positional derivatives must be easily accessible. Kinetic energy is a transparent quantity in MD, because atomic velocities are known. In an equipartition regime, each molecular degree of freedom (dof) contributes $1/2k_B T$ to the kinetic energy (see the discussion around equation 7.24). For a polyatomic system the total kinetic energy is the sum of the kinetic energies of each atom, and can be equated to the equipartition value, thus yielding a key link between molecular motion and temperature:

$$E(\text{kin}) = \sum_i 1/2 m_i v_i^2 = 1/2 k_B T (n_{\text{dof}}) = 1/2 k_B T (3 n_{\text{atoms}}) \qquad (9.6)$$

The total energy of the system is:

$$E(\text{tot}) = E(\text{pot, intra}) + E(\text{pot, inter}) + E(\text{kin}) = U - U^\circ \qquad (9.7)$$

The last equality underscores the equivalence between this microscopic definition of internal energy and the macroscopic definitions given by equation 7.22 in statistical thermodynamics or in equation 7.26 in classical thermodynamics: the internal energy that Joule was able to measure more than a century ago, and that Boltzmann extracted from the partition function, is just the sum of all molecular potential and kinetic energies – what else? Of course all the thermodynamic definitions – energy and temperature – acquire a proper meaning only by considering large ensembles of molecules, and one of the essential requirements for a successful MD simulation of these properties is the consideration of a large enough number of molecules in the computational sample.

Some fundamentals of physical chemistry can be extracted from simple considerations based on the intermolecular model potential and the molecular dynamics model. Molecules are pulled together by the attractive force field and stay together, oscillating around the equilibrium distance between their centers of mass or even sliding past one another, in a molecular dynamics analog of crystalline or liquid cohesion; the shape and curvature of the potential close to the minimum accounts for the rheological properties of the material. At some temperature, the available kinetic energy exceeds the depth of the stabilizing potential energy well, and the molecules will split apart, in a molecular dynamics picture of evaporation. The very steep repulsive wall of the potential curve accounts for the very scarce compressibility of condensed phases.

9.2.3 Pressure

Pressure can be easily understood for a gas, by the traditional didactic picture of traveling molecules that hit the container walls and transfer a certain amount of momentum to them. But the definition of pressure does not require the presence of a wall and need not invoke collisions between wall and molecules; a physical barrier is just convenient for the visualization of an arbitrary plane through the material. Pressure is flow of momentum through any plane, and can thus be defined at any place in any system of any state of aggregation.

A clearer insight may be gained by considering the statistical mechanical definition of internal pressure, valid for any state of matter:

$$P = 2/(3V)[E(\text{kin}) - \Xi] \tag{9.8}$$

$$\Xi = -1/2 \sum_k \sum_l R_{kl} F_{kl} \tag{9.9}$$

V is the volume of the system, and the quantity Ξ (greek "csi") is called the virial, a sum of products of the distance R_{kl} between any two atoms k and l in the system and of the force F_{kl} acting between them. Distances are always positive numbers so the virial is positive when forces are negative, that is attractive. For an ideal gas, $E(\text{kin}) = 3/2RT$ per mole and since forces are all zero $\Xi = 0$ and $P = RT/V$ per mole, as it should. A positive virial is something that reduces the effect of kinetic energy and hence subtracts something from the internal pressure of the ideal gas: if molecules attract one another, they will act less strongly toward the flow of momentum. Looking at the matter from another point of view, one might ask why the volume of a mole of gaseous water at an internal pressure of 1 atm is about 24 liters while the volume of a mole of liquid water in the same conditions is 0.018 liters: since $E(\text{kin})$ is the same for gas and liquid at the same temperature, an enormous internal pressure would build up in the liquid were it not for the virial, or the attractive forces between molecules.

9.2.4 NPT and NVT simulations

Consider now a computational box comprising a large number of molecules, each composed of a given number of atoms, for which all the starting dynamic variables have been properly defined and have been input to a computer. This "zero" frame can be prepared in many ways, by just a rough guess at the positional coordinates and velocities of each atom; for example, initial velocities can be assigned so that their maxwellian distribution is compatible with a selected temperature. In principle, the computational box may contain any number of molecules of any number of different chemical species; in practice, these numbers are limited by feasibility considerations, computing times, and disk space allotment. The computer will evaluate all potentials and forces, and start a trajectory calculation. During the simulation, kinetic energy and virial are easily computed, and they fluctuate along atomic trajectories.

The larger the number of molecules in the computational box, the more realistic the simulation of the properties of the system. Besides, when simulating the properties of a bulk material, surface effects must be minimized: some expedient must be adopted to reduce the difference between the potential field experienced by molecules at the boundary of the computational box and that experienced by molecules in its interior. For this purpose, periodic boundary conditions are adopted: the original computational box is surrounded by a number of identical replicas of the same, so that if in the course of the simulation one molecule is driven out of the original box in one direction, it re-enters the box from the opposite end.

Suppose that the zero box has been wrongly guessed so that its potential energy is very high, because some bonds are too stretched, or because some molecules are a bit too close and strong repulsive non-bonded interactions arise. As soon as the computer starts the dynamic simulation some atoms are pulled or pushed by strong forces and will rush to relieve the potential energy strain. Kinetic and potential energy are interchangeable, just recall the dynamics of the pendulum. In the strained system, potential energy is quickly poured into kinetic energy, velocities increase, and the computational temperature rises. But the computer can be instructed to multiply all atomic velocities in equation 9.6 by a common reducing factor, so as to rescale temperature to any specified input value, in what is the computational equivalent of the disposal of excess heat toward the surroundings. If on the other hand the molecular system is too "cold" and heat is required to reach the preset temperature, the computer will simply upgrade all atomic velocities as necessary. In both cases, after some equilibration time the system will reach a steady state at the preset temperature using the computer as the equivalent of a thermostat.

In the same conceptual manner, the computer can be instructed to keep the internal pressure at a given preset value: if for any reason the internal pressure becomes too high, all interatomic distances within the computational box, appearing in equation 9.9, are increased by the appropriate factor to readjust the virial. As a consequence, the computational box expands, in a computational equivalent of the disposal of mechanical energy towards the surroundings. If on the other hand the internal pressure becomes too low, the computer will "do work" on the system by shortening interatomic distances and squeezing molecules together a little bit, thus reducing the size of the computational box [3].

Simulations in which the total number of particles, N, the pressure P and the temperature T are kept constant are called NPT simulations. In practice, every so many simulation steps of duration Δt the temperature and pressure within the computational box are checked and readjusted to the preset values. In a similar manner, one can run NVT simulations, in which virial and pressure are not computed, and the volume of the computational box is kept constant. In an NPT simulation the density is a predicted property, while in an NVT calculation it is imposed and kept fixed.

9.2.5 *Performance and constraints in molecular dynamics simulations*

The duration of an MD run should be commensurate with the phenomenon of interest, so that a realistic picture is obtained, and the integration step should be shorter than the fastest phenomenon to be studied, so as to ensure a smooth integration path. The size of the computational box, the degree of complexity of the interaction force field and of its derivatives, and the integration time, concur in determining the amount of computing time required and hence the affordable duration of the simulation. A typical integration step in a molecular dynamics run is 0.002 to 0.005 ps, or 1 to 5 fs (femtoseconds, 10^{-15} s). This interval is a couple of orders of magnitude shorter than a typical molecular vibration time, and several orders of magnitude shorter than a typical molecular diffusion time in organic liquids. When one is interested in bulk properties of a condensed phase, molecular vibrations, especially those over bond stretching, are however scarcely interesting, so bond lengths are usually kept fixed in the simulation. This is accomplished by an algorithm called SHAKE, whose details can be found in MD package manuals.

Compared with real systems, molecular dynamics simulations use a very small number of atoms – 100,000 atoms is nowadays considered a very large number, and is still eighteen orders of magnitude short of Avogadro's number – and incredibly short times – a simulation one millisecond long is about the present limit. However, in spite of all its limitations, classical mechanics, ridiculously small samples, and short times, MD gives a detailed and reliable account of many thermodynamic and kinetic events and properties in molecular systems and during their evolution. This computational tool is almost the only window presently available on molecular-level facts, both at equilibrium and during irreversible evolution. Another appealing feature of MD is its flexibility, because the operator can specify at will the initial positions and velocities, so that some atoms can be constrained to stand still, and some others can be aimed in special interesting directions, biasing the further evolution of the system.

The typical output of an MD run consists of the following items:

1. Positional trajectories: every so many integration steps, the computer can be instructed to output the coordinates of all atoms (the so-called "frames"); the operator can thus follow the evolution in space-time of each and every atom in the computational box. This is the basic structural information, and can be used to monitor a great many different phenomena, from the occurrence of conformational changes, to the onset of molecular rotations, or of cluster formation or phase separation in liquid solutions.
2. Velocity trajectories: atomic velocity vectors are output together with atomic coordinates, and further information on correlated motion can be gained.
3. Energy trajectories: total and partitioned potential and kinetic energies can be recorded as a function of time, giving essential information on the energetics of the process under investigation. In the simulation of stationary states of condensed phases it is customary to allow an equilibration period, during which energies fluctuate significantly while the internal structure of the computational box advances to a state where a smooth and constant energy versus time plot is obtained: the latter phase of the simulation is usually called "production run", and the average properties over this period are taken as the predicted properties of the system. See some examples in Section 13.3.

9.3 The Monte Carlo (MC) method

In molecular dynamics, the computational "pull" that drives the system in its exploration of phase space is the passing of time consequent to the solution of Newton's equation of motion. This repetitive task is already very dull, but another computational technique for molecular simulation, the Monte Carlo method [4], rests upon an even duller procedure. MC postulates a force field, just as MD, but then the values of all positional variables are changed in random fashion: the ensuing change in total energy at the k-th move, ΔE_k, is the discriminating agent in a simple checking procedure (called the Metropolis algorithm):

$$\Delta E_k < 0 \quad \text{move accepted} \tag{9.10a}$$

$$\Delta E_k > 0 \quad \text{move accepted when} \exp(-\Delta E_k/k_B T) > R \tag{9.10b}$$

where R is a random number between 0 and 1.

When the move is accepted the new configuration is taken as the starting point for the next move, otherwise the old configuration is re-considered as the origin for another random move [5]. Condition 9.10b is in fact a Boltzmann probability (recall equations 7.1–7.3) so that if the random sampling is uniform and smooth, and the number of steps is sufficiently high, a Boltzmann ensemble under the assumed potential field will be generated. Time is not a variable, unless it is taken as being proportional to the number of moves, but in this case one must assume that all moves require the same amount of time.

In MC, fluctuations in box dimensions, and hence in density, are periodically attempted by scaling the coordinates of all the molecules in the box, and are judged by the usual acceptance or rejection criteria, so that the density can also be predicted. Temperature appears through equation 9.10b, as there are no velocities and hence no direct evaluation of kinetic energy. Although the final results of MC and MD are often the same, MC lacks the physical appeal of the MD representation of the true (at least, as true as the force field is) evolution of the system in time towards its equilibrium, with a view on the actual kinetics of the process. Otherwise, MC is better suited for the treatment of systems whose properties are dominated by weak intermolecular forces, because they are more uniformly distributed over the system and random sampling is more justified; very large and uneven energy changes occur when the random moves touch upon intramolecular bond stretching or bending forces. So, for example, protein folding is better studied by MD, while the equilibrium structure of liquids made of rigid molecules is better simulated by MC.

The nature and frequency of MC moves can be devised and regulated by the operator, by appropriate instructions to the computer. For an approximate order of magnitude, it may be stated that at the time of writing an upper borderline MC simulation runs on a computational box with 10^5 atoms for some 10^6 moves, but significant physicochemical information can also be obtained from much smaller systems and less moves.

9.4 Structural and dynamic descriptors for liquids

Evolutionary simulation, that is, molecular simulation that explores a large part of configurational space, like MC or MD, has an incredibly vast range of chemical applications: the optimization of conformation in single macromolecules such as proteins, or the description of their folding and unfolding mechanisms in aqueous solution; the description of the pathways and energies of molecular docking, for example in drug-receptor site recognition or in molecular complexation; the prediction of the equilibrium structure of liquids, with a description of the type and persistence of intermolecular bonding; the derivation of thermochemical quantities of condensed phases, either of neat liquids or of solutions; the behavior of complex systems such as micelles in binary phases; the study of nucleation and phase separation in liquid mixtures. Further applications to phase changes and crystal polymorphism will be described in Chapter 13. This extremely wide range of applicability stems from the simultaneous availability of positional trajectories and of intra- and intermolecular energies along the explored pathways. Were it not for the limitations in the physics of the force fields and in computing resources, molecular dynamics would be the ideal and definitive tool for the investigation of all non-reactive chemical phenomena [6].

The methods and procedures for the systematic analysis of the output of evolutionary simulations will now be reviewed. This output consists of extremely long lists of numbers, with a consequent occupation of computer disk space. Some standard procedures will be reviewed, but trajectory analysis is an enterprise that depends on the particular kind of chemical problem at hand, and specific functions for time-dependent analysis are often left to the inventive power of the operator.

9.4.1 Radial distribution functions

Structure descriptors for isolated molecules have been discussed in Chapter 1, and consist mostly of interatomic distances and bond or torsion angles. For crystals, these data must be supplemented by some information about cell dimensions and crystal symmetry operations. For liquids, the matter is more complicated because molecular motion is so widespread, and it would at first appear that there is no way of representing the features of molecular aggregation and molecular recognition in unequivocal numbers. In fact, some kind of time- or space-averaged statistical analysis must be employed. Most of the information on the structure of a liquid is conveyed by radial distribution functions (RDF). Consider first a liquid composed of n_A single atoms of atomic species A. The average atom density (sometimes called number density), assuming a random distribution of atoms within a given volume V, is given by

$$D = n_A/V \qquad (9.11)$$

Consider now an MD or MC output frame with a snapshot of the liquid structure, in which the instantaneous coordinates of all atoms are known. Consider a small spherical shell between a distance R_k and a distance $R_k + dR$ from any atom, and let the number of atom–atom distances that fall within this interval be N_k. The density of atoms within the shell is

$$D_k = N_k/(4/3\pi[(R_k + dR)^3 - R_k^3]) \qquad (9.12)$$

By construction, the RDF gives the probability that two atoms are at a given distance in the sample under study, relative to the same probability over a completely random distribution of atoms in space. In principle one could chose any atom at random within the computational box, and calculate all the distances to surrounding atoms. However, in order to get better statistics, in the actual calculation all atom–atom distances are considered for all the atoms present in the sample, so the total number of contacts must be divided by the total number of atoms. The RDF is usually constructed using a few hundred distance bins of width 0.1–0.2 Å, and the final form for distance bin k is:

$$g(R)_k^{AA} = (1/n_A)D_k/D \qquad (9.13)$$

The RDF starts at zero for $R = 0$ (no chance that two atoms may be at very short distance) and converges to 1 at infinite distance, as it should because in the limit of very large distances all intermolecular interactions vanish, any structuring disappears, and the distribution of contacts is identical to the random one. The RDF is a mirror of how intermolecular interaction within the liquid dictates cohesion: the weaker the interactions, the sooner the RDF converges to unity. Figure 9.1 gives a pictorial representation of these features of the RDF. Intramolecular distances also contribute to the RDF, but these are interesting only for non-rigid molecules, and are usually screened out when studying intermolecular cohesion.

In liquids of small and globular polyatomic molecules one could calculate by equation 9.13, for example, the RDF of the centers of mass. More interesting information is however obtained in atom–atom form. The form and the meaning of the RDF are the same if instead of atoms of the same species A, atoms of two different species A and B are considered, in a polyatomic molecule. A peak in the AB RDF means that the potential within the liquid is such that atoms of species A and B are called into close contact more frequently than they would be in a random distribution, and indicates therefore some kind of attraction between these atomic species. The RDF is thus a specific structure descriptor. All RDFs for a hypothetical perfect motionless crystalline sample would show only one infinitely sharp peak for each

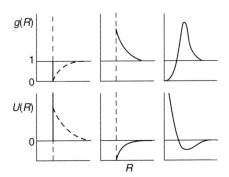

Fig. 9.1. A schematic representation of the radial distribution functions (above) and the respective potential functions (below): left, square well (full lines) and normal repulsive potential (dashed lines); center, attractive potential; right, a complete intermolecular potential function.

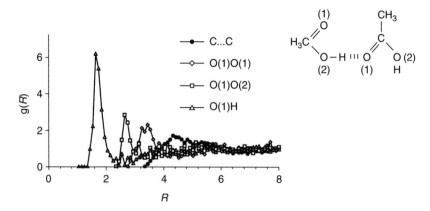

Fig. 9.2. The radial distribution functions in liquid acetic acid, from a molecular dynamics simulation.

set of interatomic distances present in the periodic crystal structure. The RDFs of liquids show broader peaks because of the structural variability induced by extensive molecular motion.

As an example, consider the intermolecular RDFs of liquid acetic acid, see Fig. 9.2. The most prominent peak pertains to the hydrogen bonding O···H distance of about 1.8 Å. This peak is accompanied by the peak for the corresponding O(carbonyl)···O(hydroxy) contacts at 2.8 Å, as expected from the previous result since the O—H distance is 1.0 Å and the O—H···O system is nearly linear. At a slightly longer distance, about 3.5 Å, there is a peak corresponding to the distance between two carbonyl oxygens across the hydrogen bond; note that this peak is broader, since the orientation of the two molecules related by the hydrogen bond may vary and there are many possible spatial arrangements for this contact, with slightly different distances. At much greater distance is a very broad peak corresponding to C···C distances. The intensity of this peak is much smaller, because the C···C interaction is not really attractive, so that the distribution of C···C distances is not much different from that for a random dispersion across the liquid medium. For the same reason, the RDFs for other atom pairs are not significantly different from unity. Integration of RDF peaks may give the total number of "bonds" over the AB interaction.

The radial distribution functions of liquids are experimentally accessible through neutron diffraction studies [7]; Fig. 9.3 shows the scattered intensity from liquid deuterated benzene. The information carried by such diffraction profiles is limited, but is sufficient for deriving some important structural conclusions, like the presence and nature of stable dimers in the liquid state. Some principles of this experiment can be gathered from a consideration of the examples in Section 5.8, but in practice, especially for complex molecules or liquid solutions, the details of the structure determination are demanding and require a complex deconvolution of the pair contributions

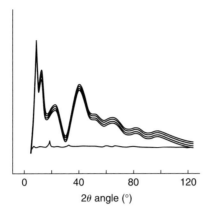

Fig. 9.3. Experimental neutron scattering intensities of liquid benzene; the different, closely spaced lines are at different temperatures. These diagrams are conceptually similar to those of Fig. 5.16. Reproduced by permission from ref. [7b].

out of the experimental diffraction profile. It is usually convenient to perform these experiments in conjunction with MD or MC simulations.

9.4.2 *Correlation functions*

In liquids, molecular arrangements change with time, and hence some functions can be devised to describe this dynamic behavior and to supplement the instantaneous or averaged information supplied by radial distribution functions. The essential idea in the construction of these correlation functions is to consider the evolving distribution of some vectorial property as a function of time, with respect to a reference state; the average value of the dot product between the actual distribution and the reference state must be close to unity for a perfect correlation, and decays to zero for complete loss of coherence. For example, consider an ensemble of N planar molecules and the unit vector, $\mathbf{u}_k$, perpendicular to the plane of the k-th molecule. The correlation function for molecular orientation is given by:

$$C(t) = \left[\sum_k \mathbf{u}_k(t) \mathbf{u}_k(0) \right] / N \qquad (9.14)$$

where t denotes current time, and state (0) is an arbitrarily chosen reference state. These functions are easily calculated from the positional trajectories output by an MD run. In a crystal or in a glass, this function should stay constant in time and nearly equal to 1 as all molecules preserve their reciprocal orientation, barring small thermal oscillations. In a proper liquid this function decays from one to zero in what is called the correlation time, the time it takes for the system to completely lose orientational "memory". This quantity depends on temperature and on the strength of cohesion forces. For example, for non-hydrogen bonded organic liquids at ordinary

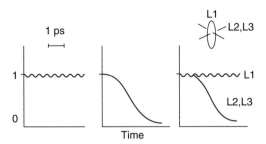

Fig. 9.4. Schematic representation of orientational correlation functions: left, a medium without rotational freedom, such as a crystal or a glass; centre, a system with full rotational freedom, a proper liquid; right, a liquid crystal in which correlation is preserved only along one molecular axis.

temperatures this time is of the order of a few ps [8]. In a liquid crystal, the correlation function for the elongation axis should stay close to unity, while those over the other two molecular axes should quickly decay to zero. These features are represented schematically in Fig. 9.4 (see more examples in Section 13.3).

More time-dependent structural correlation functions can be constructed, depending on the chemical nature of the system and of the phenomenon under investigation. For example, correlation functions over the positions and velocities of centers of mass yield information on translational diffusion. Molecular diffusion properties are often described by the self-diffusion coefficient, D, using a simple formula that involves the mean square displacement of the centers of mass [9,10]:

$$D = (1/6) < |\mathbf{r}(t + \Delta t) - \mathbf{r}(t)|^2 > /\Delta t \qquad (9.15)$$

where $\mathbf{r}(t)$ is the position of a center of mass at time t, and the average implied by the $<>$ notation is taken over all molecules and many possible choices of reference times. Figure 9.5 gives a sample plot of the mean square displacement: ideally, equation 9.15 is valid for very long time intervals, but in practice the slope of D versus Δt plots can be taken as equal to $6D$. The self-diffusion coefficient can be interpreted as the area swept by a molecule per second, and appropriate units are 10^{-8} m^2 s^{-1} = Å^2 ps^{-1}. This quantity obviously increases with temperature, and decreases with increasing molecular size and strength of the intermolecular potential. Fast-diffusing liquids like hydrocarbons have D coefficients of the order of 10^{-9} m^2 s^{-1}, while a heavy molecule like octanol has $D = 0.3 \times 10^{-10}$ m^2 s^{-1} [10]. These coefficients are measurable through NMR experiments [11]. A related quantity is the shear viscosity, which can be estimated via the calculation of fluctuations elements of the pressure tensor [12].

Simple time correlation functions can be designed to describe the lifetime of a particular bonding situation. For example, let $B_{ij}(t)$ be a function that takes the value of 1 if two molecules i and j are hydrogen-bonded, and the value zero if they are not. After defining a bond length threshold for hydrogen bonding, the set of B_{ij}s can be

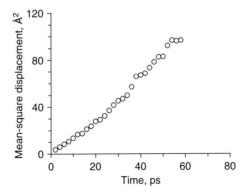

Fig. 9.5. A plot of the mean square displacement of the center of mass against time (equation 9.15) for the derivation of the diffusion coefficient. MD simulation on liquid succinic anhydride at 420 K.

calculated for a given positional trajectory and the time-autocorrelation function for the persistence of a hydrogen bond is given by [10]:

$$C(\text{HB}) = \sum B_{ij}(t)B_{ij}(t + \Delta t) / \sum B_{ij}(t) \qquad (9.16)$$

Such functions usually show an exponential decay in time. Their design is limited only by the ingenuity of the operator.

9.5 Physicochemical properties of liquids from MD or MC simulations

9.5.1 *Enthalpy, heat capacity and density*

The calculated density of the computational system is evaluated by simply dividing the mass within the computational box by its volume after equilibration at a certain temperature and pressure. The enthalpy of vaporization is a measure of the intermolecular cohesion in the liquid, and the largest contribution to this quantity naturally comes from the intermolecular potential energy. The complete expression is:

$$\Delta H(\text{vap}) = \Delta U(\text{vap}) + P\Delta V(\text{vap})$$
$$= \Delta U(\text{vap}) + PV(\text{g}) = \Delta E(\text{intra}) + \Delta E(\text{inter}) + RT \qquad (9.17)$$

where the volume of the liquid is negligible with respect to the volume of the gas, considered ideal, and the intra- and intermolecular internal energy variations are calculated from the potentials of the force field in equilibrated MD or MC simulations. Note that the difference in internal energy coincides with the difference in potential energy because the kinetic energy term is the same for the two phases at the same temperature. For rigid molecules, $\Delta E(\text{intra}) \approx 0$; the intermolecular potential in the gas is assumed to be equal to zero, and if the RT term is absorbed into the parameterization, one simply gets that $\Delta H(\text{vap})$ is equal to the negative of the intermolecular potential energy in the liquid.

The heat capacity can be estimated as [9,12]:

$$C_V = (\partial U/\partial T)_V \approx \Delta E/\Delta T + 3R + C_V(\text{vibr}) \qquad (9.18)$$

The variation in potential energy, ΔE, is estimated by the difference in average energies over NVT simulations at different temperatures. $3R$ accounts for the kinetic equipartition contribution (see Section 7.2), and $C_V(\text{vibr})$ corrects for the vibrational constraints imposed in the simulation (for example by the SHAKE algorithm), and can be estimated by the usual quantum statistical mechanics formulas (equation 7.30) if the vibrational frequencies are known.

The thermal expansion coefficient and the isothermal compressibility can be estimated as [9,12]:

$$\alpha = 1/V(\partial V/\partial T)_P = (\partial \ln V/\partial T)_P = -\ln(\rho_2/\rho_1)/\Delta T \qquad (9.19)$$

$$\kappa = -1/V(\partial V/\partial P)_T = (\partial \ln V/\partial P)_T = \ln(\rho_2/\rho_1)/\Delta P \qquad (9.20)$$

by performing NPT equilibrations at variable temperature and NVT equilibrations at variable pressure, respectively. The heat capacity at constant pressure is then:

$$C_P = C_V + Tv\alpha^2/\kappa \qquad (9.21)$$

where v is the molar volume and ρ is the density.

9.5.2 The Jorgensen school

W. L. Jorgensen and his school were the first to apply systematically the force field method in conjunction with Monte Carlo simulation for the calculation of the thermodynamic and thermophysical properties of organic liquids. In a series of papers published between 1985 and 1991, they used the united-atom approach, in which CH_3, CH_2, and CH groups were treated as a single atom to save computer time, which was still a somewhat limited resource in these days. They carried out extensive investigations of amides and peptides [13], alcohols [14], sulfur compounds [15], alkyl ethers [16], acids and esters [17]. These studies were ground-breaking in the sense that they gave the chemical community definite evidence that the properties of liquids could be reliably obtained by a simple model and by not too abstruse theoretical and computational procedures. Later on, as computers became faster and faster, the united atom approach was abandoned in favor of the all-atom approach, in which every single atom is an interaction site, in what eventually became known as the OPLS (optimized potentials for liquid simulations) all-atom force field [18]. Table 9.1 gives some typical results of these simulations, where computed data are compared with experimental measurements [19]. The accuracy of the calculations is astonishing, if one considers the small number of parameters and their transferability, and the results

PHYSICOCHEMICAL PROPERTIES OF LIQUIDS 245

Table 9.1 Thermodynamic properties of liquid organic compounds from early Monte Carlo simulations. Data at ambient conditions unless otherwise stated. Refs [13–18]

Compound	density g cm^{-3}		ΔH(vap) kJ mol^{-1}		Cp JK^{-1} mol^{-1}		Year[a]
	exp.	calc.	exp.	calc.	exp.	calc.	
N-methylacetamide	0.894	0.859	55.6	55.1	—	169	1985
dimethylformamide	0.873	0.879	43.5	43.6	162	140	1985
(CH$_3$)$_2$S	0.842	0.773	27.8	23.7	118	120	1986
ethyl alcohol	0.785	0.748	42.3	41.8	112	109	1986
2-methyl-2-propanol	0.781	0.772	46.6	46.4	220	192	1986
tetrahydrofuran	0.884	0.882	31.8	31.5	124	112	1990
tetrahydrofuran, 5000 atm	1.063	1.060	—	—	—	—	1990
acetic acid	1.044	1.041	52.3	53.5	—	—	1991
cyclohexane	0.774	0.755	32.9	32.7	156	183	1996, AA
butanone	0.800	0.805	34.5	35.8	159	156	1996, AA

[a] United-atom simulations except AA, all-atom simulations.

clearly show that the method can deal with chemical systems even in extreme temperature and pressure conditions. Some thermodynamic properties in these unusual regimes are sooner calculated than measured, because, contrary to experiment, the cost of calculation in extreme conditions is the same as that for ordinary conditions. This is one true success of theoretical chemistry in real prediction: a derivation of accurate chemical properties before or in absence of experimental measurement.

9.5.3 Crystal and liquid equations of state

The MC or MD treatment of an organic crystal proceeds along the same lines as for liquids. In fact, the computation is much less demanding, because the very limited configurational freedom of the crystalline state reduces the computing times needed for equilibration. Since energies and densities can easily be calculated as a function of temperature and pressure by MC or MD, these techniques can be applied to the calculation of complete equations of state for organic materials, including crystals and their melts.

The isobaric thermal expansion coefficients, $\beta = 1/V(dV/dT)_P$, for organic crystals are of the order of 2 K^{-1} (a more detailed discussion is in Section 11.4.1), of course much smaller than the corresponding values for liquids. Sample results of a detailed calculation of the state equation of an organic molecule by MD, carried out in conjunction with a temperature-dependent experimental determination of the crystal densities by X-ray diffraction [20], are shown in Fig. 9.6 for succinic anhydride.

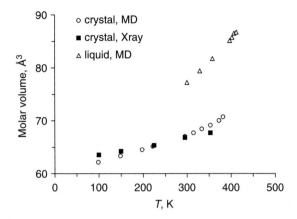

Fig. 9.6. The state equation of succinic anhydride [20]. Crystal: full squares, experimental from X-ray structure determinations; open circles, calculated by MD. Liquid: triangles, calculated by MD. The melting temperature is 393 K.

Notice that the MD simulation comfortably and reliably yields the density of the supercooled melt, a quantity not so easily accessible from the experimental side.

9.6 Polarizability and dielectric constants

Polarization (recall Section 4.4) is the action of one charge distribution on another charge distribution, resulting in a displacement of the polarized distribution with the formation of instantaneous dipoles. Typical atom–atom force fields used in MD or MC do not routinely include polarization terms. While this approximation may not be too harmful with regard to total energies, it is however a rather severe limitation when studying the fine detail of liquid structure and dynamics. The incessant molecular motion, with large reorientations, does in fact cause large time-dependent changes in the electric properties of molecules in the liquid state.

The simplest way of introducing polarizability in molecular dynamics simulations for organic liquids [21] is to use a previously developed force field in which coulombic energies had been calibrated to also include polarization; while these coulombic energies are toned down by a factor of about 0.9, a separate polarization contribution is then computed with the linear polarizability relationship (equation 4.20), using molecular or atomic polarizabilities. Even with this simplified approach, however, the computational load for molecular simulation increases significantly, because electric fields must be computed and energies are not pairwise-additive. This is a serious obstacle to the accurate treatment of polarization in liquids, and is also the reason why polarization contributions are neglected in many studies. They become crucial, for example, when studying the solvation shell around ions [22], and are indispensable for the estimation of dielectric properties [9], which depend on the correlation of instantaneous dipoles within the liquid. This correlation is estimated from the results of MD calculations by functions that are conceptually similar to the rotational correlation function, equation 9.14.

9.7 Free energy simulations

MD and MC provide a convincing description of internal energies and reliable ways for their quantitative estimation as averages over the part of phase space scanned by the simulation. It is fortunate that even a rather restricted sampling of phase space leads to sensible statistical averages for internal energy. On the other hand, there is no direct access to entropies from any of the results of evolutionary simulation, at least as described so far. Entropy cannot be obtained as a statistical average property over MC or MD runs, but rather depends on dispersion, or the amount of phase space accessible to the system, which cannot be directly established by a molecular simulation over a restricted portion of phase space.

Again, there is apparently no straightforward way of using an MD or MC simulation to mimic the integral required in equation 7.33, or to evaluate the statistical weights of configurations required in equation 7.36, so neither of the thermodynamic definitions of entropy may apply. However, consider again equations 7.20 and 7.43 together:

$$<E> = \sum P_i^* E_i = 1/Q \sum_i E_i \exp(-\beta E_i) \quad (7.20)$$

$$dU = T\,dS - P\,dV \quad (7.43)$$

When equation 7.20 is differentiated, the following result is obtained:

$$d<E> = \sum E_i dP_i^* + \sum P_i^* dE_i \quad (9.22)$$

On comparing these expressions an analogy appears: the first term in 9.22 is a change in population at constant level energy, and hence can be interpreted as an exchange of heat, as in the first term in 7.43; the second term in 9.22 is a change of quantum level energy at constant population, and hence can be interpreted as an exchange of work, as in the second term of equation 7.43. In principle, then, one could perform simulations in which temperature is raised in a stepwise fashion, drawing energy (heat) from the computational thermal bath implicit in the temperature control mechanism of the simulation, and actually carry out the Clausius integral $\Delta S < \int dq/T$ (equation 7.33, for an irreversible process), to obtain an upper bound for the entropy change over the process.

Recalling the form of the Helmholtz free energy, $A = U - TS$, the following derivations are easily carried out:

$$dA = (dU - T\,dS)_T = (dq + dw - T\,dS)_T; \quad (9.23)$$

$$(dq = T\,dS)_{\text{rev}}; (dA = dw)_{\text{rev}}; (dA < dw)_{\text{irrev}} \quad (9.24)$$

Therefore, simulations can be performed in which the force field is gradually "turned on" over the system, accumulating work from the computational work supplier, and use the fundamental relationship $w > \Delta A$ to obtain an upper boundary for the free energy variation. This way of proceeding [23] is obviously rather complicated and computationally demanding.

Alternative procedures must be found. Given the expression 7.41 for the statistical entropy, one obtains, after some algebra:

$$A - U° = -k_B T \ln Q \qquad (9.25)$$

Consider a state α described by a hamiltonian $H_\alpha = (E_{\text{pot}} + E_{\text{kin}})_\alpha$, evolving toward a second state β and a parameter λ such that for any value of this parameter the hamiltonian is given by

$$H(\lambda) = \lambda H_\beta + (1 - \lambda) H_\alpha \qquad (9.26)$$

The molar free energy difference between state α and β can be written as [24]:

$$G_\beta - G_\alpha = \sum_\lambda -RT \ln[<\exp(\Delta H'/RT)>] \qquad (9.27a)$$

$$\Delta H' = H(\lambda + d\lambda) - H(\lambda) \approx \Delta E_{\text{pot}} \qquad (9.27b)$$

where the last equation assumes that changes in the hamiltonian are changes in potential energy only. If $d\lambda$ is sufficiently small, the evaluation of the free energy difference is accurate. This is called the free energy perturbation approach, and of course ΔEs can be equally well obtained from MC or MD simulations. Alternatively, the so-called thermodynamic integration method requires an evaluation of the average ensemble $<(dH/d\lambda)>_\lambda$ at various values of λ over very small $d\lambda$ steps, and the final evaluation of the integral:

$$G_\beta - G_\alpha = \int <(dH/d\lambda)>_\lambda \, d\lambda \qquad (9.28)$$

Parameter λ can be chosen to be any pointer of variation within the system: for example, a torsional rotation to be followed in small angular steps; or, more radically, the change of a solute molecule into a solvent molecule, of a hydroxyl group into a methyl group, or the actual transmutation of a reactant into a product in a chemical reaction, provided a suitable hamiltonian is available. In principle one could chose the starting state as the ideal gas, use equation 9.25 to calculate the exact free energy by statistical mechanics, and then use 9.26 or 9.27 to turn on the intermolecular potential and obtain the value of the free energy of a condensed phase.

While in principle more revealing than enthalpy-based simulations, free energy simulations are plagued by a number of inconveniences. First, each simulation step has to be long enough to provide exhaustive equilibration over the slowest relaxation times, and since there must be many, many steps to provide acceptable accuracy, the calculations are computationally very demanding, so that very accurate calculations can be carried out only for simple systems. Secondly, the choice of the λ parameter and its sampling are critical, both to the meaningfulness and to the numerical stability of the calculation. Above all, the theoretical underpinnings of the free energy methods rely on subtle interpretations of statistical thermodynamics, so that great care and thorough understanding of the principles must be applied when planning a free energy calculation. While some well established procedures are available for the simulation of events in isolated or solvated single molecules, in applications to condensed states and phase equilibria one sees a potential danger of false steps due to some hidden incompatibility of variables, force fields, and states. The beauty of adhesion to simple physical principles and of immediate comprehension in energy-based MD and MC simulations is lost. Sometimes a sound energy-based calculation with solid interpretations and well tested potentials may be better than a tentative free energy calculation.

9.8 A theme with variations

As already mentioned, MC and MD are very versatile computational techniques. They can be elaborated upon in many ways. For example, MD simulations can be guided through conformational space by taking averages of forces over steps through fast and uninteresting molecular vibrations, to speed up the conformational search [25]. MC simulations can be carried out on dual boxes containing the gas phase and the liquid phase, with some MC steps consisting of jumps of molecules between the two phases, in the Gibbs ensemble Monte Carlo formulation [26]; the key point is the derivation of acceptance rules for these moves, which are only in part based on energy exchanges and hence cannot be evaluated by the standard Metropolis criterion. Such calculations give access to vapor pressures and phase coexistence profiles. Some of these points are taken up again in Sections 13.5 and 13.10

9.9 Water

Water is not just a chemical compound, it is in itself a chapter in natural history. Water is the universal solvent, from biological systems to industrial processes, due to its extraordinary properties and cheap availability [27]. Its molecular structure causes its condensed phases to be in fact a pure assembly of hydrogen bonds, with the unique consequences that the density of the liquid decreases with temperature in a range just above the melting point, and that the solid is less dense than the liquid; the phase diagram of water is extremely complicated, with a large number of different solid phases. The simulation of such a system is as difficult as can be, and it is no wonder that the model representation of the molecule itself and the simulation of liquid and solid water has been one of the main concerns of molecular modellers.

The literature on water modeling is accordingly too vast to suffer even a spot sampling. The simplest and most widely used models for the force field of water for

Table 9.2 Some simple force fields and calculated properties for water.[a] See Ref. [28]

	SPC	TIP3P	TIP4P	
r(OH)	1.0	0.9572	0.9572	
HOH angle	109.47	104.52	104.52	
A^b, kJ Å^{12} mol^{-1}	2633410	2435090	2510400	
C, kJ Å^6 mol^{-1}	2617	2489	2510	
q(O)	−0.82	−0.834	0	
q(H)	0.41	0.417	0.52	
q(M)	0	0	−1.04	
	calculated			*experimental*
D, g cm^{-3}	0.971	0.982	0.999	0.997
ΔH(vap), kJ mol^{-1}	45.1	43.7	44.6	44.0
C_p, JK^{-1} mol^{-1}	97.9	70.2	80.7	75.3
$10^5 \alpha$, deg^{-1}	58	41	94	25.7
$10^6 \kappa$, atm^{-1}	27	18	35	45.8

[a] Distances in Å, angles in degrees, charges in electrons. M is a supplementary pseudoatom site on the bisector of the HOH angle, 0.15 Å away from the oxygen atom.

[b] Parameters for the supplementary function $E(OO) = AR^{-12} - CR^{-6}$.

use in molecular simulations consist of a representation of the electrostatic potential by a number of point charges, plus an empirical oxygen–oxygen interaction; they are designated by the symbols SPC, TIP3P, or TIP4P [28]. Table 9.2 shows some sample results. In spite of their approximations, the performance-to-cost ratio of some of these simple models is so high that they are still on the modeling scene after so many years.

References and Notes to Chapter 9

[1] For more detail on principles and computational methods in molecular dynamics, see: van Gunsteren, W. F.; Berendsen, H. J. C. Computer simulation of molecular dynamics: methodology, applications, and perspectives in chemistry, *Angew. Chem. Int. Ed. Engl.*, 1990, **29**, 992–1023; Scott, W. R. P.; Hunenberger, P. H.; Torino, I. G.; Mark, A. E.; Billeter, S. R.; Fennen, J.; Torda, A. E.; Huber, T.; Kruger, P.; van Gunsteren, W. F. The GROMOS biomolecular simulation program package, *J. Phys. Chem.* 1999, **A103**, 3596–3607; Tuckerman, M. E.; Martyna, G. J. Understanding modern molecular dynamics: techniques and applications, *J. Phys. Chem.* 2000, **B104**, 159–178.

[2] The essential concept is easy to grasp. Consider just one atom at a distance R from the source of the potential acting on it. The numerical integration is over finite time steps Δt. Let $R°$ and $v°$ be the initial interatomic distance and velocity. The initial force is $F° = F(R°)$, and the initial acceleration is $a° = F°/m$. Then

$v' = v° + a°\Delta t$, and the new distance R' is found using an average velocity over the time interval, or $v = (v' + v°)/2 : R' = R° + v\,\Delta t$. R' and v' are then used as the starting point for the next step, and so on in a cyclic fashion: the ideal fodder for a stupid but tireless computing machine. There are of course a number of technical tricks to ensure that the integration is smooth, that energy is conserved, and so on.

[3] In practice, temperature and pressure control algorithms may be a little more complicated, to ensure a more appropriate or less obtrusive interference with the deterministic nature of the simulation. For much more detail on the practical aspects of these manipulations see the general introduction to MD given in Rapaport, D. C. *The Art of Molecular Dynamics Simulation*, 1995, Cambridge University Press, Cambridge.

[4] The name reportedly originates from a meeting of mathematicians held in Monte Carlo, in which it was realized that random sampling was an efficient method for measuring extensive properties.

[5] Statistical manipulations require here sequences of random numbers, so the quest for sources of true randomness has become an active field of mathematical research. For example, one possibility is to use the digits in the infinite sequence of decimals of π. Calculating several million decimals for π is an almost esoteric task: Carl Sagan, in his book *Contact*, tells of a scientists who finds in that sequence the proof of the existence of God. It would have been hard to imagine that the decimal digits of π would be important for the determination of the internal energy of liquid alcohols. One never knows when or where non-applied science will become applied, and forcing scientists to be applied scientists is as crazy as many politicians are.

[6] Evolutionary simulation is indeed the ideal link between molecular physics and a molecular level understanding of the thermodynamics of condensed phases, and as such is the ideal medium for didactic purposes in chemical thermodynamics. Unfortunately, this approach has not yet penetrated into even the most successful physical chemistry textbooks (e.g. Atkins, P. W., *The Elements of Physical Chemistry*, 2001, 3rd edn,, Oxford) that insist on the "ideal gas" approach to chemical thermodynamics and hardly, if at all, mention molecular simulation.

[7] For some examples see (a) Bako, I.; Radnai, T.; Belisant Funel, M. C. Investigation of structure of liquid 2,2,2-trifluoroethanol: neutron diffraction, molecular dynamics, and ab initio quantum chemical study, *J. Chem. Phys.* 2004, **121**, 12472–12480; (b) Cabaco, I.; Danten, Y.; Besnard, M.; Guissani, Y.; Guillot, B. Neutron diffraction and molecular dynamics study of liquid benzene and its fluorinated derivatives as a function of temperature, *J. Phys. Chem.* 1997, **B101**, 6977–6987. This last paper has a clear description of the derivation of pair correlation functions from experimental scattered intensities from liquids.

[8] See, for example, the discussion section in: Geerlings, J. D.; Varma, C. A. G. O.; van Hemert, M. C. Molecular dynamics studies of a dipole in liquid dioxanes, *J. Phys. Chem.* 2000, **B104**, 56–64.

[9] Walser, R.; Mark, A. E.; van Gunsteren, W. F.; Lauterbach, M.; Wipff, G. The effect of force-field parameters on the properties of liquids: parametrization of a simple three-site model for methanol, *J. Chem. Phys.* 2000, **112**, 10450–10459.

[10] DeBolt, S. E.; Kollmann, P. E. Investigation of structure, dynamics, and solvation of 1-octanol and its water-saturated solution: molecular dynamics and free-energy perturbation studies, *J. Am. Chem. Soc.* 1995, **117**, 5316–5340. This is an exemplary paper on the use of MD simulations for liquids.

[11] See for example Harris, K. R.; Newitt, P. J. Diffusion and structure in water–alcohol mixtures: water+tert-butyl alcohol, *J. Phys. Chem.* 1999, **103**, 6508–6513.

[12] Tironi, I. G.; van Gunsteren, W. F. A molecular dynamics simulation of chloroform, *Mol. Phys.* 1994, **83**, 381–403. This paper has a careful description of methods for extracting radial functions, thermodynamic properties, rotational and diffusion correlation functions, and dielectric properties from MD simulations.

[13] Jorgensen, W. L.; Swenson, C. J. Optimized intermolecular potential functions for amides and peptides. Structure and properties of liquid amides, *J. Am. Chem. Soc.* 1985, **107**, 569–578.

[14] Jorgensen, W. L. Optimized intermolecular potential functions for liquid alcohols, *J. Phys. Chem.* 1986, **90**, 1276–1284.

[15] Jorgensen, W. L. Intermolecular potential functions and Monte Carlo simulations for liquid sulfur compounds, *J. Phys. Chem.* 1986, **90**, 6379–6388.

[16] Briggs, J. M.; Matsui, T.; Jorgensen, W. L. Monte Carlo simulations of liquid alkyl ethers with the OPLS potential functions, *J. Comp. Chem.* 1990, **11**, 958–971.

[17] Briggs, J. M.; Nguyen, T. B.; Jorgensen, W. L. Monte Carlo simulations of liquid acetic acid and methyl acetate with the OPLS potential functions, *J. Phys. Chem.* 1991, **95**, 3315–3322.

[18] Kaminski, G.; Duffy, E. M.; Matsui, T.; Jorgensen, W. L. Free energies of hydration and pure liquid properties of hydrocarbons from the OPLS all-atom model, *J. Phys. Chem.* 1994, **98**, 13077–13082; Jorgensen, W. L.; Maxwell, D. S.; Tirado-Rives, J. Development and testing of the OPLS all-atom force field on conformational energetics and properties of organic liquids, *J. Am. Chem. Soc.* 1996, **118**, 11225–11236.

[19] Repertoires of heats of vaporization and methods for their estimation by group additivity have been collected: Chickos, J. S.; Hesse, D. G.; Liebman, J. F. Estimating vaporization enthalpies of organic compounds with single and multiple substitution, *J. Org. Chem.* 1989, **54**, 5250–5256. Corrections for temperature variation can be carried out by the approximate relationship $-\Delta C_p(\text{vap–liq}) = 10.58 + 0.26\, C_p(\text{liq},298\text{K})$ (ref. [7], Chapter 7).

[20] Ferretti, V.; Gilli, P.; Gavezzotti, A. X-ray diffraction and molecular simulation study of the crystalline and liquid state of succinic anhydride, *Chem. Eur. J.* 2002, **8**, 1710–1718.

[21] Caldwell, J. W.; Kollman, P. A. Structure and properties of neat liquids using nonadditive molecular dynamics: water, methanol and N-methylacetamide, *J. Phys. Chem.* 1995, **99**, 6208–6219. For the expression of polarization energies, see Dykstra, C. E. Electrostatic interaction potentials in molecular force fields, *Chem. Revs.* 1993, **93**, 2339–2353.

[22] Carignano, M. A.; Karlstrom, G.; Linse, P. Polarizable ions in polarizable water: a molecular dynamics study, *J. Phys. Chem.* 1997, **B101**, 1142–1147.

[23] Reinhardt, W. P.; Miller, M. A.; Amon, L. M. Why is it so difficult to simulate entropies, free energies, and their differences? *Acc. Chem. Res.* 2001, **34**, 607–614.

[24] See for example: Kollmann, P. Free energy calculations: applications to chemical and biochemical phenomena, *Chem. Revs.* 1993, **93**, 2395–2417; Mordasini Denti, T. Z.; Beutler, T. C.; van Gunsteren, W. F.; Diederich, F. Computation of Gibbs free energies of hydration of simple aromatic molecules: a comparative study using Monte Carlo and Molecular Dynamics computer simulation techniques, *J. Phys. Chem.* 1996, **100**, 4256–4260.

[25] Wu, X.; Wang, S. Self-guided molecular dynamics simulation for efficient conformational search, *J. Phys. Chem.* 1998, **B102**, 7238–7250.

[26] Panagiotopoulos, A. Z. Direct determination of phase coexistence properties of fluids by Monte Carlo simulation in a new ensemble, *Mol. Phys.* 1987, **61**, 813–826.

[27] At the time of writing, water resources are running out fast. The times of abundance and mindless waste of this key commodity may well be over, but politicians hardly seem to be interested – they care much more about cultivating their personality by waging nonsensical wars against purported foes, for oil, a much less indispensable resource. Without oil, the production of many of the gadgets of our opulent civilization may slow down. Without water, people die by the millions in a few days. Water shortage can easily wipe out *all* civilization on Earth.

[28] Jorgensen, W. L.; Chandrasekhar, J.; Madura, D.; Impey, R. W.; Klein, M. L. Comparison of simple potential functions for simulating liquid water, *J. Chem. Phys.* 1983, **79**, 926–935.

10

Computers

Computers ... carry out wrong and useless calculations at the same speed and with the same wholeheartedness as correct and meaningful ones.

(See Section 10.1.)

10.1 Bits and pieces

Chemical applications of force field simulations (Chapter 2), of quantum mechanics (Chapter 3), as well as X-ray data processing (Chapter 5), lattice dynamics (Chapter 6), and molecular dynamics simulations (Chapter 9) are all made possible by fast and reliable numerical computation. Therefore, electronic computers are a theoretical chemist's vital tool and very few, if any, quantitative results can be obtained without them. Computers however, fortunately have no ideas and no personal conceptions, so they have no part in the creative process that is the essential part of the development of science, and that is entirely the product of the human mind. Also, having no preferences and no critical power, they carry out wrong and useless calculations at the same speed and with the same wholeheartedness as correct and meaningful ones.

Computers handle a very large and very diversified range of tasks on a surprisingly small fundamental basis: the electric representation of only two numbers: zero and one, called binary digits (bits, in short). A zero is no current in a tiny conductor, or an open switch, a one is current through, or a closed switch. How this representation is actuated in practice is the concern of engineers, but Fig. 10.1 shows the ingenious way in which this problem was first solved, using one magnetizable ferrite ring the size of a few millimeters for each bit. Miniaturization is the key word in computer development; a modern bit is but a small bunch of electrons and a modern laptop holds billions of such bits in a few square centimeters, while the same number of ferrite rings would have taken up a surface area about the size of St Peter's square in Rome.

Computers use bits to represent numbers in binary notation: in fact, a group of n bits can represent a number up to $2^n - 1$. Consider the following conversion from the

Fig. 10.1. The IBM 1620, see the historical portrait in Chapter 5, and its ferrite ring memory.

usual base 10 notation to base 2 notation:

$$(123)_{10} = 1 \times 10^2 + 2 \times 10^1 + 3 \times 10^0$$
$$= (1111011)_2$$
$$= 1 \times 2^6 + 1 \times 2^5 + 1 \times 2^4 + 1 \times 2^3 + 0 \times 2^2 + 1 \times 2^1 + 1 \times 2^0$$

The decimal representation is a good compromise between economy of symbols for digits (10 are needed in decimal, only 2 in binary) and economy in the amount of digits needed to represent a number (in the above example, 3 in decimal, 7 in binary). Computers favor the former kind of economy and therefore adopt the binary representation. Also, it would be difficult to find simple ways of creating 10 different electrical states in any piece of matter.

A bit is also a signal, and by showing a sign with a zero or a one you could signify anything from whether or not you would like another cup of tea to whether or not a million men and ten thousand ships should proceed to invade Normandy. Groups of n bits can take 2^n different forms and hence can represent as many different symbols: a universally accepted convention uses a combination of 8 bits, called a byte, to represent a character or symbol, while a combination of 4 bytes composes a word. Thus, the usual 32-bit computer word can represent a very large number, $2^{32} - 1$ [1], or four characters. Viewed in this way, the first essential part of an electronic computer is just a container that stores an incredibly large number of bits in a core memory; this can best be mentally visualized as a string of consecutive words. Other bit sequences can be stored in memory devices such as internal hard disks or external floppy disks (by now almost forgotten), stick memories, and the like [2].

The second essential part of an electronic computer is a central processing unit (CPU), which houses the computer's electronic ego, usually called the "control" (the *geist* of a computing machine). Bit sequences are of two different kinds: program instructions and data on which programs act. Programs are sequences of commands, each of which is a coded sequence of bits stored in core memory; control scans, decodes, and executes these commands. While programs for execution must be stored in core memory, numerical data can be apportioned at will between core memory storage and hard disks or external devices. Access by a program to the data in core memory is immediate, while access to data in storage devices requires some input–output operation and is consequently slower. Commands are of many different kinds: read and write commands, providing communication with the external world; arithmetic commands, essentially addition, out of which all numerical routines are built by appropriate series; and logical commands, which cause control to execute some or other command depending on the results of previous operations. All commands can be interpreted as numbers, but not all numbers can be interpreted as commands: whenever control tries to execute a bit sequence that does not correspond to a recognized instruction, the result is an error condition. In favorable cases, a warning is issued, in less favorable cases, the result is a crash, or that sudden, total, disappointing lack of response that usually occurs when you are in the most compelling hurry to obtain some information from the machine.

In a very successful metaphoric style, all items of the computer world that have to do with programs are called software, while all the rest (electronic parts, wires, input–output devices) are called hardware. Partly countering the intrinsic semantics of these two words, in the last years software has become much more expensive than hardware. Computer elaboration is a matter of switching between electronic zero and electronic one, and therefore on the hardware side computer speed depends on the speed at which the electronic status of a solid device can be modified. This response of electrons to an electric stimulus in turn is a matter of the electronic structure of the solid (see Section 6.2). This is a technical matter. On the software side, higher speed depends on a better organization of the code. Computer speed is anyway an essential variable in the development of computational theoretical chemistry [3].

10.2 Operating systems

Just out of the factory, a computer memory is almost completely blank, except for very restricted sequences of commands that may control elementary procedures like, for example, the appearance of a logo on the screen or the subsequent loading of more complex programs (these self-loading procedures were once called "bootstraps"). Core memory is a real *tabula rasa,* like the brain of a newborn child. In the past, programs were loaded mechanically into core memory by pushing a button to activate the reading of punched cards or punched tapes, and then pushing another button that called for control to execute the program from the first instruction. Nowadays even a modest computer does many different things, and keeps track at the same time of many tasks: for example, on an ordinary working day in a chemistry lab, a desktop computer might be running a molecular orbital computation in the background, while checking periodically for the arrival of new mail, or executing a graphic task to represent a crystal packing, or checking for the arrival of incoming data from a spectrometer, all this while your colleague next door is using it to play chess. All these tasks are regulated by an operating system, which must be loaded into memory as soon as the computer is first turned on. An operating system is a very complex program that runs all the time and directs control to execute one of the many different tasks (programs) that happen to be pending. Even when the computer seems to be doing absolutely nothing, the operating system is running in a cycle and waiting for an interrupt (the click of a mouse, the hitting of a key, a signal from the digital clock built into all computers) that signifies that the user wants to run a particular program or section of a program, or that some periodic task must be performed, say refreshing the screen. The operating system then transfers control to the appropriate program, which may already be in core memory or may be waiting on some external device. Loading and unloading entire programs into and from core memory is a matter that takes insignificant amounts of time, so this can be done continuously with only minor prejudice to the running times of the background programs.

It is nowadays impossible to buy a computer that runs without an operating system [4]. The system is a program written by a computer programmer, who must foresee a very wide range of possible uses for the machine, while each user has a preferred subset of tasks. For sure, a large percent of the system activities and capabilities are never exploited by any given user, and the 99.9% of computer users who are not system analysts can never know exactly what the system is actually doing. Since commercial systems are designed by money-making companies, the suspicion that from time to time the system might be doing something for the benefit of the company, rather than of the user, is extremely well founded, especially if the computer is regularly kept online and hence open to communication with the external world. Understandably, when too many different tasks are piled one upon one another, or when a peculiar sequence of interrupts builds up, conflicts may arise and the system may break down because control is transferred to some wrong memory location, causing the computer to halt and requiring some restart procedure.

If a computer is to be used for scientific calculations, the operating system should be as simple, as reliable, and as transparent as possible, because the tasks to be carried out are relatively few and always the same: loading a program into core memory, reading the data, and then letting the program run undisturbed. Ideally, a computer devoted to scientific computation should have as little of a system as possible, in order to waste as little time as possible in useless tasks built in by the system producers; and it should be left alone while doing a calculation, without bothering it with many accessory tasks that might generate conflicts.

Viruses are short programs that sneak into core memory through external connections, and purposefully direct program control towards faulty memory locations or to the execution of detrimental commands. This is a very easy thing to do, if one considers that changing one bit out of a million may change one instruction from "add" to "subtract" and hence can completely dismember a given task. Anti-viruses are programs that try to detect viral program sequences in core memory or in other storage devices. This is much less easy. For example, a typical elusive strategy adopted by virus writers is to insert instructions that continuously move the virus itself to different memory locations and then direct control to that memory location.

For all the above reasons, a highly recommended attitude is then to use separate machines for scientific computing and for carrying out the networking, mail, and office tasks [5].

10.3 Computer programming

When a new software tool is acquired, usually in a smart and colorful box that contains a compact disk, or in a file downloaded from the Internet, or when a colleague hands over some software for doing a theoretical chemistry calculation or a molecular dynamics simulations, one actually acquires a sequence of bits that, once loaded into core memory, will perform a given task. These sequences of bits have been somehow put together by a human programmer. Of course no sane person would be prepared to write down actual strings of zeroes and ones, which for complex programs can easily run into millions, using what is traditionally called machine or base [6] language. Recourse is made to symbolic programming: programs are written in a symbolic, coded language (called the source code), more akin to the processing abilities of the human mind and sight, and then a machine language program is invoked to translate these symbolic instructions into machine language, creating what is called an executable module [7]. These translators, or meta-languages, are called compilers; they are extremely complex and very difficult to write, but once they are operational, they solve the problem once and forever for millions of users everywhere. Besides, the use of symbolic programming and compilers helps to improve the transferability of programs between different kinds and brands of machines. If a program is going to be used by people other than the program author, it makes a big difference whether the source code or the executable module is distributed: the former can be analyzed and possibly changed by an average user, while the latter is precluded from analysis, except by highly trained specialists, and is to be used as a black box.

An essential request for any scientific product is reproducibility. In this perspective, access to the source code should be as wide as possible, and its analysis, evaluation, and modification should be as easy as possible. The most widely used symbolic language in the scientific milieu is called Fortran, short for Formula Translator. Its syntax and the corresponding compiler were originally developed in the 1950s by a team of programmers at IBM [8], and, despite sometimes harsh criticism, it is still the language of choice for dealing with mathematical expressions. In Fortran, the instruction for an addition of two numbers reads $a = b+c$, algebraic subroutines are invoked by writing "exp", "sin", and the like, input–output grammar is in terms of "read" and "write" expressions, and a conditional branching reads "if ... go to". In addition, Fortran allows an easy manipulation of large indexed arrays of numbers through its do-loops, the ideal tool for matrix algebra (think of the matrices of double- and quadruple-index integrals in Box 3.1). There have been many attempts at standardizing Fortran syntax and grammar; in spite of only partial success, the use of a Fortran instruction subset, not including many of the most sophisticated options that spoil transferability and are in fact seldom indispensable, does allow cheap and efficient scientific computer programming along with reasonable portability [9].

Should all scientists, at least all theoretical chemists, be good computer programmers? True progress in theoretical chemistry comes from new ideas, not from better computers, but it is also absolutely true that ideas in theoretical chemistry travel and expand in the best way when they are given the form of computer programs, because in most cases the success or failure of new theories and procedures can be judged only if and when actual numerical calculations are carried out. A chemical theory whose principles require a thousand years for the fastest computer to carry out the necessary calculations would be in a sort of scientific limbo, until numbers could be produced to validate or disprove it. The use of standard or commercial softwares is sometimes useful, but in such cases one is restricted to always using somebody else's ideas. It can be safely stated that there can be no top theoretical chemistry or molecular simulation group without one or more good computer programmers who can modify existing softwares or develop entirely new ones. For example, a computer program whose core is the calculation of molecular orbitals develops in time with the addition of routines to calculate the electric potential, molecular multipoles, vibration frequencies, energy derivatives for optimization of molecular conformation; or by the introduction of new functional forms in DFT or of new configuration interaction approaches; and so on.

Faster computing is the result, to a large extent, of writing better programs; a good program writer is also a person who can optimize the performance of a computer by cutting down computing times and disk or memory space requirements. For the above mentioned reasons, such a person can substantially help the development of new theories. Better programs in this sense result from careful ordering and organization of the code, as well as from optimization of input–output operations, usually very much slower than central number processing. Nearly all Fortran compilers are also optimizers, that is, they analyze the sequential logic of the program and automatically arrange the instructions to the sequence for best performance [10].

On the hardware side, other speedup strategies are essentially based on the simple idea that since hardware is becoming less and less expensive, one may use many processing units instead of only one and split up the job. Parallelization and clustering strategies belong to this family: either the central processors carry out many calculations at the same time, as in parallel machines, or physically separate machines actually load different parts of the total job, as happens in clusters, which may nowadays contain up to 50–100 computers. In both cases the big problem is the synchronization of the tasks. This is in part a hardware problem, and that comes at a low monetary cost, but at a sometimes high cost in performance, because the system may break down more often. It is also in part a software problem, because it may require some substantial extra effort on the part of the programmer. In every case, a careful balance has to be struck between advantages and disadvantages.

The absolutely safest strategy is to load the same software on to separate machines and to split manually the job from the very beginning. This can be done in discontinuous jobs, such as when one has to calculate the lattice energy of 100 crystals and, with two machines, put 50 calculations on each. It cannot work in molecular dynamics simulations, where the trajectory must be continuous.

10.4 Bugs and program checking and validation

A well documented law of computer programming states that no computer program is free from errors, which in programmers' lingo are called "bugs". Program bugs are of many different types, ranging from faulty logic, to wrong instruction ordering, to just plain typos made while keying in the program source code. The most easily spotted bugs are those that produce a complete stop of program execution, or impossible operations, such as square roots of negative numbers, or patently absurd numbers such as negative temperatures, etc. For obvious reasons, much more subtle and hence more dangerous are those bugs that produce wrong results with an acceptable appearance.

Program checking and validation is an essential step in computer programming. The task is less easy than it might seem. Slow perusal of the instruction sequence by a human being is very time-consuming, and sometimes ineffective (a zero digit typed in place of an "O" character can escape human control almost forever). Ideally, a program should be checked by doing the same operations by other means, anything between running another code for the same purpose (when available) and maybe even carrying out a hand computation on some particularly simple case. Large programs usually contain a proportionally large number of different options, and it becomes proportionally more tedious or difficult to check them all. A program that has been working perfectly for years for molecules of up to 100 atoms may run amok when a molecule with 101 goes through, because of memory allocation instructions or wrong do-loop limits. Inventiveness and determination in devising program checks is as important as inventiveness and determination in code writing.

Providers of commercial or academic software may or may not be willing to analyze potential bugs brought to their attention by other users; in any case, time is a good and impartial judge, and mature software that has been in use for years with plenty of time for bugs to crop up and be corrected is certainly more reliable than newly written code. In many cases, for many standard tasks in theoretical chemistry, such as integral evaluation over gaussian orbitals, or self-consistent solutions of the Fock equations, or integration of time trajectories in molecular dynamics, benchmark calculations have been made available against which prospective programmers can test their new codes, both for correctness and for running times.

Finding and fixing program bugs may become a real nightmare, and the worst possible scenario is one in which a bug is spotted only after a consistent amount of results has been obtained from the wrong code, and worst of all, published. Cases have been reported of computer software authors who gave up publishing when they realized that they could never be 100% certain of having "shot" all the "bugs" in their programs. Is this a sensible attitude? Of course not, and the reasons are given in the next sections.

10.5 Reproducibility

Akin to bugs but much less detrimental are instabilities due to machine-dependent code. A classical case is the recursive integration of molecular dynamics equations, in which infinitesimal differences in the accuracy of the results may accumulate to the point that the simulation path actually depends on details of machine architecture like the number of bits in a word. This causes a loss of reproducibility because a calculation with the same input to the same code may produce different results. No harm is done in this particular case – a different zone of phase space is explored, that's all, but the matter is nevertheless disturbing.

This is one of the aspects of the general problem of reproducibility in computational chemistry. Once a program has been developed and tested, perhaps having spent in the process an amount of time that can be equivalent to the long hours spent by synthetic chemists over solvent distillation, recrystallization, and similar exasperating chores, a computer program can be published like any other piece of chemical information [11]. When publishing a program, or when publishing the results obtained by computer programs developed in the authors' laboratories, publication should be accompanied by as much information and supplementary material as possible, in order to facilitate the use of the code even by people with no experience. In the interest of the correct propagation of science, whose cornerstones are dissemination of methods and reproducibility of results, the following rules should be followed [12].

1. The code should be clear and linear, and comment lines should be introduced as frequently as possible to identify the particular task being carried out in each section of the code.
2. Heavily machine-, system- or compiler-dependent instructions should be avoided.
3. Variables should be given names that recall the physical quantity they represent: classical cases of traditional names are, for example, FOBS and FCALC for observed and calculated structure factors; even better, a list of variables with their meaning in the program should be given.
4. The source code should be deposited or anyway made available, possibly along with executable modules for the most common machines and operating systems.
5. A detailed description of the program structure, of its input and output, should be given in a separate document.

6. To ensure the reproducibility of the calculations, as much supplementary material as reasonably possible should be deposited: for example, three or four key sets of input data with their respective outputs, covering as many as possible of the different options available to users.
7. In particular, care should be taken in specifying all the parameters of a force field in molecular simulations.
8. Authors should be available for a reasonable amount of time to answer questions, to give advice on application, to suggest alternative uses, and to take care of newly found bugs.
9. Clear and unequivocal credits for program development should be given.
10. When using somebody else's non-commercial computer codes, the publication should be pointed out to the authors of the program, along with any unusual situations or malfunctions encountered while using the program.

Some of these prescriptions may not be easy to follow when dealing with commercial software, which often contains copyrighted parts and is always protected against unauthorized duplication. Authors of papers in which such "black box" software has played a crucial part should carefully specify what options have been taken, what points have not been checked, what defaults have been adopted, especially in such crucial issues as the formulation of force fields and their parameters. A warning should be issued against the proliferation of scarcely documented computational results, a lurking danger with today's journal proliferation and plethoric publishing policies.

10.6 "Because it's there"?

The widespread availability and common use of computers has revolutionized chemistry, as it has almost anything else in human activity. In the 1970s a new generation of chemical operators was born: computational chemists, whose tools, rather than test tubes, are electronic computers. The following applications of computers in chemistry are relevant to the subject of this book:

- the calculation of the electronic energy and of the electronic structure of organic molecules by molecular orbital or density functional theory;
- conformational analysis, or the search of the energy-minimizing structure of organic molecules either by empirical force fields or by electronic structure calculations;
- the representation and analysis of crystal structures, and the search for the best crystal structure for a given organic compound;
- the simulation of the dynamics of fluid phases and the calculation of their thermodynamic and thermophysical properties.

Computers and computing times are becoming cheaper and cheaper. Rather than a dearth of machine time, the main problem of computational chemistry is more and more that of finding suitably demanding problems. The story goes that a famous mountaineer, Sir Edmund Hillary, when asked why he wanted to climb Mount Everest, answered: "because it's there". This is an appropriate answer for a sportsman, but a computational chemist, when asked why he or she wants to do a certain calculation, cannot simply answer "because the computer is there". And yet computers are so widespread, programs are so readily available, screen colors are so bright, and the frustration of seeing an idle computer is so great, that many calculations are done for no better reason than to keep the machines busy. Add to that the fact that chemistry journals are so numerous that with a little cosmetic treatment nearly anything can nowadays get published, and one is tempted to agree with the rather cynical statement that "5% of the published papers in computational chemistry are relevant, 10% are wrong, and 85% are not even wrong" [13].

Natural sciences proceed by experiment. A good experiment can be thought of as a "question" posed to the observable world, which, if the experiment is well designed and properly carried out, answers by a nod – a yes or no, a number, a current flowing through the circuitry of the measuring machine. Theoretical chemistry has evolved into computational chemistry, which has an obligation to proceed by the same conceptual protocols as experimental chemistry. Ideally, at the origin of each computational chemistry experiment there should be a well defined question to which the operator seeks a well defined answer.

True enough, however, science is often similar to a game and throughout the history of science there are numerous examples of discoveries that have been made while toying with ideas and playing around with instruments – serendipity. It is also true, that any instrument must be thoroughly and continuously used to better understand its capabilities and its responses, and that even a publication reporting some routine calculation made on a very common machine with a standard set of programs may contain some grain of novel experience that can be useful to somebody else; and this is the strength of the modern publishing policy. The less palatable side of this policy is that sometimes it looks as if science is now proceeding more by the thrust of sheer mass than by the spark of clever ideas.

The balance, as usual, rests upon critical judgment. At one extreme, one should not boast the results of a mediocre calculation as if they were a new messianic announcement; at the other extreme, care should be applied in designing computer experiments and in sorting out the real importance of the results. This brings us back to the issue of undetected bugs: how should one deal with the fact that all computer programs are potentially flawed and hence the results are potentially wrong? As in all experiments, a computer experiment includes a technical part (the executable code, with its potential bugs) and a significance part. (the elements of new chemical knowledge that the experiment brings). A badly flawed computer program can never produce significant chemical results, but a marginally flawed program can, just as a spectrometer with a slight malfunction in its photomultipliers can still give an acceptable spectrum [14]. Conversely, a perfect program may run forever without producing any new chemical knowledge or new experience. The final success of the computational experiment should be judged on the amount of chemical information acquired, not on the technical perfection of the tool [15].

References and Notes to Chapter 10

[1] In practice, in the usual representation of decimal numbers (floating point numbers] some bits are used for the mantissa and some are used for the exponent. One bit may be used for the sign.
[2] Transfers of bits occur with astonishing reliability because of complex self-checking algorithms, also carried along in special bit sequences (parity checks). These algorithms are one of the hidden keys to the widespread success of digital information. Without that reliability, mistakes in an infinitesimal percent of the transferred bits would make the whole information completely useless.
[3] The term "electric" is reserved for macroscopic currents, e.g. a washing machine is an electrical appliance. The term "electronic" is reserved for infinitesimal amounts of charge and current. Electronic currents can be amplified to drive electrical appliances, such as when a computer signal drives the movements of an inkjet printer. Intrinsic (i.e., hardware-controlled) computer speed has been steadily doubling every year and a half, or increased by one order of magnitude every five years. Nobody knows if there is a definite physical limit to this increase.
[4] No wonder, the man who sells the most widely used computer operating system is the richest man in the world and has several times come to trial facing charges of breaking the laws of free competition.
[5] A recommendation after many years of experience: do not hesitate to unplug the net connection wire. Files can be transfered using stick or Flash memory devices.
[6] Do not confuse base language with BASIC, which is an old-fashioned symbolic language.
[7] There is in fact an intermediate stage: the compiler generates a machine language object module, while a second step, called the linker, builds the final executable module by linking together many object modules, which can also come from different compilers and different languages.
[8] The project was the first attempt at producing a transferable computer language, and took much more time than originally planned, even though the developers were among the best available programmers (David Sayre, see Section 5.9, was one of them). When it became clear that the goal was not in immediate sight, an average company might have stopped the whole thing because of the substantial expenses being incurred, but IBM had the patience and the resources to wait for its consort of geniuses to fully hatch the golden egg. Had the research been confined within time or application boundaries, there would have been a great loss to real scientific progress.

[9] The safest choice is the Fortran 77 format, which is simple, widely known, easily analyzable and modifiable, and accepted without hesitation by all compilers and all machines. Use of Fortran 90 may cause compatibility problems.

[10] A safe procedure is to check at least some of the results of calculations done with different optimization levels against the results of calculations done without optimization.

[11] Publication raises issues connected with protection of intellectual property. These subjects will not be dealt with here.

[12] Some of these rules are standards for publication, see, for example, the *Journal of Applied Crystallography*. Similar procedures were followed in the program distribution project that used to go under the name of Quantum Chemistry Program Exchange (ref. [20], Chapter 3).

[13] This statement was made in the late 1990s by a European Community officer, during a meeting of a scientific steering committee in Bruxelles. It was meant to apply to publications in the whole chemical field and has been here adapted to computational chemistry. The author of this book tends to agree, including his own papers in the list.

[14] Even more than spectrometer designers fear photodiode fading, computer programmers suffer from a neurotic fear of bugs, because they know that the consequences of even an apparently harmless imprecision in a computer code can propagate itself by unpredictable waves of ruin at any unpredictable moment.

[15] On one occasion, after completing a long sequence of calculations and publishing in a major journal, we discovered that the computer code we had been using sported a *plus* sign in front of the *attractive* part of an intramolecular non-bonded energy potential function. The results made sense anyway because intramolecular non-bonding functions are only there to prevent undue atom–atom overlap and hence the attractive part is much less important.

II

The Frontier

11

Structure-property and structure–activity relationships

...the synthetic chemist is likely to have many opportunities to encounter unusual phenomena by accident during everyday chemical work with crystalline solids...
           Curtin, D. Y.; Paul, I. C. *Chem. Revs.* 1981, **81**, 525.

There doesn't exist a category of science to which we can give the name of applied science. There are science and the applications of science, bound together as the fruit to the tree which bears it.
           Attributed to Louis Pasteur.

11.1 Fundamental research and applied technology

The study of the structure and transformations of organic condensed phases is at the forefront of modern chemistry. There are two main reasons for this, both equally valid although by far unequal on the conceptual side: one reason is that these structures and transformations are poorly understood, and hence stimulate the curiosity of researchers; the other reason is that molecular devices are a worldwide and ever expanding market with immense opportunities for money-making. Advancement in the understanding of these phenomena can be rewarding both to the intellect and to one's wallet. Very favorable as it appears, this circumstance has nevertheless several undesirable consequences. The most conspicuous one is that the officials and politicians who supervise funding agencies tend to favor those proposals that lean towards the second of the above-mentioned aspects of research, on the vulgar assumption that public money is better spent in the development of technology than in the development of new theories. When applying for funds, scientists, in turn, comply, and try to disguise themselves as manipulators of devices rather than speak out as developers of theories [1]. Aside from the fact that technology is not always good and proper, as when directed to the development of weapons or of some of the useless, energy-wasting and ultimately repulsive electronic gadgets that pester our present civilization, this pseudo-technological drift in chemical research has a serious chance of drying up the very vital life-blood of scientific thought and, by consequence, also of technological development. The reason of course is that technology cannot exist, let alone develop, in the absence of new scientific ideas. Theory is less prone than technology to the whims of the times, and is therefore much more intrinsically valuable than many of its practical, and often transient, applications.

11.2 The structure–activity dogma

A chemical can be designed and manufactured for an action that derives from its molecular structure (a pharmaceutical drug, a pesticide, a preservative) or from the structure of the material formed in its condensed phases (an optical switch, a liquid-crystal display, a plastic fiber). Understandably, a compound made for its molecular properties must interact with some receptor in a living organism, and therefore can be used either in solid form or in solution, preferably in solution for faster action. A material must function as a macroscopic object and hence will have to be in a form that either maintains its shape over time, that is a crystalline or amorphous solid, or that can be easily contained, like a semi-solid or even a high viscosity fluid. Ideally, the joint application of good science and good technology should lead in a straight path from the basic chemical unit, the molecule, to the prediction and control of the properties of all the products of chemical technology. In chemical science and technology as we understand it today, the cornerstone of progress is the concept of structure–property and structure–activity relationships: the macroscopic, active properties of a preparation or of a material will be predictable on the basis of the microscopic structures of the molecule itself and, when applicable, of the condensed aggregate. Table 11.1 has a list of molecular and macroscopic properties; Fig. 11.1 shows a diagram of the aggregation manifold, and of the relationships that may in principle exist between the microscopic and macroscopic features of a material. Figure.11.2 is a scheme showing the interplay of various disciplines in the study and interpretation of the properties of a chemical system, on the theoretical, experimental and simulation sides. Of course the structure–activity relationship is also the key to our current theories of chemical reactivity, but this subject is not dealt with in this book.

The advantages of a molecular view of chemical phenomena are evident, because chemical synthesis is a difficult and time- and money-consuming task. In the molecular view one can draw a chemical structure on the back an the envelope and be confident that the final product will conform to one's expectations, while in the absence of such a guideline one can only proceed by expensive trial and error. A natural offspring of the molecular view of the chemical world is molecular simulation, in which chemical intuition is complemented and supplemented by a computer-driven analysis, based on first principles (quantum mechanics) or expert systems (force fields), which often yields an accurate and convincing picture of matter at the microscopic level. Molecular simulation often allows even more confident predictions at the macroscopic level; it is in a way the very materialization of the structure–activity relationship. A large part of this book is henceforth dedicated to the analysis and the exploitation of the relationships between structure, properties, and activity of chemical systems, inspired

THE STRUCTURE–ACTIVITY DOGMA

Table 11.1 A list of macroscopic and microscopic properties. There is no one-to-one correspondence between entries in the right- and left-hand columns

Properties of molecules	Properties of macroscopic bodies
Molecular mass, volume and surface	Temperature, pressure
	Density
Bond distances and bond angles	Heat capacity
Molecular conformation, stereochemistry	Enthalpy, entropy, vapor pressure
	Solubility, osmotic pressure
Vibrational force constants	Surface tension
	Viscosity, rheology
	Condensation form at room conditions
Electronic structure	Polymorphic crystalline forms
Electrostatic potential	Hardness, mechanical strength
Charge distribution (multipoles)	Boiling and melting temperature
Polarizability and hyperpolarizability	
	Refractive index, optical rotatory power
	Color
	Magnetism
	Dielectric constant

by and focusing upon the theory and practice of molecular simulation for the organic condensed phases, with a critical review of existing results and an attempt at prediction of future trends.

There are three main kinds of molecular simulation:

1. Static simulation: a single structure is considered, and forces and energies are calculated with fixed nuclear positions and electron density, under the given potential. This technique samples a unique point in phase space, and therefore can be used only for periodic symmetric systems, ideally perfect crystals at zero temperature, for which static, vibration- and temperature-less lattice energies can be obtained, or harmonic vibrational frequencies can be guessed.
2. Static minimization simulation: starting from a given structure, a structure of minimum energy is searched for, under the action of the given potential. Minimization techniques may or may not require the calculation of energy derivatives, i.e. forces (slopes) and force constants (curvatures of the potential energy surface); in a true minimum, all eigenvalues of the second derivative matrix are positive. This approach follows (and samples) a convenient path in phase space towards a minimum, which, however, may or may not be the absolute minimum, and anyway does not include the effects of temperature or of molecular vibrations. It can be used for single molecules, for crystals, and is of some use also for less ordered systems or liquids, for example to provide a not too unreasonable starting point for dynamic simulation.

STRUCTURE–PROPERTY AND STRUCTURE–ACTIVITY

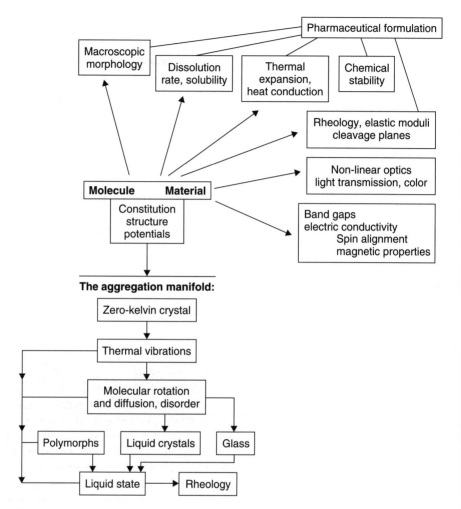

Fig. 11.1. A scheme of the interplay between the microscopic and macroscopic properties of materials.

3. Dynamic, or evolutionary simulation: random or Newtonian paths are followed to sample a significant portion of phase space, on which averages can be taken (as in ordinary molecular dynamics or Monte Carlo) or to force somehow the access to entropy (free energy simulations). Temperature and pressure can be varied.

To what extent does the structure–property–activity dogma, with its simulation–prediction sub-dogma, really work in practice? Many obstacles stand in the way. Present theories of molecular aggregation are tentative (Chapter 15), so that the prediction and control of condensation modes from molecular structure alone cannot yet

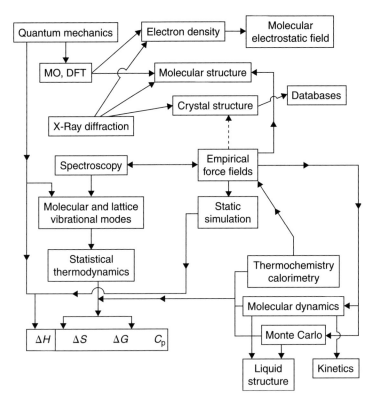

Fig. 11.2. The interplay between theories, methods and experiments in molecular science.

be achieved, and also theories on the relationships between aggregation modes and macroscopic properties are incomplete. The challenge, however, is there, and so is the future. Pharmaceutical companies, alongside *in vitro* and *in vivo*, now speak openly of *in silico*, and (although the neologism could have been better chosen) if profit-making institutions start taking computer simulation seriously, that is a good sign.

11.3 Crystal dissolution

A very old but still widely accepted model for the dissolution of a crystal [2] involves two steps: the detachment of molecules from the crystal with the formation of a saturated layer of thickness h near the solid surface, followed by diffusion of solute molecules from this layer into the bulk solution. The formation of the saturated layer, with competition between detachment of molecules from the crystal surface and re-adsorption onto it, is the structure-sensitive step; it depends on crystal structure overall, on macroscopic crystal morphology, and on the number and type of surface defects, the latter being quite often not so easily determined [3]. The structural influence on the overall solubility is, however, minor, if it is accepted that once the equilibrium is (quickly) reached, dissolution is effectively controlled by the rate of diffusion into the bulk solution, which no longer depends on the structure of the solid. Such a dissolution model can be described by the Noyes–Whitney equation, which successfully reproduces observed dissolution rates of organic compounds:

$$dB/dt = (\varepsilon [A]/h)(B^{eq} - B) \tag{11.1}$$

where B is the bulk concentration of solute, ε is the diffusion coefficient (units of length^2 time $^{-1}$), $[A]$ is the ratio of dissolution-available area to volume of the solid (units of length^{-1}), and B^{eq} is the equilibrium solubility. The integrated form is:

$$B = B^{eq} [1 - \exp(-k_{NW}t)]; \qquad k_{NW} = [A]\varepsilon/h \tag{11.2}$$

The apparent value of k_{NW} (dimension of time^{-1}, as a regular first-order kinetic constant) is easily obtained by kinetic experiments. The thickness of the saturated layer, h, and the active site concentration, $[A]$, in principle depend on crystal structure, but are not easily accessible experimentally, so the whole equation remains an essentially phenomenological one.

The simulation of dissolution rates perforce requires a dynamic treatment, although some information could be gathered by considering just the attachment energy, or the energy required to remove a molecule from a given crystal surface, which can be estimated even by a simple static calculation. The equilibrium bulk solubility, B^{eq}, can be easily measured, and its van't Hoff treatment (equation 7.60) readily yields the dissolution enthalpy, ΔH^*(diss). B^{eq} and the dissolution free energy ΔG^*(diss) $= -RT \ln(B^{eq})$ are controlled by the stabilization of the solvated complex which is, in turn related to the molecular structure, rather than to the structure of the crystal or the strength of its intermolecular forces. For all these reasons, structure–dissolution correlations are difficult to establish, involving as they do quite a number of scarcely controllable parameters, and also because experimental measurements on the surface properties of organic crystals are awkward due to their small size and scarce mechanical resistance. Differences in dissolution rates among different crystalline samples of the same substance are often ascribed to crystal polymorphism, but can equally well arise from different morphology or granulometry of the preparation, a typical case being aspirin, whose "polymorphs" turned out to be in fact just different faces of the same crystal [4]. Computer simulation of dissolution must depend on very long dynamic analysis for the kinetics, or on carefully designed and parameterized free energy calculations (Section 9.7) for the thermodynamics. The subject is treated in more detail in Section 13.9.

11.4 Thermal properties

11.4.1 *Thermal expansion coefficients*

In a crystal, displacements of atomic nuclei from equilibrium occur under the joint influence of the intramolecular and intermolecular force fields. X-ray structure analysis encodes this thermal motion information in the so-called anisotropic atomic displacement parameters (ADPs), a refinement of the simple isotropic Debye–Waller treatment (equation 5.33), whereby the isotropic parameter B is substituted by six parameters that describe a libration ellipsoid for each atom. When these ellipsoids are plotted [5], a nice representation of atomic and molecular motion is obtained at a glance (Fig. 11.3), and a collective examination sometimes suggests the characteristics of rigid-body molecular motion in the crystal, like rotation in the molecular plane for flat molecules. Lattice vibrations can be simulated by the static simulation methods of harmonic lattice dynamics described in Section 6.3, and, from them, ADPs can also be estimated [6].

Nuclear motion "drags along" the electronic cloud, so that as temperature rises, molecular envelopes oscillate more and more. If the intermolecular potential were perfectly harmonic, the overall volume effect would be nil, because the compressions and expansions would average out; but the potential is much steeper on the compression side (Fig. 4.4), so expansion is hindered less than contraction and molecules effectively occupy more and more space as mobility increases. So thermal expansion is very strictly dependent on the shape of the potential curve, that is on the strength and anisotropy of the intermolecular potential, in a typical structure–property relationship. The simple equation that defines the isobaric thermal expansion coefficient α is

$$\beta = (1/V)\,(\partial V/\partial T)_P \qquad (11.3)$$

This coefficient is obviously comparatively larger and isotropic in liquids, and smaller in crystals (Table 11.2), where the anisotropy increases with increasing intensity of internal cohesive forces. The data in Table 11.2 and Fig. 11.4(a) give a semi-quantitative picture of the phenomenon of thermal expansion in organic materials. Figure 11.4(b) shows typical plots of cell volume against temperature for some organic

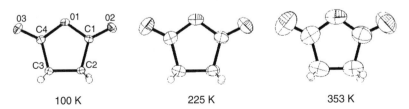

Fig. 11.3. An ORTEP [5] view of the atomic displacement parameters for succinic anhydride as a function of temperature. Ref. [20], Chapter 9.

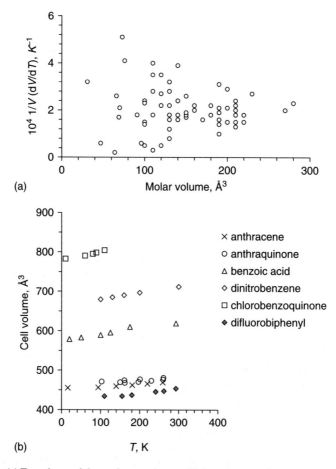

Fig. 11.4. (a) Experimental thermal expansion coefficients for organic crystals calculated from X-ray cell dimensions in a survey of temperature-dependent crystal structure determinations in the Cambridge Structural Database (Ref. [1], Chapter 8). (b) Typical plots of cell volume ($Å^3$) against temperature for some organic crystals.

crystals. One sees, however, how difficult it can be to trace simple relationships between crystal potentials and expansion, which, apart from the obvious order of magnitude differences between metallic materials and organics, is not just inversely proportional to the relative strength of cohesive forces: for example, hydrogen-bonded benzoic acid has a larger expansion coefficient than condensed aromatics. The expansion coefficient is a complicated function of intermolecular forces, molecular shape, and other less easily identifiable properties. It also increases with temperature, so that more proper comparisons should be made at constant temperature rather than on T-averaged coefficients.

Table 11.2 Thermal expansion coefficients of some materials (units of 10^{-4} K^{-1})

Compound	β, crystal[a]	β, liquid	Compound	β, crystal	β, liquid
benzene	3.6	12.1	average organic[b]	2.0	—
anthracene	1.4		succinic anhydride	2.6	10.5
naphthalene	1.6		water	1.6	2.1
anthraquinone	1.2		diamond	0.010	—
1,3-dinitrobenzene	2.3		graphite	0.07	—
chlorobenzoquinone	2.7		metals	0.08–0.15	—
difluorobiphenyl	2.5				
benzoic acid	2.6				
acetic acid	2.3	10.5			

[a] From the variation in cell volume with temperature, X-ray diffraction data.
[b] See Fig. 11.4(a).

Thermal expansion coefficients can be easily estimated by running molecular dynamics or Monte Carlo simulations, yielding molar volumes as a function of temperature, a complete equation of state for the considered material (as was shown in Fig. 9.6; notice there the significant difference in steepness between the curve for the solid and that for the liquid). Note however that for $\beta \approx 2 \times 10^{-4}$ K^{-1} the total expansion of an organic material from zero-K crystal to its melting point is as low as 5–6%, so highly refined or ad hoc potentials may be required for accurate results (ref. [22], Chapter 9).

11.4.2 *Heat capacity and heat transport*

The heat capacity, C_p, of an organic crystal can be subdivided into an internal part and an external part (recall Section 6.3 and the discussion around equation 6.19). The external part is the structure-sensitive part because it depends on the lattice vibrational frequencies and hence on the crystal structure and the packing forces within the crystal. External contributions can be estimated by lattice dynamics simulations [7], or can be derived from variable-temperature molecular dynamics simulations (recall Section 9.5 and especially equation 9.21).

Heat transport properties are also a complex function of density and external vibrational modes. Thermal conductivity can be defined as

$$\lambda = (q/t)\, L/(A\, \Delta T) \qquad \text{W m}^{-1}\text{K}^{-1} \tag{11.4}$$

that is, the flow of heat q through a length L of material in the direction normal to a surface of area A, divided by the temperature gradient. In an anisotropic medium this is again a tensor quantity. Typical values are 10^3 for diamond, 10^2 for metallic materials, 2.2 for ice, and 10^{-1} to 10^{-2} Wm^{-1}K^{-1} for organic materials.

The molecular dynamics simulation of λ is formally similar to that of other transport coefficients, such as self-diffusion coefficients (equation 9.15), but there is a conceptual difference between the actual methods for the simulation of mechanical and of thermal diffusion. These topics are too specialized to be dealt with here [8].

11.5 Strain and stress, elastic and viscous properties

Stress is a force per unit area acting on a piece of material, and strain is the resulting relative (dimensionless) deformation of the material. In a simple example, consider (Fig. 11.5(a)) a cylindrical sample under a uniaxial stress $\sigma = F/A$, undergoing a change in length ΔL with strain $\varepsilon = \Delta L/L$; the Young modulus E of the material is defined in this case as follows:

$$E = \sigma/\varepsilon \tag{11.5}$$

Typical values are 10^{11} N m^{-2} for steel and 10^9 N m^{-2} for organic materials.

More generally (Fig. 11.5(b)) consider a differential strain along a direction x, where two points $x°$ and x' experience positional changes $\delta x°$ and $\delta x'$; the generalized strain is

$$\varepsilon_x = du/dx \tag{11.6}$$

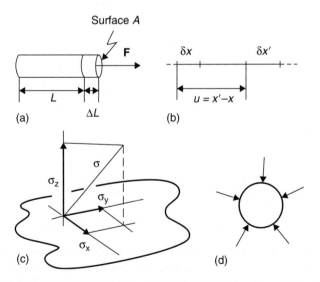

Fig. 11.5. (a,b) The definition of the elements for the calculation of Young's modulus. (c) The subdivision of a stress per unit area into its perpendicular and tangential components. (d) The elements for the description of bulk modulus; arrows are forces acting isotropically on a piece of material.

In a three-dimensional body the displacement is a vector, **u**, and if k and m label any two directions in 3D space, three-dimensional strain is defined as

$$\varepsilon_{km} = 1/2(\partial u_k/\partial x_m + \partial u_m/\partial x_k) \quad \text{or e.g.} \quad \varepsilon_{xx} = \partial u_x/\partial x \tag{11.7}$$

For a parallelepiped of original dimensions $a°$, $b°$, and $c°$, which deforms under stress from $V°$ to V' one gets:

$$V' = [a°(1+\varepsilon_{xx})][b°(1+\varepsilon_{yy})][c°(1+\varepsilon_{zz})] \approx V°(1+\varepsilon_{xx}+\varepsilon_{yy}+\varepsilon_{zz}) \tag{11.8}$$

where the last equality holds because the products of two or more εs are second-order, negligible differentials.

Consider now a generic surface of an anisotropic three-dimensional object (Fig. 11.5(c)), and a stress vector σ_k decomposed along three directions, the one perpendicular to the surface corresponding to direct stress, and the other two forming what is called shear (tangential) stress. Considering an elementary cube within the material, there is one such stress vector with its three components for each face, so there are nine stress components σ_{ij}, only six of which are independent because of the condition $\sigma_{ij} = \sigma_{ji}$ in mechanical equilibrium (no net displacement under shearing stress). The shear modulus G is defined as

$$G = \sigma_{xy}/2\varepsilon_{xy} \quad \text{N m}^{-2} \tag{11.9}$$

being again a ratio of the applied tangential stress to the resulting strain.

If the six strain components and the six stress components are arranged in column vectors, one gets:

$$\boldsymbol{\varepsilon} = \mathbf{S}\boldsymbol{\sigma} \tag{11.10}$$

where **S** is the compliance matrix, with 21 independent components. The inverse of **S**, matrix $\mathbf{C} = \mathbf{S}^{-1}$, is called the stiffness matrix. All the above treatment assumes direct proportionality between stress and strain, and is thus nothing else but a generalization of Hooke's law.

A simpler treatment applies in the isotropic case, where a material experiences a uniform load (e.g. a hydrostatic pressure, Fig. 11.5(d)), so that shear stresses are all zero and $\sigma_{xx} = \sigma_{yy} = \sigma_{zz} = \sigma$, $\varepsilon_{xx} = \varepsilon_{yy} = \varepsilon_{zz} = \varepsilon$. The bulk modulus, K, is the force per unit surface applied for a given volume variation, and is the reciprocal of the compressibility, α:

$$K = \sigma/(\Delta V/V) \text{ N m}^{-2}; \quad \alpha = 1/V(\partial V/\partial P); \quad K = 1/\alpha \tag{11.11}$$

Typical values for K (GPa) are 160 for steel, 40–50 for glass, 5–10 for elemental phosphorous or sulfur, and, in an extreme example of structure–property relationship, 600 for diamond and 33 for graphite. Bulk moduli are of the order of 10 GPa for organic solids (see below), and of 2 GPa for liquids.

Understandably, elastic moduli regulate the transfer of mechanical energy through solids. Consider a mechanical impulse applied at one end of a piece of elastic solid;

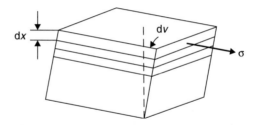

Fig. 11.6. A shear stress, σ, and the displacement of successive layers of thickness dx with relative velocity dv.

this impulse displaces some of the peripheral atoms, which in turn act on atoms in next neighbor layers, and in this way the impulse is propagated through the solid. The propagation velocity v is approximately given by

$$v \approx (E/\rho)^{1/2} \quad (11.12)$$

where E is the Young modulus in the propagation direction, and ρ is the density of the material.

Elastic behavior is that of bodies that tend to restore their original shape after stress is discontinued; viscosity is a property of bodies that irreversibly change their shape, i.e. flow, under stress. Imagine a fluid medium as composed of parallel layers (see Fig.11.6) which slide past one another under shear stress; viscosity is defined as the ratio of shear stress to differential displacement of successive layers (flow "speed"):

$$\eta = \sigma/(dv/dx) \quad (11.13)$$

Viscosity is related to the self-diffusion coefficient (equation 9.15) and can be estimated by molecular dynamics simulations [8].

The elastic moduli, as well as viscosity, depend on how the inner structure of the material responds to an external mechanical stimulus, and hence on internal cohesion forces; for example, the activation energy for viscous flow in liquids is found to be about one third of the vaporization energy. The anisotropic aspects of elastic behavior in crystalline solids depend on the details of the molecular arrangements in the crystal, with consequently different responses in different directions through the material. Anisotropy introduces conceptual and computational complications. For example, the number of independent terms in the compliance matrix depends on the symmetry of the crystal system; shearing stresses in general produce a change (lowering) of overall crystal symmetry, which has to be taken into account in the simulations, and so on.

The bulk modulus and the compressibility can be calculated by running molecular dynamics simulations to give equilibrium volumes at variable pressure. A simple method for the approximate calculation of the bulk modulus and of the elastic stiffness constants in the three main directions of a solid, using static simulation, will now be described, to illustrate how a macroscopic property like an elastic stiffness constant relates to crystal and molecular structure. A cluster representing the crystal is built by including all molecules whose centers of mass are within a given range, say 20 Å, from a reference molecule. The cluster has a globular overall shape of approximate radius $R°$, surface $A°$ and volume $V°$ [9]. The edges of the unit cell are then increased, and a new cluster is built using the new metrics, leaving the cell angles and the space group symmetry unchanged, with a radial inflation of the cluster equal to ΔR and a volume change ΔV. Calling W the intermolecular potential energy between all molecules in the cluster, $\Delta W/\Delta R$ is the finite difference ratio that approximates the radial force, which is supposed to equilibrate an external isotropic stress. The bulk modulus, K, can then be estimated as (compare with equation 11.11):

$$K = (\Delta W/\Delta R)/[A°(\Delta V/V°)] \tag{11.14}$$

K must be determined by imposing simultaneous isotropic increases in all three cell dimensions, but calculations done by increasing one cell dimension at a time may approximate the stiffness constant along each of the three directions, related to the relative strength of the intermolecular interactions in each direction. The calculated moduli decrease with increasing inflation, and should ideally be determined at zero pressure; however, the numerical calculation diverges for very small radial increments. As a compromise, data for comparison with experiment are taken at a constant 3% relative volume increment.

Table 11.3 shows that equation 11.14, with energies calculated by standard intermolecular potentials, gives answers of the correct order of magnitude. Stiffness constants for more strongly bound crystals, like hydrogen-bonded crystals, are only marginally higher than crystals bound only by dispersive interactions; in general, due to the overall weakness of the intermolecular forces in organic crystals, the differences among different directions or among crystals of different chemical constitution are rather small. As far as mechanical properties go, organic materials are all similar and all weak, and do not develop large anisotropies. The unacceptable negative values calculated for the succinic anhydride crystal show that the experimental crystal structure is calculated to be mechanically unstable, a clear (and well known) failure of the model interatomic potentials. When the crystal structure is optimized under the action of the same potentials, a significant distortion occurs, but the elastic constants become all positive. The same happens when the potentials are improved by adding a coulombic contribution over point charges, which, however, is almost irrelevant along x and decisive along the other two directions. For benzene, the calculated bulk modulus K becomes negative, as it should, on crystal structures determined at high pressure; calculated Ks to some extent follow the trend in energy density within the crystal, $W/V°$ (Table 11.3). These are all sensible results, and calculated bulk moduli can be a sensitive test of the applied potentials.

Another aspect of the influence of structure and internal forces on mechanical properties of materials is the surface tension, or the work needed to create a surface out of a piece of bulk material. In crystals, this will in general be different for different

Table 11.3 Stiffness coefficients along a, b, c and bulk modulus, K (GPa units[a]), and energy density for some organic crystals. UNI potentials

Compound	C_{11}		C_{22}		C_{33}		K	
	Calc.	Obs.	Calc.	Obs.	Calc.	Obs.	Calc.	Obs.
benzene (270 K)	13	9	15	10	13	10	13	—
naphthalene	14	15	16	15	15	20	14	—
anthracene	17	12	16	17	19	20	15	—
aspirin[b]	13	11	18	12	20	—	15	—
β-HMX[c]	23	18.41	7	14.41	13	12.44	12	9.6–10.2
succinic anhydride	6	—	−13	—	12	—	0	—
id., optimized[d]	14	—	10	—	17	—	11	—
id., with partial charges	7	—	2	—	21	—	8	—

Compound class	K	Energy density[e]
n-alkanes	6–7	0.20–0.25
condensed aromatics	13–15	0.30–0.40
oxohydrocarbons	15–18	0.30–0.40
carboxylic acids	9–15	0.40–0.65
bicarboxylic acids	16–19	0.75

[a] With energies in kJ mol^{-1}, distances in Å, K (GPa) = (10/6.022) $\Delta W/[A\ \Delta R(\Delta V/V)]$.
[b] Ref. [12b].
[c] HMX = cyclotetramethylenetetranitramine [12a].
[d] Values calculated on a structure optimized from the experimental one.
[e] Total potential energy within the cluster divided by the cluster volume (kJ mol^{-1} Å^{-3}).

surfaces. For a simple but instructive simulation of this process, the N-molecule crystal cluster constructed as described above can be rotated so that the x-direction of the reference system after rotation coincides with the perpendicular to a selected Bragg plane (Section 5.5) [10]. The coordinates of all molecules are then referred to the center of mass of the whole sphere, which is then divided into two parts, assigning the $N/2$ molecules with the lowest x-coordinates of their centers of mass to the lower hemisphere, and the other $N/2$ molecules to the upper hemisphere. This identifies through the crystal model two formal surfaces whose area, A, can be estimated from an average radius, as already described. This surface can be very rough at a molecular level (Fig. 11.7). The interaction energy W between all molecules in the upper hemisphere and all molecules in the lower hemisphere can be equated to the work that has to be expended for creating the two free surfaces. An estimate of γ the surface tension coefficient, is as follows:

$$\gamma \approx W/(2A) \tag{11.15}$$

Table 11.4 gives some typical results, again of the correct order of magnitude, remarkably for such a simple model which, among other things, neglects the possible surface reconstruction after detachment, and thus anyway yields an upper boundary

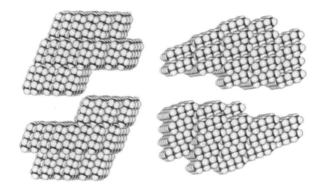

Fig. 11.7. An extreme example of different faces in the same crystal: two cuts through the *n*-hexane crystal structure, perpendicular (left) and parallel (right) to the chains.

Table 11.4 Calculated surface tension coefficient for different faces of some organic crystals. Units of N m^{-1}. UNI potentials

Compound	γ(111)	γ (100)	γ (001)	γ (110)
benzene	0.098	0.120	0.108	0.118
naphthalene	0.115	0.135	0.120	0.109
anthracene	0.130	0.105	0.146	0.123
n-hexane	0.096	0.087	0.092	0.087
aspirin	0.174	0.182	0.204	0.210
β-HMX	0.217	0.280	0.198	0.212
hydrocarbons	0.08–0.15	—	—	—
oxohydrocarbons	0.12–0.18	—	—	—
carboxylic acids	0.15–0.25	—	—	—
bicarboxylic acids	0.18–0.33	—	—	—

[a] With energies in kJ mol^{-1}, distances in Å, γ(N m^{-1}) = W/[6.022 (2A)]; γ(kJ mol^{-1} Å^{-2}) = 6.022 γ (N m^{-1}).

value of the surface energy. γ-values for hydrogen bonded crystals are significantly larger than those for dispersive crystals. γ-values are not available for organic crystals, and for a rough comparison, γ ≈ 0.07 N m^{-1} for liquid water.

The methods so far described for the estimation of elastic properties illustrate some principles, but are inaccurate due to the numerous geometric approximations and to the use of finite differences rather than true derivatives of the potential. A more ingenious method [11] introduces an external force and computes the energy expended by this force in deforming the system under investigation; minimization of the total energy, internal plus external, directly yields the equilibrium structure under external stress, from which all the resulting strains can be extracted. For example, hydrostatic pressure is simulated by a barostat whose pressure work is $P\Delta V$, where ΔV is the volume change in the system; and even external electric fields can be simulated through their effects on the internal polarization of the material.

Elastic constants of crystals can be measured by Brillouin scattering, which is sensitive to the instantaneous variations in density that occur with the propagation of mechanical energy waves [12]. The experiment requires an identification of crystal faces and an appropriate orientation of the crystal, and thus needs reasonably large and well behaved single crystal samples, not so easy to obtain from organic substances; besides, organic materials are seldom important from the viewpoint of mechanical properties, so there is little drive to overcome these difficulties. This explains why the number of complete sets of measured elastic constants for organic crystals is very small.

11.6 Optical, electric and magnetic properties

All properties so far examined – crystal solubility and dissolution rates, thermal expansion, specific heats and thermal conduction, elastic moduli, viscosity – depend upon, and have been described in terms of, the interplay between the inner structure of the condensed media, the forces exerted between ground state, closed shell molecular systems (Chapter 4), and the effects of thermal vibrations. The optical properties of materials depend on the interaction between the oscillating electric field of an incoming radiation and the electron distribution and polarizability within the material. Electric properties, mainly electric conductivity, depend on the degree of delocalization of the electron distribution over the periodic entities that form the material. Magnetic properties depend on the interactions among periodic arrangements of unpaired spins. For the said ground state, closed shell molecular systems, solid-state optical properties are mainly dependent on intrinsic molecular properties and are less sensitive to the structure of the condensed medium, while electric and magnetic properties are hard to develop, because electron and spin coupling, and in general all intermolecular interelectronic effects, are small for scarcely overlapping electron clouds. Optical, electric, and magnetic properties are instead the backbone of the solid-state physics of inorganic and metallic materials. The following section will nevertheless review some aspects of the relationships between bulk structure and bulk optical, electric, and magnetic properties that may sometimes be significant even for organic molecular crystals.

11.6.1 *Color*

When matter selectively absorbs electromagnetic radiation in the visible ($\lambda = 700$–400 nm) to ultraviolet ($\lambda = 400$–100 nm) ranges, the absorbed energy is used to excite peripheral electrons. As a first approximation, the energy corresponds to excitation from the highest occupied to the lowest unoccupied molecular orbital, and the wavelength of the absorbed radiation can be calculated from the corresponding energy gap,

Δ(HOMO–LUMO), as $\lambda = 1240/\Delta$ (eV). The sensation of color is an entirely human one, depending on the sensitivity range of the human visual apparatus; when part of the visible wavelength range is absorbed out of white radiation, the sample shows the color of the complementary visible range. Given the above energy–wavelength relationship, it takes a gap of less than 3 eV to give an absorption in the visible range; this is the case only for molecules with delocalized π-systems, or otherwise carrying groups which are, precisely for this reason, called chromophores, but for most organic molecules the frontier orbital gap corresponds to the UV wavelength range.

The Δ gap is an intrinsic molecular property, and, when it is not modified by a crystalline environment, the compound in solution has the same color as in its crystal. Color may become a crystal property when overlap between molecular electron densities is strong enough to couple the electrons in different molecules, so that a band is formed (Section 6.2) and the gap between the occupied and unoccupied bands becomes smaller, shifting the absorption wavelength from the UV to the visible region. Remembering what has been said so far, this is seldom the case for purely molecular crystals. The reverse situation, or crystal packing depleting the color effect in an intrinsically colored molecule, seems even less likely, because it seems difficult to imagine a mechanism by which intermolecular interaction can increase the energy gap between molecular frontier orbitals.

Color may arise following a major conformational change (e.g., flattening of a conjugated system), by shift of a tautomeric equilibrium, by charge transfer, for example protonation–deprotonation equilibria, or by the formation of a charge transfer complex between a donor and an acceptor molecule. Even in the latter case, color is mainly a property of the dimer, being elicited also by complexation in solution, and being only moderately affected by the crystal packing. The relationship between "molecular" color (i.e. color due to intrinsic electronic gaps in the molecule) and "crystal" color (i.e. color due to a periodic electronic effect in the crystal structure) is thus a complex one: sometimes, the color of the crystal is different from that of the solution, or the colors of crystal polymorphs are different, because of molecular transformations, rather than for genuine electron coupling effects due to crystal packing. As a first approximation, at least for conformationally rigid and electronically immutable molecules, one would guess that for a given organic compound the color of the crystal should be the same as the color in solution. There are however several other caveats: especially in very small crystal samples, as organic ones mostly are, several factors related to absorption or refraction due to internal texture, defects, dislocations, impurities, as well as to thickness, size and outer morphology of a particular crystal sample, may substantially alter the observed color, again for reasons not stemming from electron coupling. For example, pigments are crystalline organic solids used in a dispersion, and the actual observed hue effect of the overall preparation depends on subtle interplay between the micromeritics of the pigment and the properties of the dispersing medium, as well as on the thickness of the layer – all extremely complex phenomena.

Table 11.5 Color designators in the Cambridge Structural Database for various subsets (as described in Section 8.3). Assigned wavelengths are intended only as approximate labels

Color	Formal "tag" wavelength, nm	%, Z(1) database 20,209 cases	%, small database 425 cases	%, Z2 database 704 cases	%, molecular complexes
Red	700	3	2	3	9
Orange	650	4	4	3	6
Yellow	600	18	17	16	17
Brown	570	1	4	2	5
Green	550	0.6	0.5	0.8	2
Blue	500	0.4	—	1	1.5
Violet	450	0.2	0.2	0.6	1
Black	800	0.5	—	1	8
White	400	6	6	7	1
Colorless	350	67	63	66	51

In an attempt to quantify the color distribution in organic crystals, the Cambridge Structural Database has been scanned for recorded crystal colors of organic compounds. Only the "red", "orange", "yellow", "brown", "green", "blue", "violet", "white", "colorless", and "black" qualifiers have been considered, neglecting adjectives and appendages (like "-ish", "pale", etc.; in double designations, e.g. "orange-yellow", the first has been retained). To each of these colors a "tag" wavelength has been assigned, for numerical identification. Table 11.5 shows some results: in agreement with the above considerations, a large majority (about 72%) of one-component organic crystals are recorded as being "colorless" (that is, presumably, transparent) or "white", while colored crystals are mostly yellow, orange, or red (brown can be assumed as a reddish color), and only a minority are green, blue, or violet. Black is also extremely uncommon. The distribution is strictly the same in the various subset databases, and the distribution of colored crystals is the same, for instance, in centrosymmetric and non-centrosymmetric space groups. The electronic situation in colorless crystals is schematized in Fig. 11.8 (left), which shows a wide gap between very localized occupied and unoccupied crystal bands.

In an attempted analysis of the molecular color-crystal color relationship, the expected molecular absorption wavelength has been calculated from the HOMO–LUMO gap obtained by the Extended Hückel molecular orbital method. Although very approximate, this method is the only affordable one for thousands of molecules, and is also fairly reliable at least for small molecules without complex valence electronic structures: typical results are 275 nm for benzene and 630 nm for dinitrobenzene – very reasonable. A plot of tag crystal color wavelengths against estimated molecular wavelengths is shown in Fig. 11.9. A faint correlation appears, but the number of colorless crystals is large for all values of the electronic gap. This suggests that many crystals are, or appear to be colorless, even for molecules with low frontier orbital

OPTICAL, ELECTRIC AND MAGNETIC PROPERTIES 287

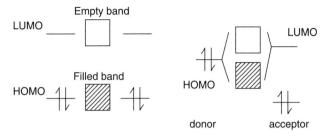

Fig. 11.8. Left: schematic of electronic interaction in a stack of aromatic molecules with large HOMO–LUMO gap and scarce overlap. Right: a donor molecule with a high-lying HOMO donates electrons to an acceptor molecule with a low-lying LUMO. If there is substantial overlap, the band spreads, although not necessary by A–D overlap – only segregated stacks are conductors.

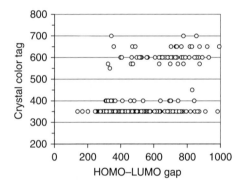

Fig. 11.9. The tag wavelengths (see text) of molecular crystals, from the colors by which they were described in the Cambridge Database (small subset), against the wavelength calculated from the Extended Huckel Δ(HOMO–LUMO) gap. nm units.

gaps, and reinforces the hypothesis that crystal size or termination effects, other than just molecular electronic transitions, are operating.

The case is significantly different in molecular complexes, where an electron-rich molecule (donor, D) with a high-lying HOMO donates electrons to an electron-deficient molecule (acceptor, A) with a high-lying LUMO; as seen in Table 11.5, white or colorless crystals are only 52%, while there is a large increase in the number of red, brown, and especially black crystals (up to 62% on a sample of 58 molecular complexes of TTF, tetrathiafulvalene). Clearly, electron exchange and donation generates a modification in the electronic structure of the constituent molecules, which in turn induces visible light absorption, in principle without the need for a solid state effect [13]. Whether part of this phenomenon is also due to crystal packing depends on the peculiarities of each single case. The increase in number of black crystals

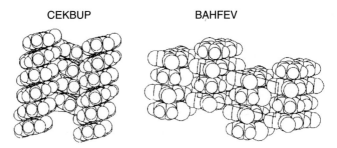

Tetrathiafulvalene-tetracyanoquinodimethane (TTFTCQ)

TCNQ TTF TCNQ TTF TCNQ

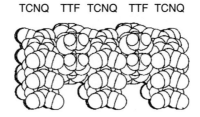

Fig. 11.10. Some examples of alternating donor–acceptor stacks in black crystals: CSD refcode CEKBUP, pyrene-naphthalenetetrone; BAHFEV, dibenzothiophene-tetracyanoquinodimethane; these materials consist of alternating D–A stacks and are semiconductors. In the TTF–TCNQ complex the material consists of segregated A and D stacks and is a conductor.

of molecular complexes is particularly revealing. For a crystal to be really black, not considering the spurious and variable physical effects described above, and excluding complete absorption due just to crystal thickness, there must be a system of delocalized electrons in a partly filled crystal band, or at least a reduced gap between occupied and unoccupied bands, so that electron mobility allows the absorption of nearly all radiation in the visible spectrum; the situation is schematized in Fig. 11.8 (right). This is the case when infinite stacks of flat aromatic molecules are formed throughout the crystal: a few typical examples of such D–A complexes forming black crystals are given in Fig. 11.10. On the other hand, crystals formed by molecules that stack but do not have a significant electronic interaction are colorless insulators (a good example is tetrafluoronaphthalene). The corresponding value of the HOMO and LUMO energies, and of the overlap integrals between molecular electron densities, should clarify the case: for the dibenzothiophene (DBZT)-tetracyanoquinodimethane (TCNQ) complex, the values are: DBZT high-lying HOMO -0.2882 au; TCNQ, low-lying LUMO -0.0667 au. On the other hand, tetrafluoronaphthalene has HOMO -0.3123, LUMO 0.0734 au, a much wider gap. The overlap integral between a stacked pair at an interplanar distance of 3.5Å is 0.0077 for the charge-transfer dimer and only 0.0047 for tetrafluoronaphthalene.

11.6.2 Optical properties and the polar axis

In crystals, the periodic symmetric arrangement of matter causes an anisotropy in the distribution of electric charge and of electric polarizability. As a consequence, when an electromagnetic wave travels through a crystalline medium a number of directionally selective interference processes occur: the light beam may change direction and speed (as described by the refraction index tensor), or may be split into a primary and a secondary beam (birefringence), or the plane of oscillation of its electric field may be constrained to certain orientations (light polarization). The intensity of these optical phenomena depends also on local or transient conditions that may alter the charge distribution within the crystal, the most obvious ones being changes in temperature or density, or the presence of stress and strain.

This is no place for a detailed description of the theory of the interaction between light and matter, in itself the subject of entire books and of countless monographs [14]. However, second harmonic generation will be briefly analyzed to give an overall impression of the proceedings. The periodic, oscillating electric field $\mathbf{E}(v)$ of an electromagnetic radiation induces a small displacement of electric charges in the crystal, that is, a charge polarization that, if there is no light absorption, oscillates along with the polarizing field. The induced polarization, $\mathbf{P}(v)$ (where v is the frequency of the radiation) can be written as a sum of a linear and a non-linear term:

$$\mathbf{P}(v) = \varepsilon_0 \, \chi^{(1)}(v)\mathbf{E}(v) + \mathbf{P}'(v) \quad (11.16a)$$

$$P'_i = \sum_{j,k} \chi_{ijk}^{(2)} \, E_j \, E_k \qquad (i,j,k=x,y,z) \quad (11.16b)$$

where $\chi^{(1)}$ and $\chi^{(2)}$ are the linear and non-linear susceptibilities; the first term of equation 11.16a is a tensor analog of equation 4.18. Intuitively, the non-linear situation is similar to that of an electric charge in an anharmonic well, whose anharmonicity vanishes in a crystal without polar axes (see below), and this is the reason why these effects appear only in non-centrosymmetric crystals. The anharmonicity effects can be Fourier-analyzed into a constant term, plus a fundamental polarization with the same frequency of the incoming field, and a second-harmonic polarization with double frequency, which is responsible for the transfer of part of the incoming radiation energy to a secondary beam with doubled frequency. The higher-order effects are always orders of magnitude smaller than first-order ones, so that if non-linear effects are to arise in efficient devices, the material must first incorporate a molecular entity of high intrinsic susceptibility, and, moreover, must withstand a very high energy field. This is no easy matter with mechanically weak, low melting, and radiation damage prone organic materials. Non-linear effects and frequency doubling have obvious applications as frequency modulators and hence processors of optical information.

Optical effects are usually much more pronounced in inorganic crystals like ionic salts, with their fully developed charge units, than in weakly polar organic materials, and the early technology of optically active materials was developed almost entirely on the inorganic side. Organic materials have become appealing due to a presumed richness and flexibility of their structural motifs with respect to the rather limited variations on the ionic theme, and, in an age of increased device miniaturization, because of their much smaller bulk and weight. Enhancing the performance of organic materials to the point where the efficiency/weight ratio becomes more favorable than that for inorganics has proved, however, to be extremely difficult, and we still do not have a fully fledged optical technology based on organics.

The optical behavior of materials is a typical example of structure–property relationship. Ideally, one should be able to have a look at the space group symbol and at the X-ray crystal structure of a given material, and to predict many if not all of its optical properties from this. In practice, the very structural complexity and flexibility that are part of the appeal of organic materials are also an obstacle to the establishment of straightforward relationships, when it is a matter of minute detail in the distribution and polarizability of delocalized electron densities, rather than of easily identifiable plus and minus charges. Besides, both for organics and for inorganics, optical properties can be extremely sensitive to many kinds of structural defects, such as partial disorder, twinning, or the presence of minor impurities, which may even go undetected under X-ray analysis. In fact, at least with respect to optical properties, the X-ray diffraction picture of a crystal is highly idealized, and the perfect infinite symmetry surreptitiously suggested by the beautifully drawn packing diagrams is, more often than not, deceiving [15].

There are, however, some gross crystal symmetry properties that really make a divide: one example is the center of symmetry versus polar axis alternative. In 1981, David Curtin and Ian Paul wrote a memorable paper [16] in which they called the chemists' attention to this remarkable fact, and into the bargain introduced the very concept of organic solid-state chemistry, giving a louder voice to the first efforts that had been made in that direction by the Schmidt–Cohen school [17] (more modestly, we followed suit not much later [18]). In a centrosymmetric crystal, any object or property at location $\mathbf{x} = (x, y, z)$ has an identical counterpart at location $-\mathbf{x} = (-x, -y, -z)$ (Section 5.2). In a non-centrosymmetric crystal this condition is not met and the polar axis makes its appearance. It is a direction in crystal space along which an intrinsic difference between "up" and "down", or "left" and "right", exists, such that a microscopic probe traveling along that direction would be able to tell the difference: consider the following scheme:

$$\cdots A\text{–}B \cdots A\text{–}B \cdots A\text{–}B \cdots A\text{–}B \cdots A\text{–}B \cdots \quad \text{polar}$$
$$\cdots A\text{–}B \cdots B\text{–}A \cdots A\text{–}B \cdots B\text{–}A \cdots A\text{–}B \cdots \quad \text{non-polar}$$

In the polar case, the probe could tell its traveling direction by recognizing whether a longer AB spacing is encountered on going from A to B or from B to A. The polar axis may or may not coincide with a crystal axis; for example, it does in space group $P2_1$, along the screw direction, as can easily be understood by looking at the sketch of a screw axis in Fig. 5.3, it does not in space group $P2_12_12_1$. The polar axis is an enhancer of directional molecular properties: any vector property (for example, molecular dipole moment) that has a non-zero component along a polar axis adds up

OPTICAL, ELECTRIC AND MAGNETIC PROPERTIES 291

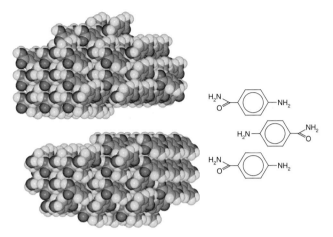

Fig. 11.11. A cut perpendicular to the polar axis in the crystal of 4-aminobenzamide (space group $P2_1$, y vertical). The amide NH_2 groups point upwards, the amide $C=O$ groups point downwards. The molecular dipole is perpendicular to the polar axis.

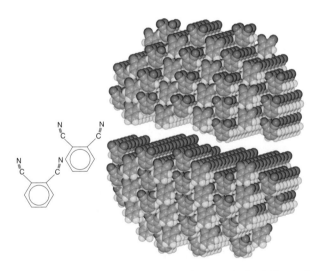

Fig. 11.12. A cut perpendicular to the polar axis in the crystal of 1,2-dicyanobenzene (space group $Pmn2_1$, z vertical): the molecular dipole is almost parallel to the polar axis.

along the polar axis direction, imparting a bulk directionality to the material. This can be visualized by inspection of appropriate cuts through sample crystal structures. Figure 11.11 shows a polar crystal in which the molecular dipole is almost exactly perpendicular to the polar axis; the resulting overall crystal polarity will be low. Figure 11.12 shows instead a structure in which the molecular dipole is almost parallel to the polar axis, with a resulting macroscopic enhancement of polarity. This is also the intuitive basis for the comprehension of the fact that only non-centrosymmetric crystals may have non-zero pyroelectric, piezoelectric, or ferroelectric effects (respectively, the change in electric response by changes in temperature, by mechanical stress, or by application of an external electric field), or can be second-harmonic generators.

A most important consequence of the polar axis is that any cut through a crystal made along a plane that intersects a polar axis must expose two different surfaces, which also means that any real crystal cut along such planes will have different "up" and "down" ends. The polar axis condition has obvious consequences on macroscopic crystal morphology, because crystal growth at the two ends is different and can even be controlled by appropriately designed additives [19], and may also affect the chemical stability of crystalline materials, because of the chemical reactivity towards external reactants. There are some obvious and some less obvious relationships between the presence of a polar axis in the crystal and the chirality of the constituting molecule and of the space group [16].

Although solid-state physics provides appropriate equations for the calculation of some optical properties of crystals, their simulation on the basis of molecular models, molecular force fields or even quantum chemical methods is more awkward than the simulation of thermodynamic or structural properties, because the interaction of electromagnetic radiation with matter is a problem in time-dependent quantum mechanics. The complex electronic structure of organic molecules and the low symmetry of organic crystals do not help along these lines. As discussed above, great progress would be made if one could develop sound methods for the prediction and control of the presence of a center of symmetry in the crystal structure, but even this limited goal is not in sight. The matter will be further taken up in Section 14.4.

11.6.3 *Electric and magnetic properties*

The electronic situation schematically depicted in Fig. 11.8 (right) leads to electric conduction (Section 6.2). About 30 years ago, in the wake of the discovery of conducting TTF–TCNQ crystals, great hopes arose for the development of a new generation of light and easily tunable organic electric conductors, especially for microelectronic applications. In a typical structure–property–activity correlation enterprise, the quest began for stronger donors and stronger acceptors, ensuring the donation of a full electron (electron transfer) as opposed to partial donation (charge transfer), and for the achievement of structural conditions appropriate for the formation of a fully-developed conduction band in the crystal: curiously enough, it was found that only crystal structures with segregated A and D stacks, like TTF–TCNQ (Fig.11.10) are conductors, while those with alternating D–A stacks are at best semiconductors. In the strive for more "polar" or "metallic" behavior, resort was made to ionic species, formed by chemical synthesis or induced by electrocrystallization, or to metalloorganic chemistry, or to appropriate doping agents, intercalating the original molecular species in crystalline, semi-crystalline or polymeric formulations.

The development of magnetic properties in a material depends in the first place on the availability of permanent unpaired spins, much more common in materials based on inorganic or metalloorganic substrates [20(a)] than in purely organic materials, apparently with the only significant exception of nitroxide compounds [20(b)]. The development of bulk ferromagnetism requires the spatial alignment of these spins without coupling; the classification and prediction of the packing arrangements required for such phenomena to occur are matters even more obscure than those connected with the development of electric conduction [21]. There seems to be very little that the structural crystal chemist or the molecular modeler can say a priori.

By definition, molecular conductors and molecular magnets are species in which a significant part of the stabilizing interaction energy is due to electron transfer and delocalization, or to spin interaction, in an unknown admixture with the ingredients of intermolecular interaction typical of organic systems, as sketched in Chapter 4 – from purely coulombic to dispersive interaction. There is little hope that molecular simulation using the empirical force fields of Section 4.9 or Section 9.2 may be successful, and crystal orbital methods (Section 6.2) should be used, with all the difficulties connected with large molecules, large unit cells, and low crystal symmetry typical of organic crystals – an organic crystallographer may consider orthorhombic as high symmetry, a solid state physicist considers tetragonal as low symmetry. For all these reasons, this is no place to describe or discuss the success of such efforts to convince the sluggish "organic" electrons to acquire a metal-like mobility or magnetism, nor for a discussion of whether the organic conductor and organic magnets projects with their applied technology implications are thriving or waning [22].

References and Notes to Chapter 11

[1] In our disgraceful times, if you must write a research proposal, you should not call it "A study of fundamentals of intermolecular forces", but "Methods in supramolecular crystal engineering for the development of nanomolecular materials, devices and biotechnological anticancer drugs". I have refereed many such abstruse documents, in which some colleagues well known to me for being fine quantum chemistry specialists boast their ability to fabricate nanotubes or to cure AIDS.

[2] Noyes, A. A.; Whitney, W. R. The rate of solution of solid substances in their own solutions, *J. Am. Chem. Soc.* 1897, **19**, 930–934; Berthoud, A. Theorie de la formation des faces d'un cristal, *J. Chim. Phys.* 1912, **10**, 624–635.

[3] Unwin, P. R.; Macpherson, J. V. New strategies for probing crystal dissolution kinetics at the microscopic level, *Chem. Soc. Revs.* 1995, **24**, 109–119.

[4] Kim, Y.; Matsumoto, M.; Machida, K. Specific surface energies and dissolution behavior of aspirin crystal, *Chem. Pharm. Bull* 1985, **33**, 4125–4131.

[5] ORTEP: C. K. Johnson, *ORTEPII: A Fortran Thermal Ellipsoids Plot Program for Crystal Structure Illustrations*, 1976, Report ORNL-5138; Oak Ridge National Laboratory, Oak Ridge.

[6] There is a rigorous theory for the extraction of rigid-body displacement tensors from ADPs (Schomaker, V.; Trueblood, K. N. *Acta Cryst.* 1968, **B24**, 63–76, as well as for the calculation of ADPs from lattice dynamics (Gramaccioli, C. M.; Filippini, G. Lattice-dynamical evaluation of temperature factors in non-rigid molecular crystals: a first application to aromatic hydrocarbons, *Acta Cryst.* 1983, **A39**, 784–791).

[7] For the lattice dynamical evaluation of external contributions to crystal heat capacities, see: Filippini, G.; Gramaccioli, C. M.; Simonetta, M.; Suffritti, G. B. Thermodynamic functions for crystals of rigid hydrocarbon molecules: a derivation via the Born–von Karman procedure, *Chem. Phys.* 1975, **8**, 136–146. Harmonic dynamics works for crystals thanks to reduced molecular mobility. By contrast, liquids exhibit so-called "instantaneous modes" (Stratt, R. M. The instantaneous normal modes of liquids, *Acc. Chem. Res.* 1995, **28**, 201–207): the eigenvalues of an instantaneous hessian for a liquid has a spectrum of imaginary frequencies, since any instantaneous frame of liquid structure is far from mechanical equilibrium because of collisions. Therefore, it is impossible to estimate heat capacities of liquids in this way, and dynamic simulation is necessary.

[8] There is no easy way of simulating the heat transport properties of an organic material in ordinary MD or MC. In principle, transport coefficients can be obtained as ensemble averages of dot products between flux vectors at reference time and at generic time, in the same form as equation 9.14. See: Evans, D. J.; Morriss, G. P. *Statistical Mechanics of Non-Equilibrium Liquids*, 1990, Academic Press, London.

[9] The volume of the spheroidal cluster, V°, is estimated by calculating the volume occupation of a single molecule, as the ratio of the intrinsic molecular volume to the space occupation factor (Sections 1.3 and 1.4), and multiplying by the number of molecules in the cluster. The expected radius R° is estimated by taking the radius of a sphere of volume V°, and the surface, A°, is then $4\pi(R^\circ)^2$.

[10] The direction of the perpendicular to the (hkl) plane is calculated by taking the direction of the maximum moment of inertia of four points belonging to that plane, e.g. points $P_1 = (1/h, 0, 0), P_2 = (0, 1/k, 0), P_3 = (0, 0, 1/l)$, and the fourth point as a vector sum of the $P_1 - P_2$ vector to point P3 (appropriate precautions are taken when some of the Miller indices are zero). The radius of the surface thus created is estimated as one half of the average largest distance between molecular centers of mass in the x- or y-directions, incremented by an average molecular cross section radius. The Fortran computer program is available in the Supplementary material (program *attach.for*). As a fringe benefit, the program produces a graphic file that shows the two hemispheres separated by a distance of 10 Å (used for Figs 11.7 and 11.11–11.12), which allows a visualization of the structure of the rough (hkl) surfaces.

[11] Busing, W. R.; Matsui, M. The application of external forces to computational models of crystals, *Acta Cryst.* 1984, **A40**, 532–538.

[12] (a) Stevens, L. L.; Eckhardt, C. J. The elastic constants and related properties of β-HMX determined by Brillouin scattering, *J. Chem. Phys.* 2005, **122**, 174-701. (b) Kim, Y.; Machida, K.; Taga, T.; Osaki, K. Structure redetermination and packing analysis of aspirin crystal, *Chem. Pharm. Bull.* 1985, **33**, 2641-2647.

[13] For a clear example of color generation by mixing molecular complexes in solution, see: DelSesto, R. E.; Botoshansky,M.; Kaftory, M.; Arif, A. M.; Miller, J. S. Charge transfer complexes of tricyanotriazine with tetrathiafulvalene and tetramethylphenylenediamine, *CrystEngComm* 2002, **4**, 117-120.

[14] For nonlinear optics see for example Zyss, J. (Ed.), *Molecular Nonlinear Optics: Materials, Physics and Devices*, 1994, Academic Press, Boston.

[15] Kahr, B.; McBride, J. M. Optically anomalous crystals, *Angew. Chem. Int. d. Engl.* 1992, **31**, 1-26.

[16] Curtin, D. Y.; Paul, I. C. Chemical consequences of the polar axis in organic solid state chemistry, *Chem. Revs.* 1981, **81**, 525-541.

[17] Schmidt, G. M. J. Photodimerization in the solid state, Pure Appl. Chem. 1971, 27, 647-678; Cohen, M. D.; Schmidt, G. M. J. Topochemistry. Part I. A survey, *J. Chem. Soc* 1964, 1966-2000.

[18] Simonetta, M.; Gavezzotti, A. Crystal chemistry in organic solids, *Chem. Revs.* 1982, **82**, 1-13.

[19] Weissbuch, I.; Lahav, M.; Leiserowitz, L. Toward stereochemical control, monitoring and understanding of crystal nucleation, *Cryst. Growth Des.* 2003, **3**, 125-150.

[20] (a) Miller, J. S. Organometallic- and organic-based magnets: new chemistry and new materials for the new millennium, *Inorg. Chem.* 2000, **39**, 4392-4408; (b) Amabilino, D. B.; Veciana, J. Nitroxide-based organic magnets, in: *Magnetism: Molecules to Materials II*, edited by Miller, J. S., Drillon, M. 2001, Wiley-VCH, Weinheim.

[21] For organic compounds, "intermolecular magnetic interactions are as yet a relatively untamed beast" (ref. [20b], page 50).

[22] Mention of optical, electric and magnetic properties of organic materials has almost vanished from the proceedings of the XVII International Conference on the Chemistry of the Organic Solid State (ICCOSS XVII) held at University of California, Los Angeles, in 2005 (proceedings to be published in a special issue of *Molecular Crystals and Liquid Crystals*). The buzzwords there are now nanostructured materials, and the polymorphism of pharmaceuticals.

12

Intermolecular bonding

Matter is a distribution of charge in real space-of pointlike nuclei embedded in the diffuse density of electronic charge, $\rho(\mathbf{r})$, defined as the expectation value of the density operator... A molecular graph, the linked network of bond paths, defines a system's molecular structure.

Bader, R. W.; Fang, D.-C. J. Chem. Theor. Comp. 2005, 1, 403.

I likewise felt several slender ligatures across my body, from my armpits to my thighs.

Swift, Gulliver's Travels, Part I, Chapter 1.

12.1 The decline of the intermolecular atom–atom bond

12.1.1 *The Feynman–Ehrenfest chemical bond*

Almost every chemist will accept a definition of an organic molecule as an aggregate of atomic nuclei and electrons, all of them contained in a recognizable region of space and distributed in that space according to certain equilibrium conditions. Besides, the molecule has a well founded basis on the ground of stoichiometry, the law of combination in integers being seldom if ever violated when H, C, N, O, F, Cl, S, P atoms join into an organic compound. Chapter 1 was dedicated to the structural and, so to speak, social aspects of the molecule concept. Yet chemists have an irresistible tendency to explain everything in terms of atoms and of bonds between pairs of them, even when atoms merge into the diffuse molecular electron density. One must then retrieve atomic imprints out of the molecular object, and somehow define the electronic effect that defines and describes the chemical bond. Neither task is trivial.

What is a chemical bond? This question is likely to haunt a chemist's life from beginning to end, as more and more chemical experience often leads to more and more uncertainty about such a fundamental concept. In qualitative models, the choice is a matter of convenience, rather than of first principles, so we may conveniently say that two atoms are joined by a chemical bond when: (1) the pristine electron densities of both atoms are significantly perturbed by the approach, (2) as a consequence, the energy of the system is lowered and a restraining force acts against separation of the nuclei, (3) hence, the nuclei are constrained at a distance which is significantly less than the sum of the radii of the bare atoms. As corollaries, one may add that (a) what other atoms in the molecule are doing is less important (a true bond is a 1 ... 2

effect, without 1 ... 3 and 1 ... 4 dependence), (b) the bond line is identified as the line joining the nuclei, and thermal vibration occurs mostly along that line, (c) when a bond is broken, chemical and structural consequences are significant, with a large destabilization, and other bonds must immediately be formed, so to speak, to repair the damage. These simple ideas account very well for the connectivity of nearly all organic molecules, with dashes being unequivocally drawn around tetravalent carbon, trivalent nitrogen, divalent oxygen and sulfur, and monovalent hydrogen and halogens. Even in this apparently strong and clear scheme there is space for uncertainties or exceptions: tautomeries, dative bonds, etc. And the model runs into troubles with conjugated π-systems, where one is forced to resort to auxiliary concepts like delocalization and resonance.

The above description of a chemical bond fits rather well with the idea of a bond electron density accumulating between nuclei, introducing the concept of a relationship between the topology of the molecular electron density and the formation of chemical bonds. The Quantum Theory of Atoms in Molecules (QTAIM), described briefly in Section 3.5.2, builds a rigorous mathematical model along this line, defining atoms within molecules on the basis of topological properties of the electron density $\rho(\mathbf{r})$. The definition is based on first principles and is not a matter of debate, although obviously the accuracy and consistency of the identification of atomic basins depends to some extent on the accuracy and consistency of the calculation of $\rho(\mathbf{r})$, or, if the electron density comes from an X-ray diffraction experiment (Section 5.6), on the accuracy of that experiment. Forces and energies are also partitioned over AIM atomic basins. Feynman forces (Section 1.1) act at nuclei, and are zero at equilibrium internuclear distance. Ehrenfest forces are of an analogous, essentially electrostatic nature, but act on entire atoms, including the electron density within the atomic basin [1]. The presence of a bond path with a bond critical point (BCP) between two nuclei implies that the atoms are bonded to one another, that the Ehrenfest force between the two atoms is attractive, and that its virial is stabilizing. The only quantum chemically legitimate description of forces and energies acting between atoms is in terms of Feynman and Ehrenfest forces, together with the virial theorem. "A molecular graph, the linked network of bond paths, defines a system's molecular structure", and the corresponding virial graph delineates "the lowering in energy associated with the formation of the structure defined by the molecular graph" [1]. The value of the density at the bond critical point to a certain extent characterizes the type of bonding. For example, a BCP density of around 0.3 au is indicative of shared (covalent) interaction.

What happens to the system's energy when a bond is formed? "The energy of formation of a molecule from separated atoms can be expressed in terms of contributions from the changes in the potential energy within each atomic basin plus a contribution from the virial of the Ehrenfest force acting on the interatomic surface between each pair of bonded atoms..."[1] Atomic basins in a molecule are thus "glued" together by Ehrenfest forces, sometimes at the expense of an elevation of the potential energy within atomic basins, so that the global bonding effect has diffuse contributions and many components, not always where one expects them on the basis of simple reasoning. The atomic basins of QTAIM have somewhat strange shapes [2], far from the familiar atomic spheres and caps of the traditional chemist; the definition and calculation of the accompanying energies and energy changes is not immediate; and in some cases, AIM concepts of bonding clash with traditional chemical thinking. For example, the AIM analysis of the biphenyl molecule shows that there is a bond path (and hence a chemical bond) between the ortho-hydrogen atoms in the two rings [3], thus overthrowing the glorious explanation of the non-planarity of the molecule in the gas phase in terms of H $\cdots$ H steric repulsion; there is a bond path between chloride ions in the NaCl crystal [4]. Since the change in total energy results from changes in Ehrenfest forces and virials all over the molecule, the appearance of a new bond path in a given structure does not in itself guarantee that the process is exothermic: for example, there are chemical bonds between the C atoms of adamantane and a He atom trapped within the adamantane cage, in spite of the huge exothermicity of the complex formation [1]. At the same time, concepts such as Coulombic attraction and repulsion between ions or polar moieties, of steric repulsion between closed-shell atoms, or even of bond energy, have no meaning in QTAIM terms. These traditional concepts stem from entirely different, arbitrary one-to-one couplings of atomic electron densities and pairing of bonding electrons, and as such are sometimes useful, sometimes misleading, simplifications.

12.1.2 *More familiar models: distance–energy analyses*

Of extreme elegance and rigor, the Atoms in Molecules theory explains in clear quantitative terms a wide range of facts about chemical bonding within molecules, and hardly needs further praise here. In some instances, however, like the above mentioned biphenyl and NaCl cases, it remains for many practicing chemists a sour mouthful to swallow.

What of the bonds between molecules, rather than between atoms in a molecule? [5] In principle, the AIM treatment makes no distinction between intra- and intermolecular facts, but in practice the predicament of the intermolecular chemist is much more awkward than that of the synthetic organic chemist. Energy-wise, intermolecular interactions are worth an order of magnitude less than intramolecular ones, and for intermolecular contacts, BCP electron densities are 0.05 au, or even less [6]. So one is faced here with the question of how robust these bond paths can be against the accuracy and consistency of the electron density determinations. Do these features survive changes in basis set, in level of the quantum chemical treatment, or even in 2θ range or in the number of spherical harmonics used to model the experimental X-ray density? And also: a saddle point in the electron density, but with insignificant weight, is a bond of zero strength – an oxymoron; and one has to be critical, especially when trying to sort out the bond paths and BCPs that are the source of intermolecular stabilization in crystals. Besides, given the diffuseness of the outer part of the electron density, the part involved in intermolecular cohesion, it is likely that in a crystal there may be bond paths and bond critical points between all vicinal atoms in different molecules, and that this abundance of atom–atom bonding indices may introduce more confusion than clarification. Perhaps, one should investigate intermolecular bonding by considering the whole network of bond paths between different molecules in some cumulative way [7], but a clear-cut procedure for doing this is not presently in sight. The use of AIM theory in intermolecular matters foresees a future of complex calculations, and even perhaps of arguing over the number of gaussians basis functions. Not a reassuring perspective.

An alternative conceptual framework, much less alarming and much more attractive to the average chemist, is the common sense link between geometry and energy: briefly, the typical structure–property relationship according to which atoms are roughly where their nuclei are, bonds are reckoned between nuclear positions, and a shorter bond is a stronger bond. Indeed, bond distances and angles are the mirror in which molecular energies take a visible shape, but relationships between geometry and energy are not always straightforward, and drawing proper conclusions on chemical stability just from structural data requires a good amount of intuition coupled with a good amount of skepticism – rather a spoonful than a pinch of salt. Nuclear positions change as a result of equilibrium conditions over the electronic energy of the whole molecule, and with large and complex organic molecules, geometry variations are distributed over the whole molecular skeleton, and are small and subtle, often merging into the noise of a few times the standard deviations of bond lengths and angles. If the matter with intramolecular energy-geometry correlation is uncertain, it becomes hopeless with the entanglement of Gulliver-like "slender ligatures" found among organic molecules in crystals: that some of them are shorter than average may depend on extrinsic factors of all sorts, without any implication on the small and evanescent bond character between any two atoms. If less alarming, the geometrical bond distance–bond energy model of intermolecular cohesion, with its reliance on fluctuations around average atomic intermolecular radii, is not consoling either, and is certainly not the promise for a future theory of solid or liquid constitution. Intermolecular cohesion is mostly a diffuse and co-operative matter. As proposed at the very beginning of this book, the term "intermolecular bonding" is much more apt than the term "intermolecular bond".

To further analyze the assumption that a chemical bond is formed through, and is defined by, the occurrence of a short separation between two nuclear positions, we consider now the distribution of atom–atom distances in organic crystals, derived from a survey of the Cambridge Structural Database. To obtain distance distribution functions (DDF), an appropriate subset of the CSD is selected, and all atom–atom distances between atomic species i and j in all crystal structures in that sample are calculated, up to a certain limit R_{max}. Let $N_k(R)^{ij}$ be the number of atom–atom distances within the k-th distance bin defined by a separation R_k, a radial increment dR, and a volume $dV = 4\pi/3[(R_k + dR)^3 - (R_k)^3]$. An inspection of the unnormalized distribution function reveals the exclusion radius $(R°)^{ij}$, or the separation below which no contacts are observed. The DDF normalization factor, F_N, can then be estimated as the total number of contacts divided by the volume of the explored space, that is, the spherical shell between the exclusion radius and the maximum allowed distance:

$$F^{ij}_N = \sum_k N_k(R)^{ij}/[4\pi/3((R_{max})^3 - (R^{oij})^3)] \quad (12.1a)$$

$$DDF = g_k(R)^{ij} = (1/F^{ij}_N)N_k(R)^{ij}/dV \quad (12.1b)$$

F_N represents the condition of uniform distribution of the observed contacts over the available contact space.

Equation 12.1b is very similar to a true radial distribution function (RDF, Section 9.4.1), but the fundamental physicochemical meaning is different: equation 9.13 refers to a distribution of many identical molecules within one molecular ensemble, like a portion of liquid, while equation 12.1b refers to sampling different molecules over many different systems, the crystal structures. The normalization of the DDF and of the RDF reflect two intrinsically different conditions, and the DDF is sensitive to the chemical nature of the selected ensemble of crystal structures. For example, the carbon–hydrogen distance distribution function would be different over a sample of saturated and unsaturated hydrocarbons, and even within aromatic hydrocarbons, the presence of electron-withdrawing or electron-donating ring substituents would change the electronic features of the carbon atoms. In other words, it is the very concept of "atom" which is at stake: variations in the molecular environment have a deep influence on the weak intermolecular bonding capability of the atomic basins, so that the distribution functions refer to "average" atomic species.

In analogy with the properties of true radial distribution functions, the appearance of a peak in the DDF corresponds to an increased frequency of contacts at the peak separation, relative to an uniform distribution, and hence one can assume as a first approximation that the peak reveals a preferential approach of the implied atomic species; to avoid semantic biases, rather than "bonding radius" we prefer to call R_P, the separation corresponding to the first peak in the DDF, the "preference radius". After the first peak, all DDFs drop to a minimum, which may be called the "recoil" distance distribution, D_R. The relevance of the atom–atom interaction is measured by the peak height, D_P, and by the "isolation ratio", $I_R = (D_P - D_R)/D_P$. As the peak merges more and more into a smooth distribution, I_R decreases, and so does the significance of the interaction. For $I_R = 1$ the peak is completely isolated from the rest of the distribution and represents a singled-out effect. All DDFs converge asymptotically to unit value at large distance, confirming that the sample is wide enough for proper statistics. Table 12.1 collects the proximity coefficients obtained from an analysis of the organic crystal structures of subset Z(1) (Section 8.3), while Figs 12.1–12.5 show some exemplary DDF plots.

Some DDFs (CC, CH, HH, CN, OO, CF) show practically no peak and rise steadily from zero to unity. Others show a distinguishable peak with increasing isolation ratio, in the order NN, CO, CS, CCl; significant peaks and larger isolation ratios are found for FF, SS, and especially the ClCl pair, whose DDF displays an almost completely developed, isolated peak. For the hydrogen bond contacts in acids, amides and alcohols, DDF peaks appear at very short distance between the H atom and the

THE DECLINE OF THE INTERMOLECULAR ATOM–ATOM BOND

Table 12.1 Proximity coefficients for intermolecular contacts found in the Cambridge Structural Database, subset Z(1). (Å) units

Contact	Exclusion radius, R°	Preference radius, R_P	Isolation ratio	Peak height
H⋯H	2.05	2.9	ripple	0.8
C⋯C	3.35	4.1	ripple	0.9
N⋯N	3.05	3.9	0.18	0.9
O⋯O	3.0	4.75	ripple	0.95
F⋯F	2.7	3.2	0.5	1.4
S⋯S	3.3	3.9	0.5	1.3
Cl⋯Cl	3.25	3.8	0.7	2.1
C⋯N	3.2	3.8	ripple	0.8
C⋯O	3.05	3.7	0.2	0.8
C⋯F	3.0	3.6	ripple	1.0
C⋯S	3.4	4.0	0.4	1.25
C⋯Cl	3.35	3.95	0.4	1.45

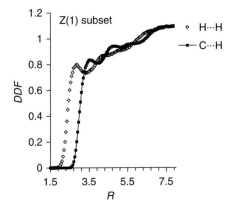

Fig. 12.1. DDF curves for H⋯H and C⋯H contacts in the Z(1) subset of the Cambridge Database, described in Section 8.3. Distances in Å.

acceptor; these peaks are well resolved, meaning that whenever the two involved atomic species have a chance of getting into contact, they will do so without hesitation – or rather, in less anthropomorphic terms, the bonding interaction is strong and driving and no spread of the bond distance toward the longer side ever occurs. The CH ⋯ O DDF shows a peak only when the oxygen is a carbonyl or nitro oxygen, while the CH ⋯ N DDF shows no peak in general, no peak with secondary amide nitrogen, and only a minor one with primary amide nitrogen. The C–H ⋯ F, C–H ⋯ Cl and C–H ⋯ S DDFs show peaks of increasing height and isolation parameter.

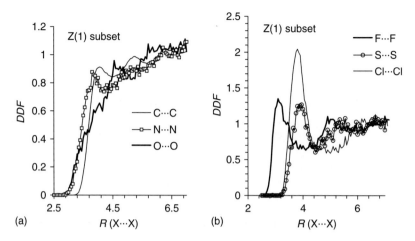

Fig. 12.2. DDF curves for contacts between non-hydrogen atoms. Distances in Å.

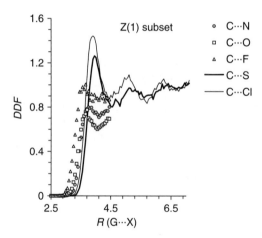

Fig. 12.3. Overall DDF curves for cross-contacts between non-hydrogen atoms. Curves over selected subsets are slightly different: for example, the C · · · O═(C) DDF has a peak of 1.3 at 3.6 Å. Distances in Å.

What is one to make of these distributions in terms of intermolecular bonds? Except for the traditional OH · · · O or NH · · · O hydrogen bond, with peak heights of 5–20 and an isolation ratio of unity, the intensity of the peaks shown in the contact distributions are less than 2 and the isolation ratios are between 0.4 and 0.7, showing that the drive to formation of a bond between these species is not compelling; therefore, the definition of a bond must rely on some arbitrary cutoff through the DDF, and hence is probably not advisable. If one is willing to propose an intermolecular bond for some

THE DECLINE OF THE INTERMOLECULAR ATOM–ATOM BOND 303

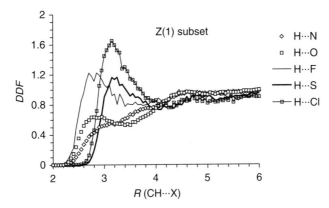

Fig. 12.4. DDF curves for contacts involving hydrogen atoms. Distances in Å.

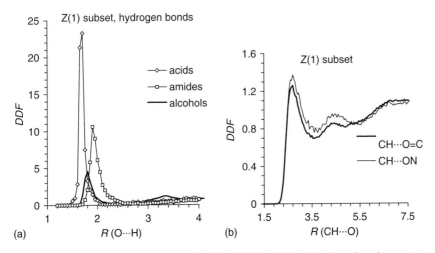

Fig. 12.5. DDF curves for (a) a hydrogen bond, and (b) the C–H··· O carbonyl or nitro contacts. Note the different vertical scale. Distances in Å.

of these distributions, for example the CH ··· O=C distribution, then one should also be prepared to accept bonds between a long list of atom pairs. In these terms it looks as if any atom is bound to nearly any other atom. Even the use of the definition of "weak bonds" may be misleading, giving a generalized assumption of a driving force while these short distances could well be the secondary result of other stronger attractions, or at best the outcome of mildly attractive behavior between diffuse molecular regions. In spite of a thriving literature on geometrically justified intermolecular bonds [8], in terms of practical use and of economy in the formulation of models, especially if the aim is a consistent and predictive theory of intermolecular bonding, there seems

Table 12.2 Proximity coefficients for intermolecular contacts involving hydrogen, found in the Cambridge Structural Database, subset Z(1)

Contact	Exclusion radius, R_0	Preference radius, R_P	Isolation ratio	Peak height
CH $\cdots$ C	2.65	3.65	ripple	0.83
CH $\cdots$ N, general	2.25	—	—	—
CH $\cdots$ NH$_2$ amide	2.3	3.15	0.3	1.1
CH $\cdots$ O, general	2.2	2.85	ripple	0.65
O carbonyl	2.15	2.7	0.45	1.3
O nitro	2.15	2.7	0.45	1.35
CH $\cdots$ F	2.25	2.8	0.35	1.2
CH $\cdots$ Cl	2.65	3.15	0.5	1.65
CH $\cdots$ S	2.7	3.2	0.3	1.2
OH $\cdots$ O acids	1.50	1.67	1.0	23
NH $\cdots$ O amides	1.75	1.92	1.0	11
OH $\cdots$ O alcohols	1.65	1.80	0.95	4.5

to be little advantage in advocating a separate bond interaction between these atomic species only on the basis of short separations.

12.2 Full density models: the SCDS–Pixel method

12.2.1 *Theory*

The philosophical transition from the atomic prejudice to a view of intermolecular interaction in terms of diffuse electron density has its proper computational counterpart in full quantum mechanical calculations, which, however, cannot at present provide complete intermolecular energies because of limitations in the treatment of electron correlation, a major ingredient of the intermolecular interaction recipe. In a different perspective, the classical atom–atom force-field approach is widely applicable but entirely parametric and of scarce adherence to physical principles. The need is felt for an extension to represent in a more realistic manner the effects of diffuse electron clouds. This is done in the so-called semi-classical density sums (SCDS) or briefly, Pixel approach [9], which will now be described. The Pixel method is based on numerical integrations over molecular electron densities, and allows a separation of the total intermolecular cohesion energy into coulombic, polarization, dispersion, and repulsion contributions.

In the Pixel formulation, the basic concept is the electron density unit, or e-pixel. Consider a molecule (molecule A) with nuclei of charge Z_j at points $(j) = [x_j\ y_j\ z_j]$. Let ρ_k be the electron density in an elementary volume V_k centered at point $(k) = [x_k\ y_k\ z_k]$. ρ_k is readily obtained through an MO wavefunction (see Chapter 3, Box 3.1); the whole Pixel theory has been developed and calibrated using valence only charge densities derived from MP2/6-31G** wavefunctions, but the use of a different level of theory, provided that a reasonable enough basis set is used, has minor consequences on the quality of the results. Each e-pixel is assigned a charge $q_k = \rho_k V_k$. With typical steps of 0.08–0.1 Å, the number of pixels in the original grid is of the order of 10^6, too large for practical use; cubes of $n \times n \times n$ original pixels are then condensed into super-pixels, n being called the condensation level. The molecule is thus represented by a set of nuclear charges and a set of e-pixel charges, whose number is around 10,000 for benzene to 25,000 for anthracene. Each pixel is assigned to a particular atom in the molecule, by the following procedure. Let p be the number of atoms for which the nucleus–pixel distance is less than the atomic radius. If $p = 1$, the charge pixel is within one atomic sphere only, and it is assigned to that atom. If $p > 1$, the pixel is assigned to the atom from which the distance is the smallest fraction of the atomic radius. If $p = 0$, the pixel is assigned to the atom whose atomic surface is nearest. This topographic assignment proves adequate also because the results are hardly sensitive to detail of the e-pixel distribution among atomic basins.

12.2.2 Coulombic energy

Consider now a second molecule, B (see Fig. 12.6) with nuclei of charge Z_m at points $(m) = [x_m\ y_m\ z_m]$, and whose e-pixels of charge $q_i = \rho_i V_i$ are at positions $(i) = [x_i\ y_i\ z_i]$. For the calculation of the intermolecular Coulombic energy, let R_{ln} be the distance between any two centers of pixels or nuclear positions l and n; the electrostatic potential generated by molecule A at point (i) of the charge density of molecule B is:

$$\Phi_i = 1/(4\pi\varepsilon_0)[\sum_k q_k/R_{ik} + \sum_j Z_j/R_{ij}] \qquad (12.2)$$

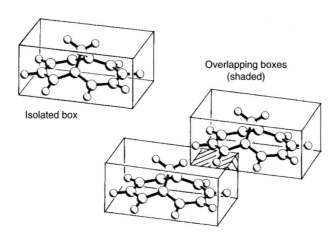

Fig. 12.6. Top: an isolated molecule surrounded by its box of electron and charge density. Below: two approaching molecules and the overlap region.

while the potential generated by molecule A at nucleus m of molecule B is

$$\Phi_m = 1/(4\pi\varepsilon_0)[\sum_k q_k/R_{km} + \sum_j Z_j/R_{jm}] \quad (12.3)$$

The total electrostatic potential energy between the two molecules is the sum of the electrostatic energies at points (i), $E_i = q_i \Phi_i$, and at points (m), $E_m = Z_m \Phi_m$:

$$E_{EL,AB} = \sum_i E_i + \sum_m E_m \quad (12.4)$$

When two molecules approach one another their electron densities overlap to some extent. In principle, this overlap should be treated as a partially bonding effect, but Pixel preserves the identity of two separate electron clouds, so that when pixels overlap, singularities in the R^{-1} dependence may result for very short pixel–pixel distances. As a practical countermeasure, all pixel–pixel distances below half the stepsize of the pixel mesh are reset at half the stepsize (the "collision avoidance" procedure).

12.2.3 Polarization energy

Let ε_i be the total electric field exerted by surrounding molecules at pixel i, α_i the polarizability at pixel i, and μ_i the dipole induced at pixel i by that field. The linear polarization energy is:

$$E_{POL,i} = -1/2\mu_i \varepsilon_i = -1/2\alpha_i \varepsilon_i^2 \quad (12.5)$$

There are obvious difficulties in the definition of the distributed polarizabilities α_i [10]. The expedient adopted in the Pixel formulation uses atomic polarizabilities taken from standard repertories [11] (Table 12.3), and takes $\alpha_i = (q_i/Z_{atom}) \alpha_{atom}$, where

Table 12.3 Physical properties of atoms needed in the Pixel formulation (see text for more explanations)

Atom	Atomic radius, Å	Atomic polarizability, Å³	Electronegativity	Ionization potential I°, au	β
H	1.10	0.39	2.1	0.500	0.4
C aliphatic	1.77	1.05	2.5	0.414	0.8
C aromatic	1.77	1.35	2.5	0.414	0.9
C aromatic bridge	1.77	1.90	2.5	0.414	1.05
N	1.64	0.95	3.0	0.534	0.5
O	1.58	0.75	3.5	0.500	0.4
F	1.46	0.50	4.0	0.640	0.0
Cl	1.76	2.30	3.0	0.477	0.35
S	1.81	3.00	2.5	0.381	0.4

Z_{atom} and α_{atom} are the atomic charge and polarizability of the atom to whose basin the pixel belongs. This procedure is at least partially justified by the fact that the sum of distributed α_i s is equal to the total volume polarizability of the molecule.

In molecular crystals, electric fields calculated by the Pixel procedure are mostly of the order of 10^{10} V m^{-1}, with a small number in the 10^{10}–10^{13} V m^{-1} range, plus a very small number of even higher ones. These high-field contributions are physically unrealistic, resulting from fortuitous short distances in the overlapping density meshes, and cause ill-conditioning in the polarization energy sums, just as happens for coulombic energies. Pixel–pixel distances are subjected to the "collision avoidance" scheme (see above); then, the damped polarization energy at pixel i is

$$E_{POL,i} = -1/2\, \alpha_i\, [\varepsilon_i\, d_i]^2 \quad \text{for } \varepsilon < \varepsilon_{max}, d_i = \exp-(\varepsilon_i/(\varepsilon_{max} - \varepsilon_i)) \quad (12.6)$$

and $E_{POL,i} = 0$ for $\varepsilon > \varepsilon_{max}$. ε_{max}, the limiting field, is an adjustable empirical parameter in the formulation.

A $\cdots$ B polarization energies are different from B $\cdots$ A energies; they are many-body energies because the total electric field depends on the simultaneous action of all surrounding polarizers (recall equation 4.33), and are not pairwise additive over atomic contributions. The total polarization energy at a molecule is the sum of the polarization energies at each of its electron density pixels, $E_{POL,TOT} = \sum E_{POL,i}$, while the total polarization energy in an ensemble of molecules is the sum of the polarization energies at each of the molecules in the ensemble.

12.2.4 *Dispersion energy*

Dispersion energies between two molecules A and B are calculated as a sum of pixel–pixel terms in a London-type expression, equation 4.24, involving the above defined distributed polarizabilities and an "oscillator strength", E_{OS}:

$$E_{DISP,AB} = (-3/4) \sum_{i,A} \sum_{j,B} E_{OS} f(R)\, \alpha_i \alpha_j / [(4\pi\varepsilon°)^2\, (R_{ij})^6] \quad (12.7)$$

Just as for coulombic and polarization terms, short distances between pixels of overlapping densities may cause singularities, so a damping function is used, $f(R) = \exp[-(D/R_{ij} - 1)^2]$ (for $R_{ij} < D$), where D is an adjustable empirical parameter.

E_{OS} was originally approximated by the molecular ionization potential, a reasonable assumption for small molecules, but a questionable one for complex polyatomic molecules, for which E_{OS} can be taken in a first approximation as the energy of the highest occupied molecular orbital, since the interacting electrons are peripheral ones and hence roughly at the HOMO energy level. This assumption is called henceforth the "constant ionization" form of the theory. A finer approach is obtained by considering each pixel as a separate oscillator, with a formal ionization potential I_i, which in turn is a function of the ionization potential, $I°$, pertaining to the atom to whose basin the pixel belongs, and of the distance between the pixel and the atomic nucleus, R_i; the pixel–pixel oscillator strength for use in equation 12.7 is then:

$$E_{OS} = (I_i I_j)^{1/2} \quad (12.8a)$$

$$I_i = I° \exp(-\beta R_i) \quad (12.8b)$$

The parameter β is a function of the atom type (Table 12.3). This "variable-ionization" form of the theory amounts, in fact, to taking different dispersion energy coefficients according to the different kinds of interacting atomic basins.

12.2.5 Repulsion energy

The repulsion energy, E_{rep}, is modeled as proportional to the intermolecular overlap. The total overlap integral between the charge densities of two molecules A and B is calculated by numerical integration over the original uncondensed densities:

$$S_{AB} = \sum_{i,A} \sum_{j,B} [\rho_i(A)\rho_j(B)]V \quad (12.9)$$

and is then subdivided into contributions from pairs of atomic species m and n, S_{mn}, using the assignment of pixels to atomic basins. For each pair of atomic species the repulsion energy is evaluated as

$$E_{REP,mn} = (K_1 - K_2 \Delta \chi_{mn}) S_{mn} \quad (12.10)$$

where $\Delta \chi_{mn}$ is the corresponding difference in Pauling electronegativity (Table 12.3). K_1 and and K_2 are positive disposable parameters. The rationale behind this approach is explained by the negative sign of K_2, because when atoms of different electronegativity meet, presumably a larger reorganization of the electron density occurs, not accounted for by the Pixel formulation which uses undeformed charge densities; the missing stabilization is then simulated through a diminution of the repulsion. The total repulsion energy is then the sum over all m–n pairs.

12.2.6 Total energies and parameters

The total intermolecular Pixel interaction energy is:

$$E_{TOT} = E_{COUL} + E_{POL} + E_{DISP} + E_{REP} \quad (12.11)$$

There are four fully disposable empirical parameters in the Pixel formulation: ε_{max}, set at 150 10^{10} V m^{-1}, D, set at 3.50 Å in the constant ionization form and 3.0 Å in the variable ionization form, K_1 and K_2, set at 4,800 and 1,200 respectively (for energies in kJ mol^{-1} with electron densities in electrons Å^{-3}; a full list of the conversion factors that apply in Pixel calculations is given in ref. [40] of Chapter 4). The β screening parameter is also discretional, and its optimal average value is around 0.4; it decreases with increasing contraction of each atomic electron cloud. The proposed values for

FULL DENSITY MODELS: THE SCDS–PIXEL METHOD

these parameters were arrived at by manual adjustment, considering: (1) the agreement between calculated lattice energies and experimental heats of sublimation for organic crystals, (2) total interaction energies between selected dimers, in comparison with the results of high quality ab initio calculations, and (3) qualitative agreement between Pixel partitioned energies and Intermolecular Perturbation Theory (IMPT) partitioned energies. While having a set of transferable parameters is certainly desirable, with very minor parameter adjustments the Pixel theory can be easily fitted to any desired thermochemical or structural property of the particular system under investigation. The gain in accuracy may compensate for the loss of generality.

The numerical data in Table 12.3 are to be considered as semi- or non-adjustable parameters in the formulation, because they are taken with only minor adjustment from different theories of atomic structure and bonding. Of course, all the techniques employed for the pixel condensation, for the collision avoidance and damping procedures, and for the numerical integrations, add to the empirical character of the formulation. Only Coulombic energies are parameter-free and are essentially as accurate as the wavefunction is [12].

12.2.7 *The generation of crystal coordinates*

For the calculation of the interaction energies within a cluster of molecules with a specified geometry, the nuclear positions and the whole array of e-pixels are repeated in space by a rigid rotation-translation procedure (see Fig. 12.6). Let x_{LG} be the coordinates of atomic nuclei and e-pixels in the standard reference frame (for example as defined in the electron density calculation by GAUSSIAN), and x_{LO} the same for any of the surrounding molecules. Then the following transformation applies:

$$x_{LO} = M_1 \, x_{LG} + t_1 \qquad (12.12)$$

where M_1 is a rotation matrix and t_1 is a translation vector. When building a crystal structure, the procedure for taking crystal symmetry into account is a bit more complicated. Let x_{FC}° be the crystal fractional coordinates of the reference molecule (in the unit cell reference system), and x_{OC}° the corresponding orthogonalized coordinates:

$$x_{OC}^\circ = P \, x_{FC}^\circ = M_2 \, x_{LO} + t_2 \qquad (12.13)$$

where P is an orthogonalization matrix. The M_2, t_2 pair are a rotation and a translation that bring the isolated molecule into the crystal structure; they can be easily calculated by separate crystallographic programs, for example by finding the transformation to inertial coordinates in the crystal structure.

Let M_s, t_s now be a matrix–vector pair representing a given symmetry operation within the crystal space group (Section 5.2). Then:

$$x_{FC}^s = M_s x_{FC}^\circ + t_s; \quad x_{OC}^s = P \, x_{FC}^s \qquad (12.14)$$

The expression for the orthogonal coordinates of the *s*-th molecule in the molecular cluster that represents the crystal, x_{OC}^s, in terms of the coordinates in the standard

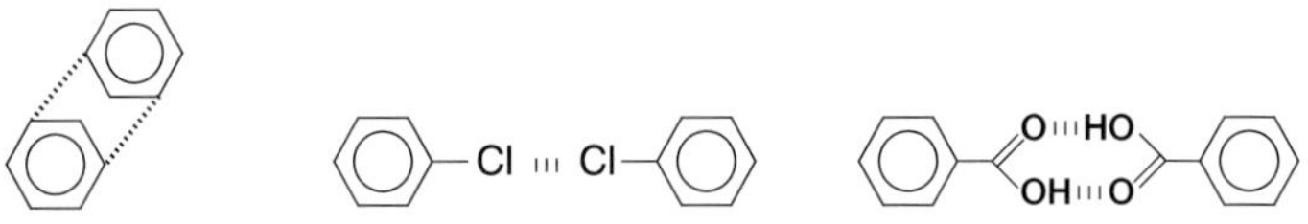

Scheme 12.1.

molecular reference system, x_{LG} is:

$$x_{OC}^s = \boldsymbol{\Omega}^s x_{LG} + \boldsymbol{\omega}^s \qquad (12.15)$$

$$\boldsymbol{\Omega}^s = \mathbf{P}\mathbf{M}_s\mathbf{P}^{-1}\mathbf{M}_2\mathbf{M}_1 \quad \boldsymbol{\omega}^s = (\mathbf{P}\mathbf{M}_s\mathbf{P}^{-1})(\mathbf{M}_2\mathbf{t}_1 + \mathbf{t}_2) + \mathbf{P}\mathbf{t}_s \qquad (12.16)$$

Many molecules surrounding the reference one in the crystal are generated by adding integer cell translations to the components of the $\mathbf{t}_s$ vectors, within a specified value of the distance between centers of mass. In this way, any molecular object specified by x_{LG} coordinates, with location and orientation in the unit cell specified by vector $\mathbf{t}_2$ and matrix $\mathbf{M}_2$, respectively, can be packed into a crystal specified by the cell dimensions appearing in matrix $\mathbf{P}$ and by space group specified by $\mathbf{M}_s, \mathbf{t}_s$ pairs.

12.2.8 Pixel calculations: General features

A calculation of the interaction energies for some prototypical dimers provides a sensitive test for a first impression of Pixel's performance. The parallel offset benzene dimer, the chlorobenzene dimer with Cl $\cdots$ Cl encounter, and the hydrogen bond formation in benzoic acid (Scheme 12.1) will be discussed.

Figure 12.7(a) shows that in the benzene dimer the main stabilizing factor is dispersion energy arising from interaction between correlated π-electron clouds, while coulombic and polarization terms are negligible in the region of minimum overall energy. Note that the coulombic energy term is destabilizing at interplanar distances greater than 3.5 Å. The equilibrium distance and the binding energy are quite reasonable and in line with the results of the best available quantum chemical calculations. Interesting, in the perspective of crystal packing, is the fact that the stabilizing interaction is still significant even at a distance of 5 Å. All these results are in line with a physically plausible interpretation of the binding between parallel aromatic rings; the stabilization for the interaction of two perpendicular benzene rings in a C–H$\cdots\pi$ approach includes a more substantial coulombic contribution, also in accord with the actual physics of the interaction.

Figure 12.7(b) shows the effects of changing the condensation level. Increase of one unit in condensation reduces the number of pixels by one half, and computing times by 75%, since computing times scale as the square of the number of pixels. The repulsion energy is calculated over the uncondensed charge distribution, and does not depend on condensation; the dispersion energy is quite insensitive to changes in condensation level; the coulombic and polarization contributions do depend on condensation, but since, in this case, they are comparatively small, good results are obtained for $n = 4$

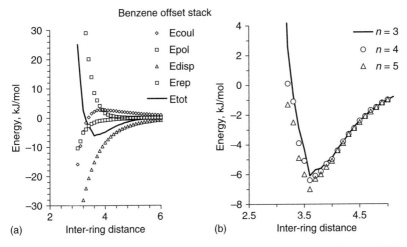

Fig. 12.7. (a) Pixel partitioned energies for a parallel offset benzene dimer, Scheme 12.1, as a function of the distance between molecular planes (condensation $n = 3$). (b) Total interaction energy as a function of condensation level, n. Computing times range from one minute to a few seconds for one point on the curve. Distances in Å.

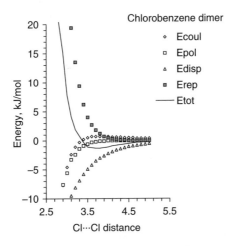

Fig. 12.8. Pixel partitioned energies for the approach of two coplanar chlorobenzene molecules, Scheme 12.1. Distances in Å.

or $n = 5$, while an acceptable result is also provided by a calculation with $n = 6$, taking about 30 seconds for the whole interaction curve.

Figure 12.8 shows the interaction picture for the chlorobenzene dimer. The dispersive interaction between the largely separated aromatic rings is negligible, and

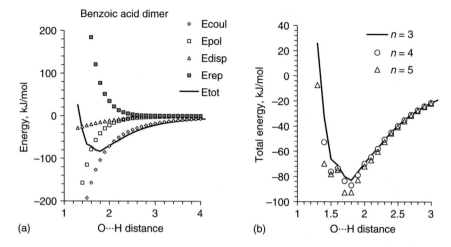

Fig. 12.9. (a) The benzoic acid dimer, Scheme 12.1; results for condensation $n = 3$. (b) The interaction curves as a function of condensation level: for very high condensation, the loss of resolution may cause severe numerical noise at short intermolecular separation. Distances in Å.

the small dispersive energy contribution between the diffuse chlorine electron densities is completely offset by mild repulsion: the Cl $\cdots$ Cl interaction is non-bonding, although not strongly repulsive. Interestingly, due to the presence of penetration terms, the coulombic interaction between the two head-on chlorine atoms is almost null at contact separations and becomes stabilizing at short separations, contrary to the results of localized point-charge models. Since in this calculation the coulombic-polarization terms are practically zero, the result is quite insensitive to the condensation level.

Figure 12.9(a) shows the case for hydrogen bonding. The dispersion term is negligible, and stabilization results almost exclusively from the coulombic-polarization terms, the former coinciding with the total energy for $R(O \cdots H) > 2$ Å. The interaction energy is in agreement with well established values for the hydrogen bond energy in carboxylic acids, while the equilibrium distance, about $R(O \cdots H) \sim 1.8$ Å, is perhaps slightly (say, 0.1 Å) longer than that found by experiment. Figure 12.9(b) shows the significant effect of changes in condensation level with this coulomb-dominated system: $n = 3$, the highest resolution, yields the most accurate energy curve, but also $n = 4$ is quite acceptable. Using higher condensation levels may cause significant discontinuities in the interaction energy curve due to numerical integration noise, for these heavily overlapping electron densities. In the minimum energy region and at short separation the coulombic energy is more and more overestimated with increasing n, but the rightmost part of the interaction curve is almost insensitive to the condensation level. Here, again, Pixel results are in line with the energetic and structural parameters of the interaction, as well as with the expected physical nature of the stabilization.

Table 12.4 Pixel lattice energies (kJ mol^{-1}) for various representative crystal structures. n is the condensation level. Constant-ionization dispersion energy model

Crystal	Number of pixels per molecule	Time, minutes	E (coul+pol)	E(disp)	E(rep)	E(total)
Naphthalene						
$n = 3$	27891	170	−26	−100	52	−73.6
$n = 4$	13122	33	−29	−100		−75.9
$n = 5$	7224	10	−32	−99		−78.7
$n = 6$	4373	4	−35	−101		−83.9
2,3-diazanaphthalene						
$n = 3$	25469	150	−52	−88	52	−88.4
$n = 4$	11991	31	−54	−88		−90.4
$n = 5$	6606	9	−57	−89		−94.8
$n = 6$	4056	4	−61	−89		−98.2
1,4-naphthoquinone						
$n = 3$	27970	142	−46	−89	55	−80.3
$n = 4$	13136	31	−49	−90		−83.5
Hexachlorobenzene						
$n = 4$	17313	39	−36	−138	91	−82.7
$n = 5$	9591	12	−41	−139		−90.1
1,4-benzene dicarboxylic acid						
$n = 3$	27266	—	−241	−102	213	−128.9
$n = 4$	12787	—	−246	−102		−134.6
$n = 5$	7084	—	−269	−102		−157.5
$n = 6$	4364	—	−276	−101		−163.6

The crystal structures of a few typical organic compounds will be considered as a further test of the performance of Pixel. The test molecules were chosen with different degrees of permanent molecular polarization: naphthalene, hexachlorobenzene, 2,3-diazanaphthalene, 2,5-naphthoquinone, and terephthalic acid. Table 12.4 collects the calculated lattice energies with variable condensation level. All naphthalene derivative crystals have about the same dispersion energy, due to the presence of the aromatic condensed core, and also the same repulsion energy. The dispersion term is larger in hexachlorobenzene, due to polarizable chlorine electrons, but also the repulsion term increases, due to increased overlap between more diffuse electron clouds. The coulombic energy of the hexachlorobenzene crystal is > 0 (destabilizing) when calculated with point charges located on atomic nuclei, and is, correctly, stabilizing in the Pixel calculation which includes the penetration contribution; this example illustrates how un-physical atomic point charges can be. The coulombic-polarization terms increase slightly on going from non-polar naphthalene to hexachlorobenzene, but are significantly larger in crystals of derivatives with permanent molecular polarization like the aza-derivative or the quinone. In the Pixel picture, the lattice energy of hydrogen-bonded terephthalic acid is dominated by the coulombic term, although dispersion is by no means negligible, again due to the presence of the aromatic ring. Note, however, that a strong coulombic stabilization does not produce an equivalent overall stabilization: on going from naphthalene to terephthalic acid, a tenfold increase in coulombic energy leads to a gain of only 20 kJ mol^{-1} in total energy, because every stabilization as molecules come closer together is paid at a high price in terms of steeply increasing repulsion. All considered, the Pixel picture of packing energies is consistent with expected physical effects of polarity and molecular constitution.

The effect of increasing condensation is, as before, significant only on the coulombic-polarization contributions, which increase by a few per cent on moving from n to $n+1$. Although the use of a low condensation level ($n = 3$) certainly leads to more accurate results, an excellent approximation with a considerable time saving is obtained by using $n = 4$, which is the regular choice for crystal calculations. Even for $n = 5$ the results are reliable, when coulombic terms are damped by the appropriate factor of 0.91–0.95. Condensation level $n = 6$ can be used for a very quick survey of the crystal potential: this gives in a few minutes an estimate of the relative importance of the various contributions to the total lattice energy, and provides an estimate of the true lattice energy, when calculated coulombic and polarization energies are multiplied by a factor of 0.9 for scarcely polar crystals or 0.8 for hydrogen-bonded crystals.

12.2.9 Pixel theory: Pros and cons

Pixel is generally applicable to charged and uncharged molecular species composed of the common atoms of organic chemistry, at the price of only a handful of truly disposable parameters. A first and essential strong point of the Pixel approach, in comparison with atom–atom ones, is that molecular electrical properties are distributed over an order of magnitude of 10^4 sites, rather than just a few nuclear positions. This large number of attractors paves the way to a more accurate, flexible, and directional description of intermolecular forces. Other favorable features are that, for coulombic energies, the scheme includes penetration effects (Section 4.3), which arise from overlapping molecular densities; these are an essential part of the interaction especially at short contact distances, and are poorly reproduced even by the most refined multipolar expansions. Besides, many-body electrostatic effects are to some extent included through the polarization term. No convergence problems appear in either coulombic or polarization energy sums, except for very peculiar cases (see Section 12.3.2). The optimization of the structure of molecular dimers and oligomers is feasible in quite reasonable computing times, and even the optimization of the lattice energy of organic crystals is within reach. In summary, it can be said that adherence to physical facts is certainly higher in Pixel than in atom–atom schemes, and that Pixel intermolecular energies are certainly more reliable than quantum chemical energies at the Hartree–Fock level. In addition, for a quite affordable computational effort, Pixel allows an estimate of the relative importance of coulombic, polarization, and dispersion contributions in intermolecular attraction; this information opens the way to a better understanding of the relationships between molecular constitution and intermolecular potentials, the key to prediction and control.

These positive aspects are counterbalanced by a number of less attractive features. The Pixel treatment of polarization leaves much to be desired, because only a linear polarization formula is used, and because the two interacting electron densities are static and no account is taken of their deformation upon polarization; and the approximation of undeformed densities with only a static polarization term becomes more and more questionable as more and more of a chemical bond is formed, as, for example, in hydrogen bonds. The London-type treatment of dispersion rests on a rather shaky physical foundation, with its empirical approach for the estimation of the oscillator strength. Repulsion calculated from overlap is also an appealing approximation, but its theoretical foundations are not entirely settled (recall Section 4.6). Pixel partitioned polarization, dispersion and repulsion energies are thus parametric, although they often compare favorably with partitioned energies derived from less empirical schemes like the Morokuma or Stone IMPT schemes (Section 4.7). This latter kind of agreement is desirable but not compulsory, because in fact each method defines its own partitioned energies, which have no absolute physical reality anyway.

The numerical integration procedures introduce other problems. In the calculation of the overlap integral, the value of the density at each volume element of the two overlapping densities is interpolated, but this causes only limited inaccuracies because of the small stepsize and the resulting large number of charge units. The electron density is poorly described close to atomic nuclei, and ideally a smaller stepsize should be used for the nuclear regions than for the outer fringes of the charge density. Moreover, when the electron densities of two molecules approach one another and start to overlap, some numerical instability may be generated especially in a colinear geometry, when entire pixel planes through the electron densities suddenly happen to fall at very short distances, causing a massive intervention of the collision avoidance procedure. Because of this computational noise, sometimes a smooth interaction energy curve shows fluctuations at certain molecule–molecule separations. Fortunately, these pathologies are nearly always easily spotted and sorted out, and are less frequent or are all absent in crystal calculations, where the electron densities are oriented along general directions in space rather than along coordinate axes.

12.3 Systematic application of the Pixel theory to intermolecular bonding

12.3.1 *A glossary of intermolecular recognition modes*

The calculation of the total interaction energy profile for molecular dimers is feasible with Pixel in reasonable computing times, so that a large number of different molecular interaction modes can be investigated with a moderate computing effort. Pixel interaction energies over a number of molecular dimers have been shown [13] to be

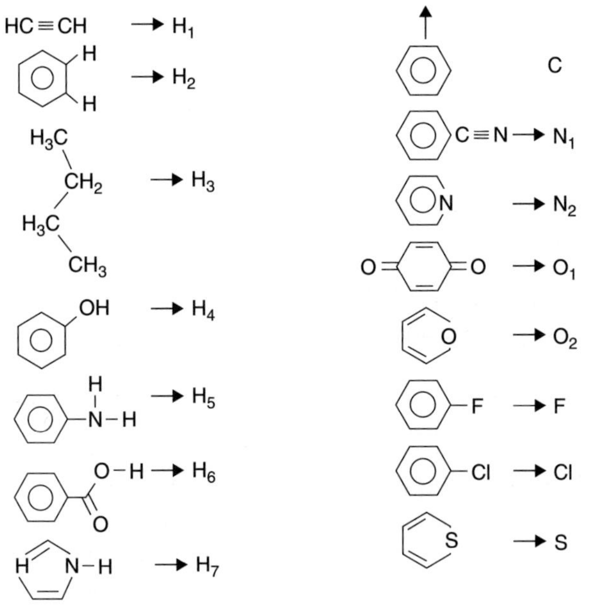

Fig. 12.10. The prototypical intermolecular interaction moieties: dimers are formed over opposing arrows.

consistently reliable, being with very few exceptions about halfway between counterpoise corrected and uncorrected MP2 interaction energies. Figure 12.10 shows a number of molecules representing atomic environments that have chances of being exposed for intermolecular contact in molecular aggregates [14]. A $\cdots$ B dimers are formed by combining pairs of molecules along the line identified by the arrow in molecule A pointing at the inverted arrow in molecule B, and most of the time the interaction energy is calculated for stepwise variation of the distance between the interacting molecules. In a few cases, where undesired very short atom–atom distances appear beside the interaction site of interest, a moderate optimization was carried out to remove the offending contacts. The calculation of the energy profiles for all the possible combinations of these prototypical systems should provide an energetic glossary of intermolecular interaction modes in organic condensed systems.

12.3.1.1 *Interactions not involving hydrogen*

Figure 12.11 and Table 12.5 show the results for interactions not involving hydrogen atoms. A steady coulombic repulsion, increasing in the order $F < O < N$, arises on approach of the most electronegative atomic species (Fig. 12.11(a)). Overlap repulsion also increases in the same order. Attractive polarization and dispersion terms are very small due to the scarce atomic polarizabilities, and cannot overcome the repulsive terms. On the whole, the interaction energies are connected in a consistent and

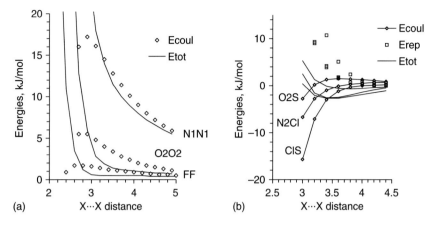

Fig. 12.11. (a) coulombic and total Pixel energies for the N1-N1, O2-O2 and F-F dimers (Fig. 12.10). (b) coulombic, repulsion and total Pixel energies for the O2-S1, N2-Cl and Cl-S1 dimers. Distances in Å.

Table 12.5 Equilibrium distances (Å) and interaction energies (kJ mol^{-1}) in dimers of molecules shown in Fig. 12.10. When no minimum exists in the interaction curve, the energy is given as positive at the distance equal to the sum of atomic radii (Table 12.3)

	N1		N2		O1		O2		F		Cl		S	
Cπ	–	+5	3.7	−2	3.35	0	3.5	−4	3.5	−2	3.8	−5	4.0	−6
N1	–	+14		+13		+9		+6		+4	3.6	−1	4.1	0
N2				+12		+8		+6		+3	3.5	−3	3.9	−1
O1						+5		+3		+2	3.5	−1	3.8	−1
O2								+2		+1	3.5	−2	3.8	−1
F										0	3.5	−1	3.7	−1
Cl											3.9	−1	3.6	−3
S													4.1	−2

understandable way with the diffuseness of the atomic electronic clouds. Cross interactions follow the same trend, so that all direct encounters among O, N, F atoms are destabilizing and unfavorable. This is reflected in the $g(R)$ plots of Fig. 12.2(a) which show no peak above unity. One can only recall that these findings are in contrast with the usual parameterizations of atom–atom force fields which include stabilizing minima in the O$\cdots$O, N$\cdots$N and similar interaction curves. This is a further reminder of the scarce adherence of atom–atom methods to physical principles.

The case is different with the approach of O or N atoms to Cl or S atoms (Fig. 12.11(b)): the latter have diffuse electron clouds, and coulombic energies, after an initial destabilizing trend, become less destabilizing or stabilizing due to penetration effects, while the higher polarizability introduces significant polarization and

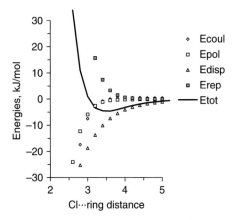

Fig. 12.12. The partitioned interaction energies in the Cπ–Cl or Cπ–S perpendicular dimers with the Cl atom of chlorobenzene or the S atom of thiophene pointing to the center of a benzene ring (Fig. 12.10). Distances in Å.

dispersion stabilizing terms. As a result, shallow minima appear in the interaction energy profiles. These effects are of course even more pronounced when two second-row atoms meet (check on moving down and to the right in Table 12.5). The trends, and their explanation, hold in the same way for the approach between ring O, N, or S atoms or chlorine atoms on one side, and the π-electrons of an aromatic ring on the other side: stabilization results from a lack of coulombic repulsion and a weak dispersion contribution (Fig. 12.12). These intermolecular approaches can hardly be classified as bonding; they occur without a substantial energy penalty, and could be better described as non-antibonding. Correspondingly, distinct peaks appear in the frequency distribution curves (Fig. 12.2–12.3), but their intensities are only slightly above unity.

12.3.1.2 C–H···X interactions

The outer rim of nearly all organic molecules consists of hydrogen atoms, so that in condensed phases H···H encounters are unavoidable. Pixel calculations show that such direct H···H interactions are slightly disfavored by electrostatic interactions: for example, the energy profile for acetylene–acetylene hydrogen pairs has a steadily rising shape due to the coulombic term. In more complex molecules, however, some dispersive contribution arising from the presence of neighbor carbon atoms can overcome this weak repulsion and accommodate for some net stabilization: this is the case for benzene and butane (H2 and H3, Fig. 12.10) hydrogen–hydrogen pairing, with a shallow interaction energy minimum of $4\,\mathrm{kJ\,mol^{-1}}$ for benzene–benzene and of $7\,\mathrm{kJ\,mol^{-1}}$ for the butane–butane case. Thus, H···H encounter is overall not so unfavorable, as demonstrated by the parallel packing of linear *n*-alkane molecules in their crystals, stabilized, according to the Pixel estimate, by about $2\,\text{kJ}\,\text{mol}^{-1}$ per carbon atom.

Table 12.6 and Fig. 12.13 show the energetic and geometrical data for the interaction between acetylenic, aromatic, and aliphatic C–H hydrogen atoms on one side, and electronegative acceptors (π-clouds or O, N, F, Cl, and S atoms) on the other, sometimes called C–H$\cdots$X "hydrogen bonds". The donor ability decreases in the order acetylenic > benzenic > aliphatic hydrogen; the best acceptor is the benzene π-cloud, thanks to a large the dispersion contribution (Fig. 12.13(a), contrary to the traditional idea that these are coulomb-sustained interactions. The stability follows in decreasing order with N and O atoms and the halogens. Acetylenic and aromatic hydrogens can pair with electronegative partners thanks to a significant coulombic stabilization and a minor dispersive contribution, while aliphatic hydrogens are less prone to doing so because of coulombic repulsion, hardly compensated by dispersive stabilization; for the latter hydrogens, the approach is not frankly repulsive but not sufficiently stabilizing either. This energetic information supplements and confirms the conclusions drawn from the parallel analysis of the frequency distribution curves for C–H$\cdots$X contacts, where only minor peaks appear at bonding distances (Figs. 12.4 and 12.5(b)).

12.3.1.3 The O–H$\cdots$X and N–H$\cdots$X hydrogen bond

Tables 12.6 and 12.7 show the Pixel estimate of the equilibrium geometries and interaction energies for the organic hydrogen bond; the dimer geometries are shown in Fig. 12.14. The nature of the hydrogen-bonding interaction, at least for the relatively weak bond formed by non-charged intermolecular partners without intramolecular electronic assistance or chain conjugation [15], is illustrated in Fig. 12.15: for the O—H$\cdots\pi$ bond, stabilization arises from equal amounts of coulombic and dispersion energy, while in the X$\cdots$H$\cdots$Y (X,Y=O,N) hydrogen bond the coulombic-polarization contribution dominates, with dispersion playing only a minor role (Table 12.8; see also Fig. 12.9(a)). The equilibrium H$\cdots$O bond distance in the O–H$\cdots$O hydrogen bond is predicted to range from 1.8 to 2.1 Å, slightly longer than the distance at which the peaks in frequency distribution curves (Fig. 12.5a) appear. Cohesive energy ranges from 24 to $36\,\text{kJ}\,\text{mol}^{-1}$; phenol is a weaker donor than benzoic acid, and the O=C oxygen acceptor power increases in the order benzoquinone < acids < amides. For the N–H$\cdots$N bond, the equilibrium H$\cdots$N distance ranges from 1.9 to 2.1 Å and the interaction energy is between 20 and $40\,\text{kJ}\,\text{mol}^{-1}$; ring N-H is a better donor than the NH_2 group, and ring N is a better acceptor than nitrile N. The strongest bond is formed between the strongest donor, COOH, and the strongest acceptor, ring nitrogen in imidazole or pyridine. The N–H$\cdots$O bond has the widest range, from $-17\,\text{kJ}\,\text{mol}^{-1}$ for the pair between the weakest donor and acceptor (benzamide–benzoquinone) to -33 for the imidazole–benzamide pair.

The Pixel results for the interaction between OH or NH groups and acceptor partners are satisfactory, also thanks to careful parameterization, but the repulsive part of the hydrogen bond curve is a particularly critical case, because of extensive overlap

Table 12.6 Equilibrium H⋯Y distances (Å) and energies (kJ mol^{-1}) for interactions involving hydrogen atoms, in dimers of molecules shown in Fig. 12.10 with linear X–H⋯Y arrangements

	Cπ		N1 Ph–CN		N2 pyridine		O1 benzo quinone		O2 furan		F Ph–F		Cl Ph–Cl		S thiophene	
HCCH (H1)	2.9	−13	2.4	−12	2.3	−17	2.3	−9	2.3	−10	2.4	−4	2.8	−2	2.5	−6
benzene H (H2)	2.9	−11	2.9	−6	2.9	−8	2.8	−6	2.8	−5	2.8	−3	3.1	−4	3.2	−4
butane H (H3)	3.2	−8	3.3	−2	3.0	−4	3.1	−2	2.9	−4	2.8	−3	3.0	−5	3.3	−3
phenol OH	2.5	−20	2.1	−30	—	—	2.1	−23	—	—	2.2	−12	2.5^a	−7	—	—
benzamide NH	2.7	−17	2.1	−20	2.0	−31	2.1	−17	2.0	−19	2.1	−9	2.5^a	−3	—	—
benzoic acid OH	2.5	−25	1.8	−38	1.8	−54	1.8	−27	1.8	−31	2.0	−12	—	—	—	—
imidazole NH	2.7	−23	2.1	−30	1.9	−37	2.0	−21	2.0	−20	2.1	−10	2.5^a	−5	—	—

^{a} The interaction curve shows minor discontinuities at short distances.

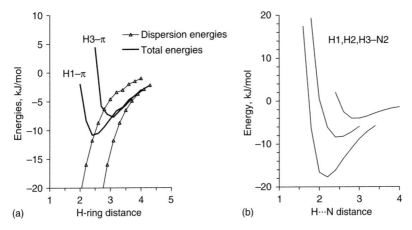

Fig. 12.13. (a) Total energies and dispersion contributions for the C—H···π confrontation. The acetylene–benzene interaction also has a non negligible coulombic contribution. (b) Shows the increasing stabilization in the C–H···N bonding with increasing proton acidity. Coulombic energies at the energy minimum: −23, −5, +1; dispersion energies, −8, −11, −6 kJ mol^{-1} for acetylene, benzene, and butane respectively. In all cases the horizontal axis is the vertical separation between the hydrogen(s) and the plane of the ring (Å).

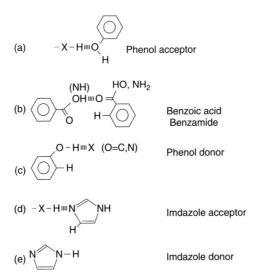

Fig. 12.14. The hydrogen bonds whose energies are shown in Tables 12.7 and 12.8. In coplanar dimers there are some unavoidable side interactions: in (b) a C–H···O interaction, in (c), a H···H contact when the acceptor is benzoic acid or benzamide; in (d), an O···HC interaction when the donor is benzoic acid or benzamide; in (e), an unavoidable H···H contact (except when the acceptor is phenol).

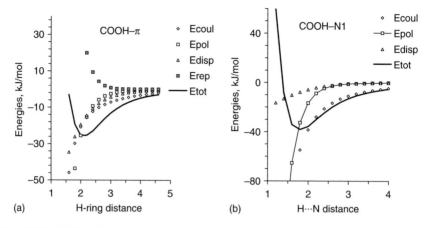

Fig. 12.15. (a) The O–H···π hydrogen bond. Note the relevant dispersion contribution. (b) The O–H···N hydrogen bond: the dispersion contribution is negligible, so that the total energy curve coincides with the coulombic energy curve at the right of the minimum. Distances in Å.

Table 12.7 Equilibrium distances (Å) and interaction energies (kJ mol^{-1}) for hydrogen bonds as in Fig. 12.14. In the diagonal entries, the second line gives the corresponding separation and peak heights in frequency distribution functions (equation 12.1)

Donor	Acceptor							
	phenol >O		benzoic acid O=C		benzamide O=C		imidazole N−H	
phenol	1.95	−24	1.9	−28	1.9	−31	2.0	−41
O−H	1.8	5						
benzoic acid	1.9[a]	−35	1.85	−36	1.8	−42	1.8	−61
O−H			1.65	23				
benzamide	1.95	−22	1.9	−21	1.9	−24	2.0	−35
N−H					1.9	11		
imidazole	1.85	−23	1.85	−28	1.85	−33	1.9	−40
N−H							2.0	2

[a] The interaction curve shows minor discontinuities at short distances.

between the interacting electron densities. In an extreme example, when the strongly polarizing OH or NH groups face the diffuse electron clouds of second row elements the interaction curves sometimes become noisy or produce false minima. This is particularly evident for the (O,N)–H···(Cl,S) interactions. These malfunctions are probably enhanced by the poor representation of the core electronic structure of second-row atoms. There is space for improvement.

Table 12.8 The Pixel partitioned energies (KJ mol^{-1}) for the hydrogen bond at a fixed bonding distance of 1.8 Å

Donor Acceptor	Bond	E(coul)	E(pol)	E(disp)	E(rep)	E(tot)
benzoic acid benzoic acid	O—H···O	−43	−17	−8	33	−36
benzamide benzamide	N—H···O	−38	−16	−10	40	−23
benzoic acid benzamide	O—H···O	−47	−20	−8	33	−42
benzamide benzoic acid	N—H···O	−36	−17	−10	41	−21
benzoic acid imidazole	O—H···N	−76	−41	−12	67	−61
imidazole imidazole	N—H···N	−73	−33	−13	79	−39
benzoic acid phenol	O—H···O	−44	−21	−9	40	−33
phenol phenol	O—H···O	−41	−18	−14	50	−22

12.3.1.4 π-*interactions*

Another ubiquitous interaction mode in organic crystals is the parallel arrangement of aromatic rings. This recognition mode was studied by the Pixel theory, using the sample molecules shown in Fig. 12.10, forming dimers where the line joining the centers of coordinates of the two molecules is perpendicular to the two parallel molecular planes. The interaction was studied as a function of the vertical inter-ring separation, without attempting optimizations with respect to in-plane sliding; extensive experiences have shown that the energetic cost or gain of such displacements is usually rather small if not altogether negligible. Tables 12.9 and 12.10 show the results. The dispersion term is dominating, and increases with increasing ring size (benzoquinone larger than furan) and with increasing polarizability of the substituent (thiophene larger than furan). The coulombic terms are attractive for antiparallel dipoles, and repulsive for non-polar pairs (e.g. benzene) or for parallel dipoles (e.g. benzoquinone). These terms are small, but in some cases act as directional pointers or are responsible for selectivity: for example, the large difference in interaction energy between the benzoquinone and benzonitrile dimers is entirely due to the modulating action of the coulombic term, and the difference in stability between homodimers and the heterodimer in the arene-perfluoroarene group is entirely due to the coulombic term. Note that the stacking interaction energy for benzoic acid is 50-80% of the energy of a hydrogen bond, demonstrating that antiparallel π-stacking of aromatic rings with polar substituents can compete with hydrogen bonding, contrary to widespread

Table 12.9 Interaction between parallel stacked aromatic rings for molecules shown in Fig. 12.10: inter-ring separation (Å) and interaction energies (kJ mol^{-1}). The two substituent groups or heteroatoms point in opposite direction (antiparallel arrangement)

	Benzo nitrile	Pyridine	Benzo quinone	Furan	Chloro benzene	Thiophene	Benzoic acid	Benzene
benzonitrile	3.6 −16	3.7 −12	3.7 −9	3.7 −8	3.7 −15	3.7 −9	3.6 −15	3.7 −10
pyridine		3.7 −11	3.7 −9	3.7 −7	3.7 −10	3.7 −11	3.7 −11	3.7 −8
benzoquinone			3.9 −4	3.7 −10	3.7 −12	3.7 −11	3.6 −15	3.7 −13
furan				3.8 −4	3.7 −8	3.8 −5	3.7 −8	3.8 −5
chlorobenzene					3.7 −13	3.9 −8	3.6 −14	3.7 −9
thiophene						3.9 −6	3.8 −8	3.9 −5
benzoic acid							3.6 −15	3.7 −10
benzene								3.8 −7

Table 12.10 The Pixel partitioned energies (kJ mol^{-1}) for the antiparallel stacking of molecules in Fig. 12.10, at a fixed ring–ring distance of 3.6 Å. The polarization energy is negligible (−1 to −2 kJ mol^{-}).

	E(coul)	E(disp)	E(rep)	E(tot)
chlorobenzene chlorobenzene	−2	−24	14	−12
benzonitrile benzonitrile	−6	−20	11	−16
benzonitrile pyridine	−3	−15	8	−10
benzoquinone benzoquinone	+6	−17	10	−3
furan furan	+2	−12	8	−4
thiophene thiophene	0	−19	17	−5
benzoic acid benzoic acid	−1	−22	10	−15
benzene–benzene	+3	−18	13	−4
benzene–C_6F_6	−7	−19	12	−17
C_6F_6–C_6F_6	+3	−21	11	−10

views that consider hydrogen bonding as a largely dominating packing factor in crystals.

12.3.2 Crystal energies

The Pixel theory can be rather comfortably applied to the calculation of the lattice energies of organic crystals. A typical calculation at condensation level $n = 4$ for a medium-size molecule (say, a 30-atom molecule) takes about 20 minutes on an ordinary personal computer, but computing times can be reduced by a factor of almost 100 if top-level computational devices are used. In a comprehensive test of the performance of the method, the lattice energies of the crystals in the Subheat database (Section 8.4) have been computed for comparison with the observed sublimation enthalpies (with the exclusion of crystals with more than one molecule in the asymmetric unit, an occurrence that the Pixel software cannot handle). Figure 12.16 shows the overall result: for the 172 crystal structures considered, comprising all combinations of the most common molecular constitutions in organic chemistry, sublimation enthalpies are well reproduced over the whole range from 30 to 160 kJ mol^{-1}, with a correlation factor of 0.79, against 0.82 for the UNI atom–atom potential scheme, which includes about 100 empirical parameters and provides no insight into the physics of the interactions [16].

Pixel calculations on crystals also provide a closer picture of the molecular packing, by showing the intensity and the physical nature (coulombic, polarization, dispersion) of the cohesive energy, both in an overall sense and subdivided into pairwise interactions in the crystal. This approach amounts to replacing atom–atom energies, which are undefined quantities, by molecule–molecule energies, which have a definite physical meaning. This will constitute the starting point for the theories of crystal packing exposed in Chapters 14 and 15.

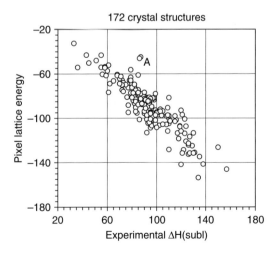

Fig. 12.16. The plot of calculated lattice energies against experimental sublimation enthalpies for 172 organic crystals of the Subheat database (Section 8.3). The least-squares straight line through the data points is $y = 0.96x (R^2 = 0.79)$.

12.4 Directed bonds versus diffuse bonding

Perhaps even more important than a consideration of the total cohesive energies – that is, the strength of the bond as represented by the depths of the bonding wells – is the consideration of the force constants for the stretching of the intermolecular liaisons, – that is, the flexibility of these bonds as represented by the widths of the corresponding potential energy wells. Table 12.11 is a pictorial view of the overall landscape of intermolecular bonding.

The joint consideration of the atom–atom distance distribution curves and of the quantitative analysis of intermolecular interaction energies and force constants produced by the Pixel theory strongly points to the conclusion that intermolecular approach, recognition and cohesion are better judged in the perspective of interactions between diffuse electron clouds than of directed atom–atom bonds, with the possible exceptions of some of the strongest of the "weak" hydrogen bonds [15], i.e. those O$\cdots$H or N$\cdots$H hydrogen bonds whose energy is in excess of 20 kJ mol^{-1} and whose well widths do not exceed 0.3 Å. The rest of the interactions commonly seen in the static picture provided by organic crystal structures, and presumably constituting some privileged recognition modes also in liquids, is a matter of diffuse adhesion between molecular charge densities which merges into a stabilization continuum between 0 and -20 kJ mol^{-1}, where the trading of a few kJ mol^{-1} to cross over between different molecular arrangements is a very likely occurrence. This is the source of the intrinsic difficulty in the prediction and control of organic crystal structures. If the analysis and the conclusions presented in this chapter are accepted, then

Table 12.11 The Pixel intermolecular energy landscape. A, B are interacting moieties; the energy range is in kJ mol^{-1}, the well width is in Å.

A	B	Binding energy range	Well width	$g(R)$ peak intensity
N, O, F	N, O, F	repulsive	—	no peaks
N, O, F	Cl, S, π	1 to 6	shallow	1.2 to 1.5
C—H	N, O, Cl, S, π			
acetylene		2 to 17	0.5 to 1.0	1.2 to 1.5
benzene		3 to 11		
aliphatic		2 to 8		
aliphatic chains		1.5/C atom	shallow	—
π-ring	π-ring	3 to 16	0.3 to 0.5	—
π-ring	hexafluorobenzene	20	0.4	—
(O,N)—H	O, N, π	17 to 61	0.2 to 0.4	5 to 23
benzamide NH	any acceptor	17 to 35		
phenol OH	any acceptor	20 to 41		
imidazole NH	any acceptor	20 to 40		
benzoic acid OH	any acceptor	27 to 61		

the path to progress in intermolecular science does not go through a theory of atomic bonds but through a theory of charge-density bonding. The Pixel theory naturally adheres to this new principle.

References and Notes to Chapter 12

[1] Bader, R. F. W.; Fang, D.-C. Properties of atoms in molecules: caged atoms and the Ehrenfest force, *J. Chem. Theory Comp.* 2005, **1**, 403–414.
[2] See Rafat, M.; Devereux, M.; Popelier, P.L.A. Rendering of quantum topological atoms and bonds, *J. Mol. Graphics Modeling* 2005, **24**, 111–120.
[3] Matta, C. F.; Hernandez-Trujillo, J.; Tang, T.-H.; Bader, R. F. W. Hydrogen–hydrogen bonding: a stabilizing interaction in molecules and crystals, *Chem. Eur. J.* 2003, **9**, 1940–1951. The stabilization occurring upon the formation of the bond path between these hydrogens may be overcompensated by other destabilizations around the molecule, so the H–H bonding argument does not imply that planar biphenyl must be more stable than twisted biphenyl. The possible lack of immediate response between bond path presence and molecular stabilization, especially in non-covalent cases, is one of the least appealing facets of AIM theory (see the ensuing discussion and ref. [6]).
[4] Abramov, Y. A. Secondary interactions and bond critical points in ionic crystals, *J. Phys. Chem.* 1997, **A101**, 5725–5728.
[5] Dunitz, J. D.; Gavezzotti, A. Molecular recognition in organic crystals: directed intermolecular bonds or nolocalized bonding? *Angew. Chem. Int. Ed. Engl.* 2005, **44**, 1766–1787.
[6] For a thoughtful review of the topological approach to chemical bonding see: Popelier, P. L. A. Quantum chemical topology: on bonds and potentials, in *Intermolecular forces and clusters*, vol. 1, edited by D. J. Wales, *Structure and Bonding* Vol. 115, pp. 1–56, 2005, Springer-Verlag, Berlin. On p.13 it is said: "...the magnitudes of the electron density at a bond critical points in a representative covalent bond, a hydrogen bond, and a "van der Waals bond' approximately follow the ratio 100:10:1".
[7] In a similar manner, Lemuel Gulliver, washed ashore on Lilliput island, finds himself firmly tied by weak bonds: "I attempted to rise, but was not able to stir ... I found my arms and legs were strongly fastened on each side to the ground ... I likewise felt several slender ligatures across my body, from my armpits to my thighs." (Swift, J. *Gulliver's Travels*, Part I, Chapter 1).
[8] For a few random examples see: Munro O. Q.; Mariah, L. *Acta Crystallogr.* 2004, B60, 598; Balamurugan, V.; Hundal M. S.; Mukherjee, R. *Chem. Eur. J.* 2004, **10**, 1683; Caronna, T.; Liantonio, R.; Logothetis, T. A.; Petrongolo, P.; Pilati, T.; Resnati, G. *J. Am. Chem. Soc.* 2004, **126**, 4500; Nguyen, H. L.; Horton, P. N.; Hursthouse, M. B.; Legon A. C.; Bruce, D. W. *J. Am. Chem. Soc.* 2004, **126**, 16. Textbooks in the subject are G. R.Desiraju, *Crystal Engineering, the Design of Organic Solids*, 1989, Elsevier, Amsterdam; Desiraju G. R., T. Steiner,

The Weak Hydrogen Bond, 1999, IUCr Book Series, Oxford University Press, Oxford.

[9] (a) Gavezzotti, A. The calculation of intermolecular interaction energies by direct numerical integration over electron densities. I. Electrostatic and polarization energies in molecular crystals, *J. Phys. Chem.* 2002, **B106**, 4145–4154; (b) Gavezzotti, A. Calculation of intermolecular interaction energies by direct numerical integration over electron densities. 2. An improved polarization model and the evaluation of dispersion and repulsion energies, *J. Phys. Chem.* 2003, **B107**, 2344–2353; (c) Gavezzotti, A. Towards a realistic model for the quantitative evaluation of intermolecular potentials and for the rationalization of organic crystal structures. I. Philosophy, *CrystEngComm*, 2003, **5**, 429–438; (d) Gavezzotti, A. Towards a realistic model for the quantitative evaluation of intermolecular potentials and for the rationalization of organic crystal structures. II. Crystal energy landscapes, *CrystEngComm*, 2003, **5**, 439–446; (e) Gavezzotti, A. Hierarchies of Intermolecular Potentials and Forces: Progress towards a Quantitative Evaluation, *Struct. Chem.* 2005, **16**, 177–185; (f) Gavezzotti, A. Calculation of lattice energies of organic crystals: the PIXEL integration method in comparison with more traditional methods, *Z. Krist.* 2005, **220**, 499–510.

[10] Rigorous methods for polarizability distribution have been proposed: Williams, G. J.; Stone, A. J. Distributed dispersion: a new approach, *J. Chem. Phys.* 2003, **119**, 4620–4628.

[11] Miller, K. J. Additivity methods in molecular polarizability, *J. Am. Chem. Soc.* 1990, **112**, 8533.

[12] Spackman, M. A. The use of the promolecular charge density to approximate the penetration contribution to intermolecular electrostatic energies, *Chem. Phys. Letters* 2005, **418**, 154–158; Volkov, A.; Coppens, P. *J. Comp. Chem.* 2004, **25**, 921–934.

[13] Gavezzotti, A. Quantitative ranking of crystal packing modes by systematic calculations on potential energies and vibrational amplitudes of molecular dimers, *J. Chem. Theor. Comp.* 2005, **1**, 834–840.

[14] Molecular geometries have been prepared without optimization, using standard bond lengths: regular, flat benzene rings with C–C 1.40Å, all angles 120°; geometries of substituent groups taken from representative crystal structures. The coordinates of all the molecules used in this investigation are available as supplementary material (file *dimer.oeh*).

[15] The "strong" hydrogen bond is between charged groups, with interaction energies of up to $100\,\text{kJ}\,\text{mol}^{-1}$; see e.g. Meot-Ner (Mautner), M. The ionic hydrogen bond and ion solvation, *J. Am. Chem. Soc.* 1984, **106**, 1257–1272.

[16] A detailed analysis of the deviations in Fig. 12.16 would be very tedious and scarcely revealing: as discussed in Section 7.6 and further in Sections 8.7 and 8.8 discrepancies may arise because of failure of the theory, but also due to the intrinsic difficulties in the comparison between lattice energies and heats of sublimation: experimental fluctuations, neglect of intramolecular rearrangement energies, etc. One case is particularly interesting, however: the lattice

energy of 1,2-dicyanobenzene is calculated to be some $40\,\mathrm{kJ\,mol^{-1}}$ below the experimental heat of sublimation (see point A, Fig. 12.16), while the calculated lattice energy of the 1,4-isomer matches the experiment well. The 1,2-isomer has a strong molecular dipole moment and packs in a polar space group with all dipoles parallel and aligned along a cell axis (Fig. 11.12), while the centrosymmetric 1,4-isomer has obviously no dipole moment and packs in P1-space group. The 1,2-dicyanobenzene crystal is a special case in which the coulombic and polarization energy sums fail to converge: an estimate of the correction for convergence to the coulombic energy by the van Eijck-Kroon method (ref. [22], Chapter 8) is $-26\,\mathrm{kJ\,mol^{-1}}$ per cell contents. A similar term is expected for the polarization energy. These corrections would bring the calculated energy very close to that of the experiment.

13

Phase equilibria, phase changes, and mesophases: Analysis and simulation

The conclusion we have reached is a special case of one of the most elegant results of chemical thermodynamics. The phase rule was derived by Gibbs and states that, for a system at equilibrium, $F = C - P + 2$.

Atkins, P. W. *The Elements of Physical Chemistry*, 2001, 3rd ed,
Oxford University Press, p.195; as elegant as useless.

13.1 Things and molecules

The story, or the myth, goes that when somebody asked Enrico Fermi if he knew the thermal conductivity of some metal, or if he could remember the mathematical equation for a given phenomenon, he would say "no", and then proceed to derive the requested information from first principles. Now there is no question that Fermi mastered the principles of physics in the same way as that Maurizio Pollini masters the keyboard of his piano, but there is also no question that there must be a certain amount of exaggeration in the story.

A normal person wakes up in the morning, yawns, gets out of bed, has a cup of coffee, chews a doughnut, lathers his chin to shave, dresses, climbs into his car, and drives to work. A chemist has his brain triggered to attention by the release of several neurotransmitter molecules, including serotonine and acetylcholine, sends an electric signal to his muscle activators to compress the lung tissue in order to pull in a 20% oxygen–nitrogen gas mixture and to stretch his legs, stimulates his taste receptors with a water solution made of caffeine, sucrose and an enormous amount of trace aromatic molecules, uses the saliva enzymes to degrade a piece of polysaccharide, fills the pores of his chin with a gas/liquid suspension of surfactant molecules, then protects his skin with a 60–40% coating of silk and polymer fibers, impregnated with appropriate dyes to make them attractive, and finally ignites a gasoline/air mixture to generate heat which forces a gas to expand, pumping some pistons up and down and chug-chugging his car along. What a complicated life. But like Enrico Fermi, chemists refuse to take objects at their face value and prefer to work their way from atoms and molecules to the shape and properties of objects as we see them. The structure–activity principle does the rest, and this attitude makes the chemist the master of the world. This is admittedly a very cavalier attitude and no doubt frustration is much more common than success, but the molecular point of view is certainly the cast of mind of the chemistry of the years and centuries to come.

All matter, from the simplest fluid such as gaseous helium to the most complex system like a biological cell, is made of electrons and nuclei. Electric potentials tend to glue the nuclei together, while kinetic energy, connected to atomic (nuclear) masses moving with given velocities, tends to pull them apart. It is this eternal struggle between electricity and temperature that ultimately gives rise to the entire world as we see it, with its properties and its changes. This chapter examines a portion of this extremely wide and complicated landscape, within the following limitations.

1. Chemical bonding is not considered, i.e. molecules are taken as they are and do not react.
2. The analysis is limited to compounds made of the usual elements of organic chemistry, with the exclusion of organometallics, rocks, silicates, and metals.
3. Molecules are considered only up to a size of some 1000 da, with the exclusion of polymers and biological macromolecules.
4. Only pure substances or binary mixtures are considered.

Within the above-mentioned restrictions, a phase is a piece of matter containing one or two chemical species in a distinguishable state of aggregation. Admittedly, this oversimplifies the real world, where even one of the purest liquids, drinking water, is a solution of hundreds of components, and most chemical systems consist of many different chemicals in different and perhaps variable states of aggregation.

13.2 Basic thermodynamic functions

The basic thermodynamic functions enthalpy, heat capacity, and entropy are strongly dependent on the state of aggregation. Enthalpy and heat capacity are higher in condensed states because of the significant intermolecular potential energy terms, and increase with the number of vibrational energy "pockets" available (recall equations 7.25 and 7.30); for example in a periodic solid the partition function includes sums over vibrational states in the whole Brillouin zone (see Section 6.3). Entropy is proportional to the amount of phase space which the system can explore, and the higher dispersion in space for gases largely overweighs the dispersion of energy among available states, which are more numerous in condensed systems. Hence the familiar rule:

$$H_{cr} > H_l > H_g; \quad S_g > S_l > S_{cr}; \quad C_{p,l} >\sim C_{p,cr} > C_{p,g}$$

The thermodynamic treatment of equilibrium is in terms of chemical potentials. Phase equilibria and phase changes are dictated by the leverage between enthalpic and entropic terms implicit in equations such as 7.49 and 7.55. Such equations, however, hold exactly for infinite and homogenous systems, but in real systems the influence of size, termination, and defects cannot be neglected. The microscopic texture of the system may then become of paramount importance, and it must be said at once that this is the case for most of the chemical phenomena and transformations in phase transitions. Size and heterogeneity can be incorporated in the thermodynamic treatment through surface free energies, but if the measurement of thermodynamic functions is not easy for homogeneous and continuous systems, it is orders of magnitude more awkward for surface-dependent systems. The very concept of chemical equilibrium may dissolve across a maze of semi- or non-equilibrium states, and the derived equations show an unpleasant dependence from poorly characterized boundary conditions, or from the chemical constitution of each system, if not altogether from the particular working conditions of each experiment, with a loss of generality that does much harm to the development of fundamental knowledge. One often hears of a crucial role played by "activation free energies", quantities whose very name casts a shadow of uncertainty by its admixture of thermodynamic and kinetic flavors. There is very little that can be done against this state of affairs.

13.3 Melting

Consider as a first example a piece of pure crystalline solid in equilibrium with its pure liquid. at the melting temperature. Assume that the crystals are large enough that surface effects can be neglected (as will become clear later, this is quite an assumption). If an amount of heat that is less than the latent heat corresponding to the crystal mass is supplied to the system, some crystalline material will melt at equilibrium without change in temperature, but as soon as more heat is poured into the system, after all the solid mass has melted the heat can be used to raise the temperature, the chemical potential of the crystal exceeds that of the liquid, and the crystal becomes thermodynamically unstable. The reasoning seems neat and simple, but if one tries to work it backwards, and expect crystallization after subtracting the appropriate amount of heat, kinetics sets in and supercooling is the rule rather than the exception, with the sample staying liquid for a long time well below the melting temperature, especially if the cooling rate is very slow. The reason is of course that it takes a comparatively long time for molecules to find their way through phase space and lock into the proper reciprocal orientation for crystallization. Equilibrium thermodynamics compares different states in an abstract sense, and not the different states of the parts of a system that are actually undergoing a given chemical process.

The thermodynamic law that governs the solid–liquid equilibrium is equation 7.57, so that the equilibrium melting temperature is a ratio of melting enthalpy to melting entropy. T_m and ΔH_m can be measured fairly easily (Section 7.6), so that equilibrium melting entropies can also be obtained. Is there a relationship between the melting temperature and the molecular structure or crystal forces? Intuitively, one would expect easier melting for less cohesive crystals, and, indeed, there seems to be some correlation between melting temperatures and sublimation enthalpy – surprisingly, a steadier correlation than with melting enthalpies [1]. A further view on this problem is provided by a statistical analysis of melting entropies. They are about 13 J K^{-1} mol^{-1} for small spherical molecules (the translational contribution) and about 57 J K^{-1} mol^{-1}

for rigid molecules (Walden's rule). But there are fluctuations; according to one hypothesis [2], the more symmetrical a molecule, the higher its probability of being in the correct orientation to be incorporated into the crystal lattice, with a decrease in the difference between periodic crystal and randomly oriented liquid molecules, and a lowering of the melting entropy. ΔS_m then consists of the translational contribution plus a rotational contribution of $57 - 13 = 44$ kJ mol^{-1}, further corrected by a term in the logarithm of the rotational symmetry number, σ [2]:

$$\Delta S_m = 13 + [44 - R\ln(\sigma)] \quad (13.1)$$

With the assumption that melting enthalpies are much less affected by molecular symmetries, symmetric molecules must have higher melting points and lower ideal solubilities [3]. Although to some extent revealing, such correlations are mostly qualitative and hardly comprehensive [3,4] Entropy from conformational flexibility, blocked in the crystal and free in the liquid, is not considered.

Molecular shape also influences crystal properties and the thermodynamic melting parameters, sometimes in a simple and understandable way. Globular molecules tend to form plastic crystals, in which extensive reorientation of parts of the molecule or even of the entire molecule occurs, without substantial translational diffusion; the crystal is made of a periodic array of rotators. Clearly, the rotational contribution to the melting entropy is lower, as the crystal already possesses some of the rotational freedom of the liquid, so that ΔS_m is depressed and the melting point is abnormally high. An example is provided by *tert*-butyl derivatives [5], whose crystals show a phase transition from a low-temperature, ordered phase II to a high-temperature, rotationally activated phase I, which then evolves into the liquid. Melting entropies are particularly low, 5–15 J K^{-1}mol^{-1}, while the sum of the melting entropy and of ΔS for the II–I transition is around 40 J K^{-1} mol^{-1}, not far from the boundaries of equation 13.1. Rotational activation in the solid is also the common explanation of the unusually high melting temperature of planar, disk-like aromatic molecules, benzene being a classical example. All things considered, however, the melting temperature of an organic compound is one of the least predictable thermodynamic properties.

Some aspects of the melting process can be studied by evolutionary simulation methods like molecular dynamics [6]. A crystalline computational box is set up, and the system is gently driven up in computational temperature until evolution into the liquid state occurs. As an example of the output of such a computational experiment [7], Fig. 13.1 shows the evolution of the most important quantities (intermolecular potential energies and density) throughout the whole process for the benzene molecule. After sequential runs with increasing temperature, decreasing density, and decreasing cohesive energy, when the computational box is brought to 300 K, which is some 20 K above the experimental melting temperature, the melting catastrophe is almost instantaneous, and is revealed (recall Section 9.2) by a sudden drop in density and rise in potential energy. Note that the kinetic energy does not change because the temperature is promptly kept constant by a supply of computational heat from the T-coupling mechanism. The enthalpy of melting can be calculated from the difference in total energy between the equilibrated crystal and the equilibrated liquid.

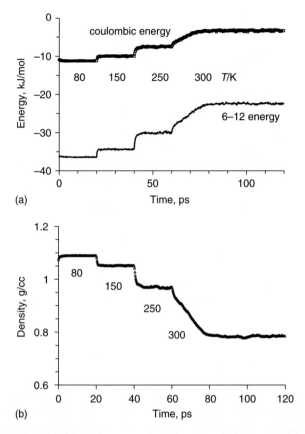

Fig. 13.1. (a) Time evolution of the 6–12 and coulombic potential energies during the simulation that goes from crystalline benzene at 80 K to liquid benzene at 300 K. The inset shows the temperatures of the simulations. The 60–80 ps section corresponds to the melting run. (b) Density profile along the same trajectories as in (a).

In the 60–80 ps section of Fig. 13.1 the benzene molecules are literally pushing the crystal lattice apart under the effect of the suddenly available kinetic energy. Diffusional and rotational freedom results and the crystal collapses to the liquid, which is then normally simulated at 300 K in the 80–120 ps section of the run. Since the part of the simulation where melting occurs is a non-equilibrium simulation, one cannot draw any conclusions from averages, nor can one claim to have simulated the actual solid–liquid equilibrium or to have predicted the melting temperature. Nevertheless, such dynamic runs offer a window over the evolution of the internal structure of the system as it goes from crystal to liquid; an example taken from a study of acetic acid is shown in Fig. 13.2. One cannot say for sure that this picture is a representation of the true structural changes that occur when the acetic acid crystal melts; molecular

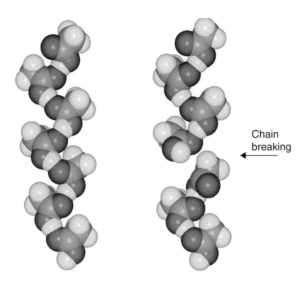

Fig. 13.2. Snapshots drawn from pre-melting trajectories for the acetic acid crystal. A regular chain of hydrogen-bonded molecules is seen on the left. On the right, one molecule has flipped around, breaking the chain and cross-linking to form a hydrogen bond to a neighboring chain (the arrow indicates the oxygen atom that is no longer engaged in intra-chain hydrogen bonding). After ref. [6a].

dynamics is only a simulation, and its results depend on the details of the potentials and of the computational setup; and yet, it offers a gallery of possible events, that become more and more realistic with increasing accuracy of the potentials and length of the simulation, and can be further examined by other experimental techniques. The role of molecular dynamics is often similar to that of a police officer who shows a gallery of photographs of possible culprits to the witness of a crime scene.

It has been shown [8] that in MD melting runs there is a no-return point, such that computational frames before that point revert to the crystal upon cooling, while frames after that point cannot be prevented from collapsing into the liquid. What this means in terms of structure is far from clear, and it is not even obvious that this crossover point is reproducible, as the effect must be to some extent dependent on the particular track through phase space taken by the simulation. In any case, computational experiments with the introduction of a small number of vacancies in the crystal lattice [6a] or on the melting of computational polymorphs with variable density [8] show that any decrease in the compactness and close packing of the crystalline material, even a very minor one, has huge consequences on the propensity towards melting, as expected.

There are many more ways in which molecular dynamics can be used to study the molecular events as a crystal structure approaches the melting point. Figure 13.3(a) shows the radial distribution function (see in Section 9.4.1) for the molecular centers of mass in the benzene crystal. As temperature rises, the peaks shift to longer separations

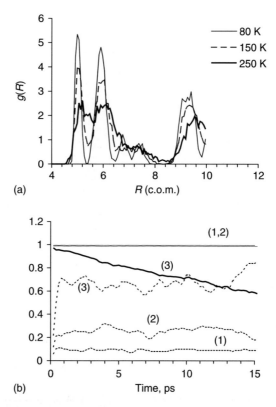

Fig. 13.3. (a) Radial distribution functions for the molecular centers of mass in the benzene crystal, from an MD simulation. (b) Same simulation: dashed lines, mean center of mass displacements; full lines: rotational correlation function for in-plane oscillation. (1), (2) and (3) at 80, 150 and 250 K respectively. The rotational function at 250 K shows the onset of molecular rotation.

and become flatter and wider, as a consequence of the increased displacement freedom due to the decrease in density (not unexpectedly; to pursue the previous criminal analogy further, presenting a picture of Al Capone to mafia prosecutors). Rotational correlation functions provide further information: Fig. 13.3(b) shows that at 250 K, even while the crystal remains integer, as testified by constant density and steady mean square displacements of the center of mass, there is a loss of orientational correlation for in-plane rotation, as the benzene crystal moves to a rotator phase by a second-order phase transition. This in-plane rotation of the molecules in the benzene crystal, so clearly pointed out by molecular dynamics, is confirmed by NMR relaxation experiments (recall Table 1.6 and the related discussion). Note that the loss of orientational correlation in the crystal is partial, and is on a much longer timescale (15 ps for a 0.4 decrease) than in the liquid state (Fig. 13.4, full loss in

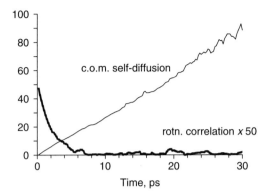

Fig. 13.4. Center of mass displacement and rotational correlation function for liquid benzene at 300 K. 80–120 ps section of the simulation shown in Fig. 13.1.

6 ps), demonstrating the directional constraint and the comparatively high activation barrier for the rotation in the crystal.

No wonder, the atomic details of the mechanism of melting are far from clear. Intuitively, one may recall that as temperature rises, the anharmonicity of the interatomic potential requires a decrease in density, so that at some point the molecules in the solid find enough space around them to become free to diffuse or rotate, with a consequent collapse of long range structural periodicity and a catastrophic evolution into the liquid state. The number (concentration) of these defects is a reproducible quantity at a given temperature. This so-called Lindemann criterion [9] sets a limiting value for the volume increase that the solid can withstand; this rule apparently holds for crystals made of simple spherical particles, which melt as soon as the vibrational amplitude exceeds 15% of the cubic cell edge. This is also the order of magnitude for the average volume increase on going from a very low temperature to the melting temperature for organic crystals (recall Fig. 11.4), but the formulation of a theory for this behavior would require at the very least an anisotropic version of the Lindemann rule, somehow taking into account also the wide variety of molecular shapes and interaction potentials – a very difficult path to follow.

A different mechanism, verified by careful experiments, is surface melting [10]: below the melting point, solid particles become coated with a liquid-like film of increasing thickness, in thermodynamic equilibrium with the solid, and it is this liquid-like portion of the system that functions as a trigger for the subsequent bulk melting. At the normal melting temperature, the melt front rushes in further, transforming the entire solid into a liquid. If this mechanism were always true and necessary, melting would be a surface-dependent phenomenon and an infinite bulk solid would never melt. In fact this mechanism certainly applies to systems made of very simple particles, like Ar atoms, but it is doubtful that it may be at work in complex molecular systems. Surface melting is a sort of extreme surface reconstruction, a phenomenon which may not be pervasive or even frequent for molecules with large and articulated anchoring points. A logical conjecture is to propose that melting may initiate at any kind of defect (grain boundaries, dislocations, etc.), the surface being the most obvious "defect" in any real solid, but without the need for a real surface liquid layer. Anyway, the Lindemann criterion and surface melting are not mutually exclusive, and neither precludes the sharpness and reproducibility of the melting points of organic compounds, as they both result from equilibrium conditions that are met at well specified temperatures.

Molecules in crystals undergo collective oscillations (Section 6.3) whose frequency and associated librational energy are sharp functions of the boundary conditions imposed by the surrounding lattice. As density decreases with rising temperature, some of these conditions may in some cases become much less imposing or may eventually vanish, and the corresponding vibrations become "soft modes". These soft modes lead to collective structural rearrangements and to crystal–crystal phase transitions with almost zero cost in enthalpy [11]. Is melting the result of some mode softening or even of a simultaneous softening of all lattice modes, or, as an alternative, of a chaotic loss of correlation at single defect points? No one can tell for sure, but the most disquieting side of this question is that there may not be a unique answer, except, once again, for crystals made of very simple objects such as hard spheres, disks, or cylinders. For example, the instabilities in n-alkane crystals above a certain temperature have been studied by molecular dynamics [12], leading to the conclusion that "the instabilities ... correspond to the softening of long-wavelength vibrational modes associated with rigid motion of the chains ... the instability then propagates to smaller wavelengths". Conceivably, many different things may happen in the almost infinite variety of shapes and potentials represented in large and flexible organic molecules, and it may be extremely difficult if not altogether impossible to find general rules. The solid-state chemist is here in the position of the physicist who (so the story goes) when asked if she could provide a general equation for the motion of a horse, answered, yes, but only assuming a spherical horse.

13.4 Solid–liquid equilibrium and nucleation from the melt

If melting is difficult to characterize in molecular terms, nucleation and growth of crystalline particles from the melt is an even more elusive phenomenon. Given the extreme difficulty of obtaining molecular level information, phenomenological, macroscopic nucleation theories have been formulated [13] before and aside from numerical molecular simulation. These theories constitute an almost completely parallel approach to the matter and their description does not belong in this book, although points of contact with molecular level simulations have been explored [14].

The following treatment of nucleation in classical nucleation theory (CNT) provides a glimpse of the methods and reasoning of classical theories. A supersaturated solution is assumed to contain spherical nuclei of radius R and with n_c particles, and the Gibbs free energy change on formation of these nuclei includes a term in the difference between the free energies of the solid and the liquid, $\Delta\mu$, and a term that represents the surface free energy. In supersaturation conditions the first term is negative (drive towards the solid) and the second term is positive and responsible for the free energy activation barrier to nucleation:

$$\Delta G = 4/3\pi R^3 \rho_s \Delta\mu + 4\pi R^2 \gamma \quad (13.2)$$

where ρ_s is the particle density of the bulk solid, and γ is a surface free energy density. This function has a maximum ΔG^* for a given R_{crit}, defining the size of the critical nucleus and also the probability of its formation, which is taken to be proportional to a Boltzmann-like factor in ΔG^*. In addition to these thermodynamic factors, the theory includes some kinetic factors that take into account the obvious facts that (1) there must be a finite frequency of attachment for molecules docking onto the nucleus, and (2) the free energy barrier can be climbed from both sides. Using other simplifying assumptions, and simple geometrical factors for attachment, one ends up with the following expression for the nucleation rate per unit volume:

$$rate = (\Delta\mu/6\pi k_B n_c T)^{1/2} \rho_{\text{liq}} (24 D_s n_c^{2/3}/\lambda^2) \exp[-(16\pi/3k_B T)(\gamma^3/\rho_s \Delta\mu^2)] \quad (13.3)$$

λ is a "typical" diffusion distance, D_s is the self-diffusion coefficient, and ρ_{liq} is the density of the liquid. Expressions of this kind have a rather unpleasant look and require funny numerical coefficients with lots of πs. More seriously, they are problematic because their overall reliability is the product of the reliability factors of all the embedded assumptions and approximations, a product that is nearly zero at the third assumption with 0.5 probability of being realistic. Models can be improved, phenomenological equations can be elaborated upon, densities can be taken as variable instead of constant, but one never gets past the basic stumbling points of a model, which is in all likelihood inadequate for the description of nucleation of crystals of complex organic molecules, for which the crucial quantities R, λ, and γ are unknown or possibly even undefined.

There are other problems. Homogeneous nucleation, or nucleation in a continuous environment, is a tempting assumption in theorization [15], but known facts seem to indicate that it is just wishful thinking, at least for organic materials. Molecular aggregation is extremely sensitive to the presence of even minimal perturbing factors like tiny impurities or even cracks in the container's surface, and every practicing chemist has experienced the effects of supplying mechanical energy by scratching the walls of a crystallization vessel. The sensitivity of nucleation processes to experimental, transient conditions is alarming from the standpoint of the construction of a unified theory of crystal nucleation. For example, stirring the melt has been reported to induce chiral symmetry breaking in the crystallization of 1,1'-binaphthyl [16].

At the other extreme, the formulation of molecular level theories is difficult if not impossible due to the intrinsic complexity of the molecular shape and potential in organic compounds. Molecular simulation studies of solid–liquid transformations concern mainly the Lennard-Jones fluid (the "spherical horse"), a hypothetical system composed of spheres interacting by some sort of sphere-sphere empirical potential [17], or simple molecules like the alkanes, usually modeled in the united-atom approach (each methyl or methylene group being a single interaction site) [18], or small globular molecules [19].

As mentioned before, a single MD run through a melting process is not an equilibrium simulation. If a proper description of the equilibrium is desired, the simulation should be carried out a large number of times forwards and backwards through melting and crystallization – a fantastic task for even the most optimistic believer in the exponential growth of computing power. Alternatively, one may calculate the free energy of the crystal and of the liquid as a function of temperature, using free energy simulations (Section 9.7). One such calculation has been carried out for n-octane, using an all-atom model and allowing for the full flexibility of the chain [20]. The thermodynamic integration requires a reference state for which the free energy can be computed exactly: for the crystal, this is the so-called Einstein crystal, an ensemble of non-interacting particles each of which oscillates around its lattice site according to a given force constant, with total internal energy U_{Ein}; for the liquid, the reference state is the ideal fluid at zero density, the ideal gas. An effective potential U_{eff} is then employed along with a step variable, λ, which carries the system from the reference state to the actual desired state. For the crystal, for example, one has:

$$U_{\text{eff}} = (1 - \lambda)U + \lambda U_{\text{Ein}} \qquad (13.4)$$

$$\Delta F = F(\lambda = 0) - F(\lambda = 1) = -\int d\lambda < (dU_{\text{eff}}/d\lambda) > \qquad (13.5)$$

A free energy difference becomes then an integral requiring only the knowledge of the internal energy U. The main problem of such approaches is that computing times scale with an unknown but for sure alarmingly high power of the number of atoms in the molecule, so the description of the molecule has to be restricted to a small number of sites and to very simple potentials. The return for such a huge computing investment rests on the accuracy of the potentials employed; for complex organic molecules where empirical force fields are often tentative and difficult to optimize, there are serious chances of using a big gun to shoot rubber bullets.

The ultimate challenge is a detailed description at a molecular level of the microscopic events that occur on the path from liquid to solid, both in structural and energetic aspects. In a brute force approach, one prepares a liquid computational box and applies standard molecular dynamics by lowering the temperature in steps, waiting for a crystallization trajectory to spring out of the computer. Success requires a big investment in time and effort, and a significant bit of luck, since MD trajectories are by definition unpredictable – if one wishes, chaotic, in the sense that small variations in initial conditions are unpredictably amplified as the simulation proceeds. The choice of temperature plays a crucial role, since too high a temperature may prevent molecules from sticking into place, and too low a temperature may deprive crystallizing molecules of the kinetic energy needed to swim through the liquid and reach their docking positions.

No wonder, successful studies of this kind can be counted on the fingers of one hand. An example is a much emphasized study of water freezing [21]: a computational box of 512 water molecules was simulated at 230 K, in "many" (one wonders exactly how many) trajectory calculations, each longer than 1 μs , or a very long time for an MD simulation. During a time lag of about 200 ns, subsequent snapshots along the trajectory clearly show the formation of an hexagonal proto-structure of hydrogen bonds, followed by a formation of the complete hexagonal network typical of solid water.

In a different approach, free energy Monte Carlo calculations have been applied for the ice nucleation process through the "umbrella sampling" formalism [22]. In this procedure, the free energy is written as a function of some order parameters, connected with the geometry of the hydrogen bonding coordination sphere; the operator knows which values of these order parameters correspond to the solid and which to the liquid. The system is then "pulled" through phase space by a systematic variation of one or two of these order parameters, the leading parameters, while the free energy is minimized with respect to the others (the "lagging" parameters); in this way, a minimum free energy path is deliberately mapped for the nucleation process (in the same manner, umbrella sampling can be applied to any kind of chemical evolution, if the proper parameters can be chosen). The calculation reproduces the latent heat of melting of water, and estimates a free energy barrier to nucleation of 80 kT. Pre-crystallization nuclei are said to be dynamic in character, and the size of the critical nucleus is estimated at 210–260 molecules. These considerations and these numbers look like distant voices coming from an unexplored, far-away planet [23].

13.5 Vapor–liquid and vapor–solid equilibrium

The most striking news that one learns when studying vapor–liquid phenomena is that not only does the vapor need to nucleate a liquid droplet to condense, but that also the liquid needs to nucleate a gas bubble to evaporate [24]. On the theoretical side, the simulation is made easier because the vapor is relatively simple to handle, on the experimental side, vapor pressure measurements in vapor–liquid equilibrium are fairly easy to perform. The Gibbs ensemble Monte Carlo method (Section 9.8) can be applied to the vapor–liquid equilibrium with considerable success: vapor pressure curves, second virial coefficients, and other equilibrium properties can be calculated by molecular simulation, and, remarkably, good results can apparently be obtained by highly accurate *ab initio* quantum mechanical potentials [25a] or by simple empirical potentials [25b].

Crystal growth by sublimation is extensively used for obtaining high purity crystals. The nucleation and growth of crystals from the vapor invariably occurs by grafting to some solid support, usually in practice some cold spot on the walls of a container. A likely hypothesis is then that the solid surface somehow acts as a nucleation catalyst. In principle, nothing forbids the simulation of the vapor–solid equilibrium along the same lines as for vapor–liquid equilibria.

13.6 Glasses

"In the past few years the glass transition phenomenon has won general recognition as one of the outstanding unsolved problems in condensed matter physics" [26a]. These not exactly encouraging words appear at the top of a paper by one of the pioneers in the study of the glassy state. Glasses are solids without diffusional freedom but without long-range structural periodicity. To some extent, they can be assimilated to supercooled liquids, but in glass forming systems the viscosity has a sharp increase close to the so-called glass transition temperature until diffusion is entirely frozen out.

The glassy state can be induced by a rapid cooling of the liquid below the melting point; it is not known whether all substances must vitrify under given conditions, or only some substances with peculiar properties can be vitrified – and, if the latter is true, it is not know what these properties should be. An empirical rule states that the glass-forming ability of a molecular liquid is the ratio of the boiling point to melting point, which is generally above 2 for substances that vitrify easily [26b]; this ratio should reflect the sluggishness of the crystallization process due to poor packing efficiency. No wonder the rule has been questioned by the accumulation of more experimental data, as invariably happens with shape-aggregation relationships for organic compounds. The glassy state can be studied at varying temperature and pressure by many techniques, including thermal analysis and all methods that probe the relaxation times within the system, plus all usual spectroscopic techniques. The structure factor can be obtained by neutron scattering measurements, yielding radial density distribution curves (recall Section 5.8).

Two diagrams may be proposed to illustrate some of the basic properties of the glassy state: the heat capacity trace (thermogram) and the viscosity–temperature diagram. Schematic examples are shown in Figs. 13.5 and 13.6. The salient features of the thermogram, left to right, are: (1) there is a residual C_p difference between glass and crystal, showing that the glassy state has some extra energy "pockets"; (2) on heating, the glass goes through an endothermic bump that corresponds to some activated process for devitrification, followed by an exotherm that indicates evolution to a crystalline state; (3) after this, the crystal behaves in the usual manner and melts at the normal melting temperature. The viscosity diagram shows the difference between materials classified as "strong" where the plot is linear and follows an Arrhenius activation law:

$$\ln(\eta) = \ln(A) - (E^*/RT_g)(T_g/T) \tag{13.6}$$

and materials classified as "fragile", where the viscosity breaks down much more sharply on increasing temperature, hence the name. These characteristic features must depend on the strength of the cohesion within the material, and, for example, hydrogen-bonding substances are usually "stronger" than "fragile" hydrocarbons, but exceptions are the rule in this kind of correlation.

There apparently is no doubt that the glass transition is reproducible, and hence can be discussed in thermodynamic terms rather than in evanescent kinetic terms.

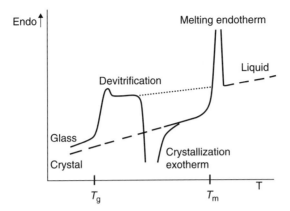

Fig. 13.5. Solid line: thermogram for the heating of a rapidly quenched liquid that has gone through the glass formation process. Dashed line: normal thermogram for the same material that has not gone through the vitrification process. Sample values for the glass transition temperature T_g and the melting temperature T_m are 117.5 and 178.15 for toluene. Adapted from ideas exposed in ref. [26].

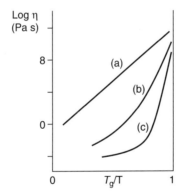

Fig. 13.6. Plots of viscosity against the glass transition to temperature ratio. Curve (a), "strong" material, Arrhenius law behavior; curves (b) and (c) increasingly "fragile" materials, where internal breakdown occurs as soon as temperature rises above T_g.

But what is the level of molecular organization within the glassy state? There certainly is a short range association effect quite similar to that observed in normal liquids, observable by peaks in the radial distribution functions, which reflects the main intermolecular bonding abilities of the compound; for example, some structuring arises from interactions between aromatic rings in benzene derivatives, or typically by hydrogen bonding. Depending on the strength of the intermolecular cohesion, there may also be some intermediate range ordering, possible due to a structuring among clusters of molecules, rather than among molecules themselves, which is revealed by a "pre-peak" in the static structure factor at very short θ or distances of 10–30 Å, as measured by neutron scattering [26,27].

The temperature and pressure evolution of glass-forming systems can be modeled by evolutionary molecular simulations like Monte Carlo or molecular dynamics. A good sample case is m-toluidine (1-methyl-3-aminobenzene), which has been extensively studied by a combination of neutron scattering and Monte Carlo simulation [27]. An analysis of the temperature, pressure, and isotopic substitution dependence allows an assignment and tentative rationalization of the features in the static structure factor: those features which depend on pressure are ascribed to the exclusion volume of the aromatic rings (a repulsive effect) while temperature-dependent features are associated with bond stretching phenomena, ideally hydrogen bonding. The pre-peak intensity increases on lowering the temperature down to the glassy state: interpreted as due to clustering phenomena induced by hydrogen bonding, this intermediate-range order introduces a fascinating, but disturbing, character of at least partial structural inhomogeneity into the glassy state and, possibly, into supercooled liquids in general [28]. A more detailed analysis [29] using molecular dynamics simulation allows a breakdown of the structure factor intensities over separate atomic contributions: it shows that the pre-peak results mainly from the contributions of nitrogen atoms involved in hydrogen-bonded molecular clusters. In addition, the simulation mimics almost perfectly the C_p anomaly at the glass transition (Fig. 13.7) and allows a molecular level interpretation of the structural rearrangement at the transition temperature, which involves a change in the reciprocal orientation between hydrogen bonds and inter-ring interactions. This example illustrates the efficiency and the richness of microscopic detail afforded by accurate molecular simulation in the analysis and interpretation of experimental observations – it goes without saying that the simulations also perfectly reproduce all the properties of the corresponding normal liquid.

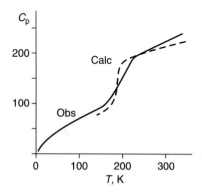

Fig. 13.7. Simulated (dashed, from ref. [29]) and experimental (full line, from ref. [26b]) evolution of the heat capacity of liquid toluidine through the glass transition. The agreement is impressive.

13.7 Liquid crystals

This beautiful oxymoron refers to a particular state of matter, encountered with organic compounds whose overall envelope presents a particularly simple shape, most often an ellipsoid or a discoid. These molecules organize themselves with a varying degree of structural periodicity (order), in one or two dimensions, intermediate between the liquid and the crystal, hence the name of mesophases and mesogenic compounds. Since the structural periodicity is to some extent commensurate with the molecular shape, the orientation of the elongation axis or of the main plane are easily recognizable geometric descriptors of the intermolecular structure of these particular condensed phases. The preparation of mesogenic compounds and the experimental analysis of the corresponding phases are the subject of entire books in themselves and will not be reviewed here. The same applies to the evolutionary simulation of mesophases [30]; only a flavor of these simulations will be given here.

Although the complete atomistic simulation of ensembles of mesogenic molecules is within reach of present computational facilities, the traditional treatment of liquid crystals in molecular dynamics or Monte Carlo simulations makes use of the Gay–Berne potential, an ingenious computational machine whose aspects deserve to be described here for their epistemological implications. An ordinary Lennard-Jones (LJ) potential, equation 4.38 or 4.40, can be written as a function of the distance between two particles, R_{ij}, the well depth ε and the equilibrium separation σ. An ellipsoidal object is identified by the position of its centroid and by an orientation unit vector $\mathbf{u}$, and the Gay–Berne (GB) potential is a modified LJ that takes into account the anisotropy of the ellipsoid, both in energy and equilibrium separation:

LJ: $\quad U_{LJ} = U(R_{ij}, \varepsilon, \sigma) \quad (= 4.38) \quad (13.7)$

GB: $\quad U_{GB} = U[R_{ij},\ \varepsilon(\mathbf{u}_i, \mathbf{u}_j, R_{ij}),\ \sigma(\mathbf{u}_i, \mathbf{u}_j, R_{ij}), \mu, \nu] \quad (13.8)$

In the above formulation R_{ij} is the distance between centroids, while the overall distance dependence is still of the 12-6 type. Both well depth and equilibrium separation depend on the distance and relative orientation of the two ellipsoids, and two extra parameters, μ and ν, add flexibility to the model by exponential scaling of the implied dot products between orientation vectors. All in all, a GB potential depends on four parameters, the length/breadth ratios for ε and σ, plus μ and ν. Figure 13.8 shows the nice result: an orientation dependence of the potential. In addition, the ellipsoids or the discoids can be made electrically active by imposing some central multipoles along the molecular envelope; the calculation of the multipole energy then is easily accomplished by standard electrostatic formulas (see Section 4.2 and ref. [30]). For two dipoles $\mathbf{d}_i$ and $\mathbf{d}_j$ at a distance $\mathbf{R}$, for example, the total simulation energy of the liquid crystalline sample then becomes:

$$U(\text{dipole}) = (d_i d_j / R^3)[\mathbf{d}_i \cdot \mathbf{d}_j - 3(\mathbf{d}_i \cdot \mathbf{R})(\mathbf{d}_j \cdot \mathbf{R})] \quad (13.9)$$

$$E_{TOT} = U_{GB} + U(\text{dipole}) \quad (13.10)$$

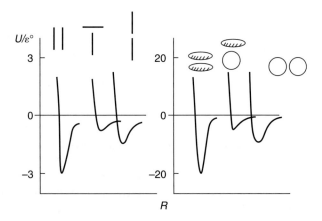

Fig. 13.8. Gay–Berne potentials: limiting cases of reciprocal orientation between ellipsoids or discoids. ε° is a reference well depth. In discotics, the length/breadth ratio is substituted by a thickness/diameter ratio. Adapted (with permission) from ref. [30].

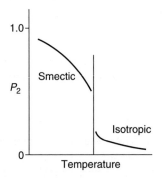

Fig. 13.9. Typical result of a simulation for a liquid crystalline ensemble: as the clearing temperature is reached, the order parameter P_2, equation 13.11, suddenly drops to zero.

In standard modeling, the parameters of the force field are optimized to reproduce experimental properties of real substances, e.g. heats of sublimation or of melting, and the parametric force field is then used to predict unknown values of the same and other physical quantities. In the study of liquid crystals by the above formalism the logical path is sometimes different: one asks for certain properties, and then tries to find the parameters that would be needed to produce such a result. Consider, for example, the clearing point between a smectic phase and the liquid; the transition is easily characterized by the evolution in temperature of the second moment of the distribution of molecular orientation vector **u** with respect to a principal direction **e** [30]:

$$P_2 = 3/2 \ <(\mathbf{u} \cdot \mathbf{e})^2> -1/2 \qquad (13.11)$$

A schematic illustration of this situation is shown in Figure 13.9. Clearly the position of the clearing point is tunable according to the values assigned to the parameters in the GB potential and to the molecular multipoles. The properties of the real molecule that would exhibit the required property are then inferred from the values of the parameters, e.g. a certain length/breadth ratio and certain values of the molecular multipoles.

13.8 Nucleation and growth from solution: Experiments

To many people, chemistry is a science of colors; a typically impressive experiment carried out by elementary school teachers is the mixing of two colorless solutions to obtain a colored product. It is easy to understand how awe-inspiring it can be, along the same line, the mixing of two transparent solutions to yield a turbid medium, or the precipitation of solid matter with a clearly distinguishable external morphology, faces, and dihedrals [31]. This is as close to a miracle as the fantasy of the layman can go, and there is no doubt that crystallization has exerted a special fascination throughout the centuries, with the birth and spread of crystal fads, some of which survive today in the selling of healing crystals, miracle pyramids, and the like. It must be said that the precipitation of a nice and clean crystalline material perhaps after a long synthetic effort is a source of special satisfaction and reward even to an experienced chemist [32], just as the stubborn refusal of some solutes to coagulate into their solids is a source of frustration and a major obstacle to further purification.

13.8.1 *Overview*

The basic tenets of the science of crystal development from solution are few and very broad: (1) solute molecules associate in solution until the newly formed aggregate grows to a certain size; this stage is usually called "nucleation"; (2) some of the associative nuclei then further develop into larger crystals; this stage is usually called "growth" – but notice that the verb "to grow" had also to be used in the definition of nucleation.

In macroscopic phenomenological terms, the formation of an elementary aggregate involves a change in free energy due to aggregation itself, and a change in free energy due to interface tensions, as already outlined in equation 13.2. The development of the theory of nucleation from solution follows a path that is very similar to that for nucleation from the melt, with the obvious additional complication that the system is now a two-component (if not a many-component) one. However, as soon as the nucleation heterogeneity appears, the basic concepts of thermodynamics are put in question. In fact, free energy must be a state function, but in a heterogeneous system the state-defining variables may not be just the traditional ones, temperature, pressure, volume, chemical compositions, but must be supplemented by some descriptor of the amount of heterogeneity (order? size distribution?), and the recognition and measurement of these supplementary variables is nearly always extremely difficult, because very few if any of the traditional chemical analysis techniques can be made sensitive enough to probe the intimate structure of these microscopic and fluctuating aggregates. Especially for large and flexible organic molecules, the very concept of phase equilibrium becomes questionable as most of the time the experimental reproducibility and identification of states and phases are difficult and one does not know, literally, what state or phase one is speaking of.

Then the nuclei grow into macroscopic crystals, but there is no way of properly defining a transition from the nucleation to the growth stage. There is no doubt that the growth mechanics must be directed by the template provided by the pre-existing stable structure of the nuclei. Phenomenological theories of crystal growth use schematic models of the growing surfaces, which are postulated to be very flat and regular at low temperatures and to become "rough" at higher temperature; this roughening is due to the presence of kinks, steps, and ledges, whose efficiency in promoting growth is evident on the basis of simple geometrical intuition, because a molecule attached at a kink site interacts with the bulk in two or three directions, while a molecule on a flat surface interacts along one direction only. Again, while working reasonably well for simple particles, such models are clearly inadequate to treat complex molecules. Given such premises, it is not surprising to learn that nucleation and growth theories for real organic molecules are presently in an unsatisfactory state.

Nucleation, growth, and transformations among nucleating and growing species have been and are being studied extensively, taking full advantage of the recent large improvements in analytical techniques. The overview is at the same time an extremely stimulating and outrightly frustrating one: the dynamics of these processes is seen to depend on nearly anything: temperature, chemical nature of the solvent, solute solubility, pH, stirring rate, total volume of the sample [33]; chemical additives (and hence impurities in nanoconcentrations) [34]; sonication (treatment with ultrasonic waves) [35]; confinement in microporous solids, emulsification [36]; epitaxy [37], and even cross-fertilization of various polymorphic nuclei [38]. Apparently, an external electric field or the extent and direction of polarization of an external light source can influence in stereoselective ways the nucleation and crystallization of highly polarized molecules [39].

13.8.2 Light scattering, calorimetry

So the hunt for this elusive entity, the growing nucleus, is open. Useful analytical techniques for the study of crystal nucleation and growth must be sensitive to the size of the evolving clusters, and, ideally, also to the detailed intermolecular structure of these clusters. The first requirement seems more within reach than the second.

When aggregates grow, the solution becomes turbid because of light scattering, and accurate measurements of the photon autocorrelation function provide a direct access to the radius of the scattering nuclei. This technique can be comfortably used for large nuclei formed by crystallizing protein molecules: in a typical study on lysozyme [40], cluster radii were seen to grow from 20 to 50 Å in about one hour, with the number of different clusters decreasing from 1000 to 10 over the same period. For ordinary organic molecules, practical difficulties arise when using this technique, due to the smallness of the clusters, to multiple scattering, and to the requirement that the suspension be of non-interacting monodisperse clusters (all nuclei of the same size). Nevertheless, light scattering techniques are employed to study the formation of liquid droplets from water vapor [41], allowing the measurement of nucleation rates (number of nuclei formed per unit of volume and time) and of cluster composition, which for water range from 10^5 to $10^9 \text{ cm}^{-3} \text{ s}^{-1}$, and from 20 to 30 molecules per cluster, respectively.

Similarly, the particle size and number concentration in crystallizing systems can be counted directly through the obscuration of a laser beam by particles flowing between the beam and a photodiode [42], or via the attenuation of a sound wave traveling through the system [43], with the added advantage in the latter case that the measurement is not hampered by optical saturation. Treating the experimental data by classical nucleation theories, nucleation and growth rates can be obtained, with orders of magnitude of $10^{12} \text{ m}^{-3} \text{ s}^{-1}$ and 10^8 m s^{-1}, respectively. Particle sizes, apparently, can vary from 5 [42] (a puzzling, astonishingly small number) to 500 [43].

In an alternative to counting the number of condensed particles, one can try to monitor the concentration of remaining solute [44], or some other thermodynamic property that evolves with cluster formation: for example, microcalorimetry can measure the heat absorbed as crystallization occurs, and thus can draw a de-supersaturation curve [45]. Needless to say, none of the above directly or indirectly touches upon the question of the internal structure of the generated clusters.

13.8.3 *Chemical spectroscopy*

Why not use standard chemical spectroscopy to study intermolecular association and nucleation? The main obstacles are (1) that one must find signals that change upon association, and this is not easy within the weak force regime proper of intermolecular bonding, except perhaps for hydrogen bonds, and (2) that one must also have a clear dependence on cluster size, which is even more difficult to achieve, with the added complication that (3) many clusters of different size may be present in the nucleating solution. For example, the existence of hydrogen-bonded aggregates of the size of a few molecules could be ascertained in the benzyl alcohol–carbon tetrachloride system [46], but only by a multivariate analysis of an entire spectral region (3,100–3,700 cm^{-1}), because there is no obvious way of clearly separating signals from monomers and from oligomers. In a similar manner, a principal component analysis (recall Section 8.10.2) of the 1,600–1,680 cm^{-1} spectral region carried out on a crystallizing progesterone solution could distinguish between spectral features of the solute in solution and the solute crystals [47], and Raman spectroscopy can distinguish the relative amounts of two polymorphic phases in a polymorphic transformation by quantitative calibration of peak heights over the whole 200–1400 cm^{-1} region [48]. A subtle change in spectral features in the carbonyl absorption region between tetrolic acid in chloroform and in dioxane spots the formation of a heterogeneous hydrogen bond that leads to the crystallization of a dioxane solvate [49]. Here we are here close enough to gleaning what molecules are doing in solution prior to the crystallization event.

NMR spectroscopy is another obvious candidate: if, after all, it is possible to determine the structures of proteins in solution, why should it not be possible to determine the structure of aggregating molecules? In a relatively straightforward experiment, the disappearance of monomeric lysozyme macromolecules from solution after supersaturation was recorded [50], but for small-molecule clusters, the problem is always the same; scarce structural sensitivity. In a test study, the NMR spectra of dimerizing amide compounds were analyzed in terms of the formation of dimers and chains, by varying the solute concentration to find association-dependent bands [51]. A correlation between dimer structures found in solution or found in the crystal by X-ray diffraction was found, again, opening a window on solution behavior, possibly on the way to more complete association studies, although even here the determinations had to rely on chemical shift changes as small as 0.1–0.2 ppm or even less. One cannot imagine what should be measured for weakly bound van der Waals complexes.

13.8.4 X-ray scattering and diffraction

The routine availability of high-power synchrotron radiation is likely to play a major role in the development of in situ analysis of the nucleation event. The high-brilliance, entirely wavelength-tunable radiation can probe particle sizes in the 30–1000 Å range by scattering, or in favorable cases even the ordering at atomic level within the clusters, in the 2–10 Å range, just as in ordinary powder diffraction experiments. The high output radiation intensity allows data collection on the timescale of milliseconds and thus permits time-resolved crystallization studies. The nucleation of glycine was studied by X-ray scattering [52]: the relationship between scattered intensity and particle radii and shape goes through a phenomenological model involving fractals, which, if rather obscure to the chemist, is apparently reliable enough to draw vital conclusions about the nucleation mechanism: glycine exists in solution as a mixture of monomers and dimers, which at increasing supersaturation coalesce into liquid-like particles, which then reorganize into crystalline entities. These conclusions are enough to dispose in one single stroke of classical nucleation theory which proceeds from differences in chemical potential between solution and crystal.

In diffraction experiments, the diffraction pattern is directly obtained from the crystallizing supersaturated solution, and when the contribution from the diffuse scattering due to the solvent is filtered out, the data allow a time-resolved study of crystal formation. Solution patterns can be compared with the pattern from dry powders, and the growth process can be monitored; for example, a nice coincidence of solution and dry powder diffraction patterns was observed for benzamide and dibromoaniline [53]. Such techniques are widely used in nucleation and growth studies for inorganic materials [54].

13.9 Crystal growth and morphology

The morphology of minerals and inorganic crystals is usually evident, and is actually impressive, as one may realize by thinking of gems. The strong forces, mainly of coulombic nature, that hold together these materials impart to them a strong mechanical resistance and a strong anisotropy. Growing single crystals of weakly bound organic materials is much more problematic: samples are usually very small, and their outer morphology is much less well defined or recognizable. For an impression of actual morphology considerations in the field of organic crystals, Table 13.1 collects the recorded qualifiers of sample shape in a large set of crystal structures in the Cambridge Structural Database; these qualifiers are perforce approximate, if one recalls that samples for X-ray analysis may be as small as fractions of a millimeter, but nonetheless one can see there a predominance of growth forms that lead to an overall globular shape, rather than plate or acicular. The sampling is not unbiased because it refers to crystals selected for X-ray analysis, and an expert crystallographer will always select for that purpose an individual as similar as possible to a sphere, even when the crystal batch contains individuals of different morphologies.

13.9.1 *Crystal faces, attachments energies, and morphology prediction*

After the early nucleation and accretion events, large aggregates of unmistakable crystalline nature eventually result, and further attachment of molecules from a surrounding fluid phase leads to what is known as crystal growth proper, an intrinsically epitaxial, two-dimensional phenomenon occurring on well-developed crystal faces,

Table 13.1 Number of qualifiers of crystal morphology in the Z(1) database (see Chapter 8): 8,519 total entries. 3D, 2D, and 1D approximately label crystals grown in globular, plate or acicular form, respectively.

Descriptor	
Prism	2946
Block	1343
Parallelepiped	238
Cube, box	140
Total 3D	55%
Plate	1546
Slab	43
Total 2D	19%
Needle	1169
Cylinder, rod	226
Total 1D	16%

as opposed to the three-dimensional shaping of previous aggregates, which can be either in crystalline, liquid, or-semi-structured states.

Each face of a macroscopic crystal must be parallel to a Bragg (*hkl*) plane (see Fig. 5.13); when cell parameters are known, it is a relatively easy matter to assign Bragg indices to each prominent face observed on a macroscopic crystal sample, and vice versa. Early crystallography used planes and angles between them to determine cell parameters – at least, when large enough crystals for visual observation can be grown. If the crystal structure has been determined by X-ray diffraction, an appropriate projection can be performed so as to have a view of the internal crystal structure down the perpendicular to each face. What, then, should one "see" on a given (*hkl*) face? It is important to realize from the very beginning that the surface population on a given face can be uniquely defined only in terms of lattice points; but when considering the actual molecular structure with its electron density envelope, and the organization of matter around lattice points, including the symmetry operations within the cell, what one "sees" depends on where one "cuts" (Fig. 13.10). In other words, a static definition of surfaces and slices through the crystal structure, a vital link between macroscopic morphology and molecular level structure on the way to the molecular modeling of surface properties, depends on some subjective choices. For simple cubic crystals made of atomic spheres (again, the "spherical horse") things may be easier, because atomic positions mostly coincide with lattice points, and surfaces can be classified as compact, flat, or rough. For organic crystals made of complex, flexible molecules, these definitions lose much of their uniqueness.

On the other hand, Fig. 13.11 shows a sketch of the actual situation at the boundary between the growing crystal face and the solution. Solute molecules approach the surface in conformational freedom, amidst solvent molecules, some of which are also adsorbed on the crystal surface. There is little doubt that the structure of the bulk solution is different from that of the solution close to the surface, which limits molecular mobility and induces an electrostatic potential onto the solute molecules,

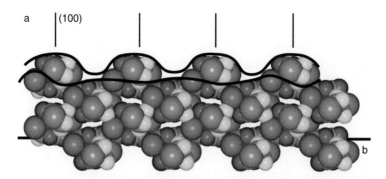

Fig. 13.10. Shows a cut perpendicular to the *a* direction (perpendicular to the plane of the page): the two wavy lines correspond to a different choice of surface molecules, and hence to a different surface structure, roughness, etc.

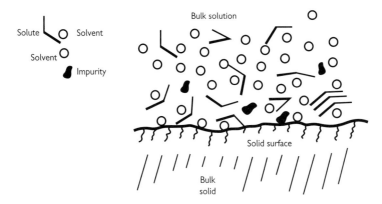

Fig. 13.11. A schematic representation of the complex situation near a crystal–solution interface. Shows solute molecules with variable conformation in the bulk solution, partial structuring of solute molecules near the interface, and solute, solvent and impurity particles adsorbed on the solid surface. The structure of the solid at the surface may be different from the bulk solid structure.

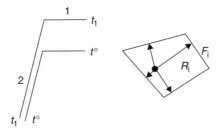

Fig. 13.12. Left: explanation of the relative growth of two faces: between time t° and t_1, surface 1 is displaced (grows) faster, surface 2 grows more slowly, so that at time t_1 surface 2 has become much larger than surface 1. Right: the construction of morphology, shown in two dimensions: the lengths of vectors R_i out of a selected origin (black circle) are proportional to the growth rate, and the external morphology results from the envelope of faces F_i perpendicular to these vectors.

and hence is likely to induce a partial solute structuring. Impurities also play a substantial although hardly predictable role, being active at a level of the order of 10^{-9} M [55]. The solid itself is likely to be sensitive to the presence of the fluid phase, with a partially reconstructed surface structure that may differ from the bulk structure. As the crystal grows, the addition of new molecules to the various possible faces occurs in a highly anisotropic fashion, because the energetic path to attachment is obviously different for different surface aspects. As a result, some faces grow faster than others, and the macroscopic specimen takes up its experimental morphology, as explained schematically in Fig. 13.12.

With a simple construction, the predicted crystal morphology is established as explained in Fig. 13.12 (right), using expansion vectors whose direction is determined by the face indices and whose modulus depends on the growth rate. The question is, how does growth rate depend on surface structure? Early phenomenological theories assume flat surfaces in the first place, which grow only by virtue of molecular attachment to some dislocation. Along these lines, the growth rates may be taken just as inversely proportional to the $d(hkl)$ spacing (Fig. 5.13); the smaller the spacing, the stronger the cohesion; or, in a more appropriate way, as proportional to the attachment energy, which is the energy released when a new layer of molecules is deposited on the surface. The attachment energy can be calculated by systematically removing surface molecules, although with some reference-dependent assumption, as discussed above, and computing the related energies by atom–atom empirical potentials. In this way, computer programs for the tentative prediction of crystal morphology can be set up [56].

As a help in the identification of the relevant surface networks, the periodic bond chain concept is sometimes used: a molecule within the network is represented by a point, and the network is kept together by "bonds" that are, in essence, molecule–molecule energies calculated by the applied intermolecular potentials [57] (see the "structure fingerprints" of Section 14.2.3). On such flat surfaces, the molecules are assumed to be closely packed in regular arrays more or less like spheres, until some external event (typically, a rise in temperature) induces a so called "roughening" transition, or the creation of non-specified surface irregularities that create molecular niches into which the incoming molecules may dock more promptly. If one considers the complex structure of a molecular surface (Fig. 13.10), it is doubtful that such simplified schemes may lead to a consistent routine prediction of crystal morphology. They may perhaps be applied more confidently to the prediction of crystal growth by sublimation, where the complications due to the presence of the solvent are removed.

The simulation of morphology using attachment energies only is static in nature. As an improvement, in an attempt to include surface roughening, the Monte Carlo method is used for the simulation of the attachment process over the crystal graph, to fill the gap due to the neglect of the effects of supersaturation and temperature; differently from free atomistic simulation, however, in this approach a number of kinetic and thermodynamic constraints (crystal network, supersaturation and kinetic parameter) are input to and enforced throughout the calculation [58]. Not surprisingly, while static attachment methods show little sensitivity to the force field employed, the Monte Carlo approach is extremely sensitive to small variations in the force field, due to the stochastic nature of the phenomenon and to the importance of molecular detail in the anisotropy of surface roughness. In an extensive comparative predictive study of paracetamol morphology, different results were obtained with Dreiding, distributed multipole and Compass force fields (see Chapter 2), none of which gave good agreement with the experimental morphology [59]. One sees here a harbinger of the complex interplay between the scope of the simulation method and the accuracy of the force field that also appears when dealing with crystal structure prediction (Section 14.4).

Crystal morphology is well known to be extremely solvent-dependent. If predictions other than in vacuo are desired, the influence of the solvent must be taken into account, and the simplest way of doing this is to consider an effective attachment energy, $E^*(\text{att}) = E(\text{att, vacuo}) - E(\text{att, solv})$, where $E(\text{att, solv})$ is the energy cost for removing the solvent from the surface. This quantity can be obtained by molecular dynamics simulations in which a crystal surface is equilibrated in the presence of a solvent layer, and by partitioning the total configurational energy into crystal–crystal, solvent–solvent and crystal–solvent terms, or by appropriate manipulations of the total energies of the separate crystal and liquid phases against the energy of the heterogenous system. The modified attachment energy $E^*(\text{att})$ is then introduced in the usual static scheme for the prediction of the crystal morphology [60].

13.9.2 *Electron micrography and atomic force microscopy (AFM)*

Well grown faces of organic crystals can be studied in some favorable cases by ordinary optical microscopy, to a detail of perhaps some fractions of a millimeter. For higher resolution, use is made of electron micrography, an imaging technique that relies upon the interaction of an electron beam with the electron density of the surface atoms. Atomic level resolution can be obtained by the atomic force microscope, a unique tool for revealing molecular and atomic detail on surfaces. Its setup (Fig. 13.13) relies on atomic level technology but is disarmingly simple in concept. A sensing probe, consisting of a tip made of an inert material (e.g. SiN) about 100 Å or less in diameter protrudes from the end of a flexible cantilever, and is brought at atomic distance from the surface to be explored by a piezoelectric actuator (z-actuator). The tip interacts with the surface and is pulled by the intermolecular force; as the surface is moved below the tip by x–y piezoelectric actuators, the z-actuator moves the tip up and down to maintain a constant force between the surface and the tip. These up and down movements are translated into angular deflections of a laser beam reflected off the end of the cantilever towards a photodetector. The response of the photodetector is converted to a topographical image of the surface by image processing techniques. The dimensions of the apparatus are amazing – the tip-surface distance is of the

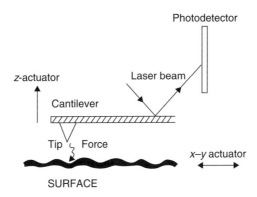

Fig. 13.13. Schematic drawing of an AFM instrument.

order of 2 Å, the cantilever spring constant is of the order of $0.1\,\mathrm{N\,m^{-1}}$, and actuators work at the rate of 1 Å displacement per volt. The images thus obtained have a captivatingly realistic appearance, showing hills and valleys of atomic dimensions. One should never forget, however, that what one is actually seeing is not really nuclei and electrons, but a reconstructed image of up and down cantilever displacements for constant-force condition. AFM can be applied to non-conducting materials, differently from STM (scanning tunneling microscopy), and, even more important for the structural chemistry of crystal formation, it can be applied to surfaces immersed in a solvent and under solvent flow conditions, so that the degree of saturation of the solution can be controlled [61].

AFM probes a crystal surface of nanoscopic dimensions at an advanced stage of crystallinity, and hence is the ideal tool for studying crystal growth and dissolution. The list of phenomena that can be studied by in situ AFM on organic crystals in solution is long and is limited only by the investigator's imagination: surface topography, dissolution and growth of faces, advancement of growth terraces; etching by different solvents; the effect of different degrees of saturation or of additives; epitaxial growth; and all these phenomena can be fairly easily correlated with the surface structure at atomic resolution, if known by X-ray diffraction on single crystals. In addition, the results of AFM experiments can be compared with and supplemented by the results of static or dynamic simulations of attachment energies or of the actual crystal–liquid interface.

13.10 Evolutionary molecular simulation

There is little hope of studying theoretically the complex intermolecular situation sketched in Fig. 13.11 with even modest accuracy, unless evolutionary molecular simulation is employed. Recourse is thus made to the classical arsenal of such simulations, broadly subdivided into Monte Carlo (MC) and Molecular Dynamics (MD). The conceptual foundations of these two methods, and their computational feasibility, are now reviewed again from the perspective of their use for nucleation and growth studies.

In MD, time is a clearly singled out variable in a deterministic simulation based on a postulated force field and on the classical equations of motion. For the simulation of an evolving crystal aggregate, MD has the obvious advantage that the kinetics of the process is transparent, as accretion rates can be immediately described as a function of computational time, although the rate of any molecular process is obviously dependent on the postulated force model. In contrast, there is no apparent time variable in an MC simulation, because evolution steps are random and may randomly affect molecular evolutions which in reality happen on different timescales. If, as is often the case, time in MC is taken as proportional to the number of moves, one is implicitly assuming that all molecular moves occur on the same timescale, perhaps not a very severe approximation in studies of molecular aggregates bound by nearly isotropic van der Waals forces. In a variant of the MC formulation, called kinetic Monte Carlo (KMC)

[62], moves are accepted or rejected according to a comparison between a random number $0 < r < 1$, and a probability of the performed move, estimated according to preselected rate rules; compare this criterion with the standard MC energy criterion of equation 9.10b. The time spacing between successive MC moves is thus calculable, but only at the price of having to formulate sensible rate rules.

MD simulations of crystal growth can be readily set up, at least in principle. One may think of using a computational box made of a crystal slab in contact with a liquid phase; this would have the obvious problem that at any non-zero temperature part of the solvent would be lost to the overhanging vacuum, and that the crystal slab surrounded by vacuum would distort under the action of extreme surface forces. These problems may be avoided by periodic boundary conditions, using a computational box in which a crystalline lamella is enclosed between two solvent slabs so that on applying the periodicity conditions, the continuity of the two phases is preserved. The MD approach has the usual shortcomings: the first is the timescale problem, which is especially acute in this case because nucleation and growth occur on the scale of seconds, at the shortest, while MD simulation may last milliseconds at best. Put another way, typical nucleation rates may be of the order of the appearance of one particle every 10^{14} seconds in a cube with a side of 10 Å. The second is the size problem, because 10^3–10^4 molecules are considered instead of 10^{23}. Such simulations are therefore restricted to the study of some elementary events in the nucleation or growth path, such as instantaneous attachment or detachment of solute and solvent particles, or the different structuring of solvent layers close to or far from the crystal surface [63]. Alternatively, extreme supersaturations must be used, to convince solute molecules to aggregate out of the solution, with the obvious consequence that calculated nucleation rates may be fourteen orders of magnitude larger than experimental ones [64]. Better results are obtained with small, globular and rigid molecules such as SeF_6 [65], or when the aggregation forces are very strong, as for covalent C—C bonds [66].

Both in MD and in MC, long simulation periods are spent in observing evolutions that are irrelevant to the process under study, such as internal molecular vibrations, or small amplitude oscillations around molecular positions in the liquid, while the process-driving steps like, for example, diffusion from the solute to the crystal surface, occur very seldom over extremely large numbers of MC moves or MD steps. In this respect, MC has an advantage over MD because in MC a "move" can be anything the operators want it to be, provided that reasonable acceptance criteria are selected. Thus, one has MC "smart move" [67], in which the smart moves are the change of identity between a solute and a solvent particle, or the swapping of the position of a solute and a solvent molecule in close contact. These smart moves greatly enhance the essential transport phenomena in the computational box. In the so-called aggregation-volume-bias variant of the Monte Carlo scheme, [68] the computational box space is explicitly divided into aggregating and non-aggregating zones, and MC moves are swaps between the two zones; in essence, the particle is "picked up" from a bulk solution region and "stuck" into an aggregating zone, thus elegantly straddling the long and uncertain diffusion path that would result from the usual Metropolis-type

MC in which moves are only as described in equation 9.10. In MD, umbrella sampling can be used to drive the system to the desired path through configurational space.

References and Notes to Chapter 13

[1] Westwell, M. S.; Searle, M. S.; Williams, D. H. The "n" effect in molecular recognition, *J. Mol. Recognition* 1996, **9**, 88–94. The title of this paper stems from an analysis of the sharpness of the melting transition, in spite of the flatness of the free energy landscape (free energy changes in the surroundings of the equilibrium point are very small: for water, for $\Delta T = 10\,\text{K}$ from equilibrium ΔG_{melt} is only $0.2\,\text{kJ}\,\text{mol}^{-1}$). The root of the problem is in the effects of size and of cooperativity in extended arrays of weak interactions.

[2] Dannenfelser, R. M.; Surendran, N.; Yalkowsky, S. H. Molecular symmetry and related properties, in *SAR and QSAR in Environmental Research*, 1993, Vol. 1, pp. 273–292, Gordon and Breach. This paper includes a large collection of melting entropies.

[3] For a more detailed discussion see Pinal, R. Effect of molecular symmetry on melting temperature and solubility, *Org. Biomol. Chem.* 2004, **2**, 2692–2599. This paper also gives the formal connection between melting entropy and ideal solubility.

[4] See Gavezzotti, A. Molecular symmetry, melting temperatures and melting enthalpies of substituted benzenes and naphthalenes, *J. Chem. Soc. Perkin 2* 1995, 1399–1404. This paper apparently shows that the effect is at work also for asymmetrically substituted compounds, so the relationship to symmetry number becomes less strict, and that more symmetric molecules also have higher melting enthalpies. Recall also Section 1.4. See also Brown, R. J. C.; Brown, R. F. C. Melting point and molecular symmetry, *J. Chem. Ed.* 2000, **77**, 724–731, quoting an original statement by Thomas Carnelley (1882) more or less to the same effect: "of two or more isomeric compounds, those whose atoms are more symmetrically or more compactly arranged melt higher...". A related paper is Yu, L. Inferring thermodynamic stability relationships of polymorphs from melting data. *J. Pharm. Sci.* 1995, **84**, 966–974.

[5] Reuter, J.; Buesing, D.; Tamarit, J. L.; Würflinger, A. High-pressure differential thermal analysis of the phase behavior in some *tert*-butyl compounds, *J. Mater. Chem.* 1997, **7**, 41–46. This paper also discusses some phenomenological theories of crystal melting (like the Pople-Karasz theory, 1961) that take into account orientational disorder in the solid, based on ideas presented in 1939 by Lennard-Jones and Devonshire. The essential parameter of such a theory is the ratio of the barriers to reorientation and to diffusion, which is also a measure for the anisotropy in molecular shape.

[6] (a) Gavezzotti, A. A molecular dynamics view of some kinetic and structural aspects of melting in the acetic acid crystal, *J. Mol. Struct.* 1999, **486**, 485–499; (b) Ferretti, V.; Gilli, P.; Gavezzotti, A. X-ray diffraction and molecular

simulation study of the crystalline and liquid state of succinic anhydride, *Chem. Eur. J.* 2002, **8**, 1710–1718.

[7] Standard NPT-MD, GROMOS package, computational box of 400 benzene molecules ($5 \times 4 \times 5$ unit cells), periodic boundary conditions, temperature, and isotropic pressure control, force field from a slight modification of the Williams potentials (ref. 44, Chapter 4), other working conditions as described in ref. [6].

[8] Gavezzotti, A. A molecular dynamics test of the different stability of crystal polymorphs under thermal strain, *J. Am. Chem. Soc.* 2000, **122**, 10724–10725.

[9] For a simple calculation based on the Lindemann criterion and comparison with experiment see for example Tabor, D. *Gases, Liquids and Solids*, 1969, Penguin Books, Baltimore, p. 207. The calculation assumes harmonic vibration of the particles, against Young's modulus. Melting temperatures are calculated with reasonable accuracy for metals and even for quartz, but not for organic molecules.

[10] See for a description Frenken, J. W. M. *Surface melting*, Endeavour, New Series, 1990, **14**, 2; Dash, J. G. Surface melting, *Contemporary physics* 1989, **30**, 89. The occurrence of surface melting is the commonly accepted wisdom for metals and crystals of things like Ar and Ne. The presence of a liquid film on the surface of ice is given as an explanation of ice's slipperiness, but not without challenge.

[11] A soft mode is a lattice mode whose frequency becomes imaginary at some value of the density (or external pressure). See e.g. Dove, M. T.; Rae, A. I. M. Structural phase transitions in malononitrile, *Faraday Disc.* 1980, **69**, 98–106. The whole discussion, entitled "Phase transitions in molecular solids", is extremely instructive reading for the theory, experiment, and simulation of second-order phase transitions, operatively defined as those transitions that occur without a major discontinuity in enthalpy and heat capacity: the onset of molecular rotation is an example, as opposed to first-order transitions like polymorphic transitions or melting.

[12] McGann, M. R.; Lacks, D. J. Chain length effects on the thermodynamic properties of *n*-alkane crystals, *J. Phys. Chem. B* 1999, **103**, 2796–2802.

[13] See for example Oxtoby, D. W. Nucleation of first-order phase transitions, *Acc. Chem. Res.* 1998, **31**, 91–97, and references therein. The classical nucleation theory (CNT) is challenged by extended theories, one of which was unfortunately called "density functional theory" because free energy is a functional of the average density profile of the nucleating site, generating some confusion, at least for theoretical chemists, with quantum chemical density functional theory.

[14] See a critical analysis of the problem in Auer, S.; Frenkel, D. Quantitative prediction of crystal-nucleation rates for spherical colloids: a computational approach, *Annu. Rev. Phys. Chem.* 2004, **55**, 333–361. See also Turner, G. W.; Bartell, L. S. On the probability of nucleation at the surface of freezing drops, *J. Phys. Chem.* A2005, **109**, 6877–6879, and references therein. These papers rely on a statistical analysis of simulated nucleation trajectories of clusters of SeF_6 molecules to estimate nucleation rates.

[15] See, for example Wales, D. J.; Berry, R. S. Freezing, melting, spinodals and clusters, *J. Chem. Phys.* 1990, **92**, 4473–4482; the theory exposed there relies on the assumption of solid-like and liquid-like clusters coexisting in equilibrium at a given temperature, a rather curious assumption, at least for a molecular crystal chemist with a view on the complex kinetic interplay in molecular aggregations. In Hettema, H.; McFeaters, J. S. The direct Monte Carlo method applied to the homogeneous nucleation problem, *J. Chem. Phys.* 1996, **105**, 2816–2827, one reads that "the traditional definition of a 'phase' has limited meaning with respect to clusters". If nucleation clusters are the key entities in solid–liquid equilibrium, we are led to the paradox of studying a phase equilibrium without a clear definition of the concept of phase.

[16] Sainz-Diaz, C. I.; Martin-Islan, A. P.; Cartwright, J. H. E. Chiral symmetry breaking and polymorphism in 1,1'-binaphthyl melt crystallization, *J. Phys. Chem. B* 2005, **109**, 18758–18764.

[17] For example: Yang, J.; Gould, H.; Klein, W.; Mountain, R. D. Molecular dynamics investigation of deeply quenched liquids, *J. Chem. Phys.* 1990, **93**, 711–723; van Duijneveldt, J. S.; Frenkel, D. Computer simulation study of free energy barriers to crystal nucleation, *J. Chem. Phys.* 1992, **96**, 4655–4668; Swope, W. C.; Andersen, H. C. 10^6-particle molecular-dynamics study of homogeneous nucleation of crystals in a supercooled atomic liquid, *Phys. Rev.* 1990, **B41**, 7042–7054; Agrawal, R.; Kofke, D. A. Thermodynamic and structural properties of model systems at solid-fluid coaxistence. II Melting and sublimation of the Lennard-Jones system, *Mol.Phys.* 1995, **85**, 43–59; Huitema, H. E. A.; Vlot, M. J.; van der Eerden, J. P. Simulations of crystal growth from Lennard-Jones melt: detailed measurements of the interface structure, *J. Chem. Phys.* 1999, **111**, 4714–4723; ten Wolde, P. R.; Ruiz-Montero, M.; Frenkel, D. Numerical calculation of the rate of crystal nucleation in a Lennard-Jones system at moderate undercooling, *J. Chem. Phys.* 1996, **104**, 9932–9947.

[18] Brodka, A.; Zerda, T. W. Molecular dynamics simulation of liquid–solid phase transition of cyclohexane, *J. Chem. Phys.* 1992, **97**, 5669–5675; Esselink, K.; Hilbers,P. A. J.; van Beest, B. W. H. Molecular dynamics study of nucleation and melting of *n*-alkanes, *J. Chem. Phys.* 1994, **101**, 9033–9041. In this last paper, the crystallization of 156 model chains representing *n*-nonane is captured in a simulation, with an energy and density landscape that is exactly specular to the one shown in Fig. 13.1.

[19] In a series of elegant experiments using electron diffraction on nuclei generated in a supersonic nozzle, together with molecular dynamics simulations, L. S.Bartell and coworkers have studied the nucleation rates of small globular molecules: see later discussion and ref. [65].

[20] Polson, J. M.; Frenkel, D. Numerical prediction of the melting curve of *n*-octane, *J. Chem. Phys.* 1999, **111**, 1501–1510. The last sentence of this paper is: "with the rapid increase in computing power, calculations that are barely feasible now should be standard in a few years time".

[21] Matsumoto, M.; Saito, S.; Ohmine, I. Molecular dynamics simulation of the ice nucleation and growth process leading to water freezing, Nature 2002, **416**, 409–413. For typical *Nature* hype, see Sastry, S. Sculpting ice out of water, Nature 2002, **416**, 376–377.

[22] Radhakrishnan,R.; Trout, B. L. Nucleation of hexagonal ice in liquid water, *J. Am. Chem. Soc.* 2003, **125**, 7743–7747.

[23] Interestingly, the freezing process can be much more easily simulated in the presence of an external electric field: Svishchev, I. M.; Kusalik, P. G. Electrofreezing of liquid water: a microscopic perspective, *J. Am. Chem. Soc.* 1996, **118**, 649–654. The timescale for the transformation is calculated to be a few hundred picoseconds, or three orders of magnitude faster than the timescale found in the study of ref. [21].

[24] Talanquer, V.; Oxtoby, D. W. Nucleation in molecular and dipolar fluids: interaction site model, *J. Chem. Phys.* 1995, **103**, 3686–3695. The phenomenological density functional theory is applied in conjunction with a simulation model including spherical objects interacting by Lennard-Jones and coulombic potentials.

[25] See for example: (a) Sum, A. K.; Sandler, S. I.; Bukowski, R.; Szalewicz, K. Prediction of the phase behavior of acetonitrile and methanol with ab initio pair potentials, *J. Chem. Phys.* 2002, **116**, 7627–7636; (b) Lago, S.; Garzon, B.; Vega, C. Accurate simulations of the vapor-liquid equilibrium of important organic solvents and other diatomics, *J. Phys. Chem. B* 1997, **101**, 6763–6771. For empirical correlations between boiling points and molecular descriptors related to shape and hydrogen-bonding ability see Katritzky, A. R.; Mu, L.; Lobanov, V. S.; Karelson, M. Correlation of boiling points with molecular structure, *J. Phys. Chem.* 1996, **100**, 10400–10407.

[26] (a) Alba-Simionesco, C.; Fan, J.; Angell, C. A. Thermodynamic aspects of the glass transition phenomenon. II. Molecular liquids with variable interactions, *J. Chem. Phys.* 1999, **110**, 5262–5272. (b) Alba, C.; Busse, L. E.; List, D. J.; Angell, C. A. Thermodynamic aspects of the vitrification of toluene and xylene isomers, and the fragility of liquid hydrocarbons, *J. Chem. Phys.* 1990, **92**, 617–624.

[27] Morineau, D.; Alba-Simionesco, C. Hydrogen-bond-induced clustering in the fragile glass-forming liquid *m*-toluidine: experiments and simulations, *J. Chem. Phys.* 1998, **109**, 8494–8503.

[28] Are some or all of these materials made of submicroscopic domains? The structural inhomogeneity is apparently parallel to dynamic inhomogeneity: "molecular rotation and translation may occur significantly faster in one part of the sample than in another part a few nanometers away". Cicerone, M. T.; Ediger, M. D. Enhanced translation of probe molecules in supercooled o-terphenyl: signature of spatially heterogeneous dynamics? *J. Chem. Phys.* 1996, **104**, 7210–7218.

[29] Chelli, R.; Cardini, G.; Procacci, P.; Righini, R.; Califano, S. Molecular dynamics of glass-forming liquids: structure and dynamics of liquid metatoluidine, *J. Chem. Phys.* 2002, **116**, 6205–6215.

[30] Zannoni, C. Molecular design and computer simulations of novel mesophases, *J. Mater. Chem.* 2001, **11**, 2637–2646. Note that equation (3) in this reference should read: $x = (\sigma_e^2 - \sigma_s^2)/(\sigma_e^2 + \sigma_s^2)$. Edited by Pasini, P.; Zannoni, C. *Advances in the Computer Simulation of Liquid Crystals*, NATO Science Series C, Mathematical and physical sciences, Vol.545, 2000, Kluwer, Dordrecht.

[31] Hulliger, J. Chemistry and crystal growth, *Angew. Chem. Int. Ed. Engl.* 1994, **33**, 143–162.

[32] One such crystal addict was W. Koerner, who in the 1920s synthesized thousands of benzene derivatives and took a special pleasure in growing large crystals: see Demartin, F.; Filippini, G.; Gavezzotti, A.; Rizzato, S. X-ray diffraction and packing analysis on vintage crystals: Wilhelm Koerner's nitrobenzene derivatives from the School of Agricultural Sciences in Milano, *Acta Cryst.* 2004, **B60**, 609–620.

[33] Ferrari, E. S.; Davey, R. J.; Cross, W. I.; Gillon, A. L.; Towler, C. S. Crystallization in polymorphic systems: the solution-mediated transformation of β to α glycine, *Cryst. Growth Des.* 2003, **3**, 53–60 (optical microscopy, XRD, FT-IR); Cashell, C.; Corcoran, D.; Hodnett, B. K. Secondary nucleation of the β-polymorph of L-glutamic acid on the surface of α-form crystals, *ChemComm* 2003, 374–375 (SEM imaging); Blagden, N.; Davey, R. J.; Lieberman, H. F.; Williams, L.; Payne, R.; Roberts, R.; Rowe, R.; Docherty, R. Crystal chemistry and solvent effects in polymorphic systems: sulfathiazole, *J. Chem. Soc. Faraday Trans.* 1998, **94**, 1035–1044 (morphology, XRD); Towler, C. S.; Davey, R. J.; Lancaster, R. W.; Price, C. J. Impact of molecular speciation on crystal nucleation in polymorphic systems, *J. Am. Chem. Soc.* 2004, **126**, 13347–53 (pH on zwitterionic aminoacid molecules).

[34] Weissbuch, I.; Lahav, M.; Leiserowitz, L. Toward stereochemical control, monitoring and understanding crystal nucleation, *Cryst. Growth Des.* 2003, **3**, 125–150 and Weissbuch, I.; Leiserowitz, L; Lahav, M.. "Tailor-made" and charge-transfer auxiliaries for the control of the crystal polymorphism of glycine, *Adv. Mater.* 1994, **6**, 952–956 (additives; small angle XRD); Pino-Garcia, O.; Rasmuson, A. C. Influence of additives on nucleation of vanillin: experiments and introductory molecular simulations, *Cryst. Growth Des.* 2004, **4**, 1025–1037 (multicell crystallization with visual inspection).

[35] Gracin, S.; Usi-Penttila, M.; Rasmuson, A. C. Influence of ultrasound on the nucleation of polymorphs of *p*-aminobenzoic acid, *Cryst. Growth Des.* 2005, **5**, 1787–1794, and Devarakonda, S.; Evans, J. M. B.; Myerson, A. S. Impact of ultrasonic energy on the crystallization of dextrose monohydrate, *Cryst. Growth Des.* 2003, **3**, 741–746 (increase in nucleation rate by ultrasound).

[36] Lee, A. Y.; Lee, I. S.; Dette, S. S.; Boerner, J.; Myerson, A. S. Crystallization on confined engineered surfaces: a method to control crystal size and generate different polymorphs, *J. Am. Chem. Soc.* 2005, **127**, 14982–14983 (glycine on gold islands); Ha, J.-M.; Wolf, J. H.; Hillmyer, M. A.; Ward, M. D. Polymorph selectivity under nanoscopic confinement, *J. Am. Chem. Soc.* 2004, **126**, 3382–3383 (anthranilic acid on porous glass); Allen, K.; Davey, R. J.; Ferrari,

E.; Towler, C.; Tiddy, G. J. The crystallization of glycine polymorphs from emulsions, microemulsions and lamellar phases, *Cryst. Growth Des.* 2002, **2**, 523–527 (optical microscopy and SEM).

[37] Boerrigter, S. X. M.; van den Hoogenhof, C. J. M.; Meekes, H.; Bennema, P.; Vlieg, E.; van Hoof, P. J. C. M. In situ observation of epitaxial polymorphic nucleation of the model steroid methyl analogue 17-norethindrone, *J. Phys. Chem.* B2002, **106**, 4725–4731; Bonafede, S. J.; Ward, M. D. Selective nucleation and growth of an organic polymorph by ledge-directed epitaxy on a molecular crystal substrate, *J. Am. Chem. Soc.* 1995, **117**, 7853-7861. (spectroscopy, microscopy).

[38] Chen, S.; Xi, H.; Yu, L. Cross-nucleation between ROY polymorphs, *J. Am. Chem. Soc.* 2005, **127**, 17439–17444.

[39] Aber, J. E.; Arnold, S.; Garetz, B. A.; Myerson, A. S. Strong DC electric field applied to supersaturated aqueous glycine solution induces nucleation of the γ polymorph, *Phys. Rev. Letters* 2005, **94**, 145503; Garetz, B. A.; Matic, J.; Myerson, A. S. Polarization switching of crystal structure in the nonphotochemical light-induced nucleation of supersaturated aqueous glycine solutions, *Phys. Rev. Letters* 2002, **89**, 175501.

[40] Peters, R.; Georgalis, Y.; Saenger, W. Accessing lysozyme nucleation with a novel dynamic light scattering detector, *Acta Cryst.* 1998, **D54**, 873–877.

[41] Wolk, J.; Strey, R. Homogeneous nucleation of H_2O and D_2O in comparison: the isotope effect, *J. Phys. Chem.* B2001, **105**, 11683–11701.

[42] Roelands, C. P. M.; Roestenberg, R. R. W.; ter Horst, J. H.; Kramer, H. J. M.; Jansens, P. J. *Cryst. Growth Des.* 2004, **4**, 921–928.

[43] Mougin, P.; Wilkinson, D.; Roberts K. J. In situ measurement of particle size during the crystallization of L-glutamic acid under two polymorphic forms: influence of crystal habit on ultrasonic attenuation measurements, *Cryst. Growth Des.* 2002, **2**, 227–234.

[44] Groen, H.; Roberts, K. J. Nucleation, growth, and pseudo-polymorphic behavior of citric acid as monitored in situ by attenuated total reflection Fourier transform infrared spectroscopy, *J. Phys. Chem.* B 2001, **105**, 10723–10730. Coupled with turbidometric measurements, the results indicate a spontaneous liquid-phase separation prior to crystallization ("oiling-out"). See also Bonnett, P. E.; Carpenter, K. J.; Dawson, S.; Davey, R. J. Solution crystallisation via a submerged liquid–liquid phase boundary: oiling out, *ChemComm* 2003, 698–699.

[45] Mohan, R.; Boateng, K. A.; Myerson, A. S. Estimation of crystal growth kinetics using differential scanning calorimetry, *J. Cryst. Growth* 2000, **212**, 489–499. See in this paper a brief review of some key references to analysis and control of crystallization processes in industrial applications.

[46] Forland, G. M.; Liang, Y.; Kvalheim, O. M.; Hoiland, H.; Chazy, A. Associative behavior of benzyl alcohol in carbon tetrachloride solution, *J. Phys. Chem.* B 1997, **101**, 6960–6969.

[47] Falcon, J. A.; Berglund, K. A. In situ monitoring of antisolvent addition crystallization with principal component analysis of Raman spectra, *Cryst. Growth Des.* 2004, **4**, 457–463.

[48] Ono, T.; ter Horst, J. H.; Jansens, P. J. Quantitative measurement of the polymorphic transformation of L-glutamic acid using in-situ Raman spectroscopy, *Cryst. Growth Des.* 2004, **4**, 465–469.

[49] Parveen, S.; Davey, R. J.; Dent, G.; Pritchard, R. G. Linking solution chemistry to crystal nucleation: the case of tetrolic acid, *ChemComm* 2005, **12**, 1531–1533.

[50] Drenth, J.; Haas, C. Nucleation in protein crystallization, *Acta Cryst.* D1998, **54**, 867–872.

[51] Spitaleri, A.; Hunter, C. A.; McCabe, J. F.; Packer, M. J.; Cockroft, S. L. A 1H NMR study of crystal nucleation in solution, *CrystEngComm* 2004, **6**, 489–493.

[52] Chattopadhyay, S.; Erdemir, D.; Evans, J. M. B.; Ilavsky, J.; Amenitsch, H.; Segre, C. U.; Myerson, A. S. SAXS study of the nucleation of glycine crystals froma supersaturated solution, *Cryst. Growth Des.* 2005, **5**, 523–527.

[53] Quayle, M. J.; Davey, R. J.; McDermott, A. J.; Tiddy, G. J. T.; Clarke, D. T.; Jones, G. R. In situ monitoring of rapid crystallization processes using synchrotron X-ray diffraction and a stopped-flow cell, *Phys. Chem. Chem. Phys.* 2002, **4**, 416–418.

[54] See e.g. Watson, J. N.; Iton, L. E.; Keir, R. I.; Thomas, J. C.; Dowling, T. L.; White, J. W. TPA-silicalite crystallization from homogeneous solution: kinetics and mechanism of nucleation and growth, *J. Phys. Chem. B* 1997, **101**, 10094–10104.

[55] Berkovitch-Yellin, Z. Toward an *ab initio* derivation of crystal morphology, *J. Am. Chem. Soc.* 1985, **107**, 8239–8253. The statement, coming from one of the founding members of the Weizmann school on crystal growth control, presumably refers to tailor-made impurities, but nevertheless sounds as a mourning bell for all theories of crystal formation based only on static energies and classical thermodynamics of pure systems.

[56] See, for the basic ideas, the paper cited in ref. [55]; Clydesdale, G.; Docherty, R.; Roberts, K. J. HABIT-a program for predicting the morphology of molecular crystals, *Comp. Phys. Commun.* 1991, **64**, 311. For a discussion of the intrinsic shortcomings of these methods, see Roberts, K. J.; Docherty, R.; Bennema, P.; Jetten, L. A. M. The importance of considering growth-induced conformational change in predicting the morphology of benzophenone, *J.Phys. D Appl. Phys.* 1993, **26**, B7–B21.

[57] See Grimbergen, R. F. P.; Meekes, H.; Bennema, P.; Strom, C. S.; Vogels, L. J. P. On the prediction of crystal morphology. I. The Hartman–Perdok theory revisited, *Acta Cryst.* 1998, **A54**, 491–500, and references therein. Periodic bond chains (PBC) define "flat" surfaces, as those surfaces with a compact structure that require energy for the formation of growth-promoting steps, while faces without PBCs are supposed to have a zero energy for step formation in some direction, and hence to be able to grow rough. These ideas have a certain amount of subjectivity, and such models, in their original formulation, neglect

some of the structural and energetic features typical of complex and flexible organic molecules.

Other variations on the theme of attachment energies include the use of quantum mechanical methods based on the crystal electron density for the calculation of attachment energies (Docherty, R.; Roberts, K. J.; Saunders, V.; Black, S.; Davey, R. J. Theoretical analysis of the polar morphology and absolute polarity of crystalline urea, *Faraday Discuss.* 1993, **95**, 11–25), or the modification of attachment energies by the inclusion of energy temrs to represent the influence of impurities incorporated in the crystal lattice: Mougin, P.; Clydesdale, G.; Hammond, R. B.; Roberts, K. J. Molecular and solid state modeling of the crystal purity and morphology of caprolactam in the presence of synthesis impurities and the imino-tautomeric species caprolactim, *J. Phys. Chem.* B 2003, **107**, 13262–13272.

[58] Boerrigter, S. X. M.; Hollander, F. F. A.; van der Streek, J.; Bennema, P.; Meekes, H. Explanation for the needle morphology of crystals applied to a triacylglycerol, *Cryst. Growth Des.* 2002, **2**, 51–54, with an approach to the calculation of the energies for step formation; Boerrigter, S. X. M.; Josten, G. P. H.; van der Streek, J.; Hollander, F. F. A.; Los, J.; Cuppen, H. M.; Bennema, P.; Meekes, H. MONTY: Monte Carlo crystal growth on any crystallographic orientation: application to fats, *J. Phys. Chem.* A 2004, **108**, 5894–5902; Deij, M. A.; Aret, E.; Boerrigter, S. X. M.; van Meervelt, L.; Deroover, G.; Meekes, H.; Vlieg, E. Experimental and computational growth morphology of two polymorphs of a yellow isoxazolone dye, *Langmuir* 2005, **21**, 3831–3837.

[59] Cuppen, H. M.; Day, G. M.; Verwer, P.; Meekes, H. Sensitivity of morphology prediction to the force field: paracetamol as an example, *Cryst. Growth Des.* 2004, **4**, 1341–1349.

[60] ter Horst, J. H.; Geertman, R. M.; van Rosmalen, G. M. The effect of solvent on crystal morphology, *J. Cryst. Growth* 2001, **230**, 277–284.

[61] For an introduction see Palmore, G. T. R.; Luo, T. J.; Martin, T. L.; McBride-Wieser, M. T.; Voong, N. T.; Land, T. A.; De Yoreo, J. J. Using the atomic force microscope to study the assembly of molecular solids, *American Crystallographic Association Transactions* 1998, **33**, 45–57. For applications see Malkin, A. J.; Kuznetsov, Y. G.; Glantz, W.; McPherson, A. Atomic force microscopy studies of surface morphology and growth kinetics in thaumatin crystallization, *J. Phys. Chem.* 1996, **100**, 11736–11743; Wen, H.; Li, T.; Morris, K. R.; Park, K. Dissolution study on aspirin and α-glycine crystals, *J. Phys. Chem. B* 2004, **108**, 11219–11227; Luo, T.-J.; MacDonald, J. C.; Palmore, G. T. R. Fabrication of complex crystals using kinetic control, chemical additives, and epitaxial growth, *Chem. Mater.* 2004, **16**, 4916–4927; Abendan, R. S.; Swift, J. A. Dissolution on cholesterol monohydrate single-crystal surfaces monitored by in situ atomic force microscopy, *Cryst. Growth Des.* 2005, **5**, 2146–2153.

[62] See, for a clear review of these problems, Kotrla, M. Numerical theories in the simulation of crystal growth, *Comp. Phys. Commun.* 1996, **97**, 82–100.

[63] Boek, E. S.; Briels, W. J.; Feil, D. Interfaces between a saturated aqueous urea solution and crystalline urea: a molecular dynamics study, *J. Phys. Chem.* 1994, **98**, 1674–1681; Hussain, M.; Anwar, J. The riddle of resorcinol crystal growth revisited: molecular dynamics simulation of α-resorcinol crystal-water interface, *J. Am. Chem. Soc.* 1999, **121**, 8583–8591.

[64] Datta, S.; Grant, D. J. W. Computing the relative nucleation rate of phenylbutazone and sulfamerazine in various solvents, *Cryst. Growth Des.* 2005, **5**, 1351–1357. This study uses a few solute molecules and 20–100 solvent molecules.

[65] Chushak, Y.; Bartell, L. S. Crystal nucleation and growth in large clusters of SeF_6 from molecular dynamics simulations, *J. Phys. Chem. A* 2000, **104**, 9328–9336. The study required nanosecond simulation for 10–12 clusters of up to 2,085 molecules at many temperatures, using a 7-site atom–atom potential and rigid-body molecular dynamics, in which the equations of motion are solved in three positional and three orientational coordinates for each molecule. Nucleation and crystallization were clearly observed.

[66] Ding, F.; Bolton, K.; Rosen, A. Nucleation and growth of single-walled carbon nanotubes: a molecular dynamics study, *J. Phys. Chem. B* 2004, **108**, 17369–17377.

[67] Huitema, H. E. A.; van Hengstum, B.; ven der Eerden, J. P. Simulation of crystal growth from Lennard-Jones solutions, *J. Chem. Phys.* 1999, **111**, 10248–10260.

[68] For applications to vapor–liquid nucleation see McKenzie, M. E.; Chen, B. Unravelling the peculiar nucleation mechanism for non-ideal binary mixtures with atomistic simulations, *J. Phys. Chem. B* 2006, **110**, 3511-3516; Chen, B.; Siepmann, J. I.; Klein, M. L. Simulating the nucleation of water/ethanol and water/nonane mixtures, mutual enhancement and two-pathway mechanism, *J. Am. Chem. Soc.* 2003, **125**, 3113–3118. The latter paper has a reference and discussion to experimental measurements using a supersonic nozzle expansion technique.

14

Crystal polymorphism and crystal structure prediction

...because high speed computers are now available, the solution of crystal structures without diffraction data ... is an interesting challenge.
Rabinovitch, D.; Schmidt, G.M.J. *Nature* 1966, **211**, 1391–1393.

We seem to be some time away from being able to control or even to predict with real assurance the packing a compound will adopt when it crystallizes, or the relationship of the crystal symmetry to molecular structure.
Paul, I.C.; Curtin, D.Y. *Chem. Revs.* 1981, **81**, 525–541.

14.1 A fundamental fact

Organic molecules can provide an almost infinite variability in shape and electrical polarization, and the solid state has a high degree of structuring and anisotropy. These are the reasons why much is expected from structure–activity relationships in solid-state chemistry, with promise for constructing a wide range of diverse materials. Moreover, since molecules in solids are blocked into a fixed conformation, and rotation and translation are forbidden, one may expect to be able to construct many different buildings with the same bricks, and one may fancy that the same organic compound can make a piece of rubbish in one crystal and a touchstone in another. As a consequence, great effort is presently devoted toward a predictive theory of the organic solid state. Two pieces of undisputable evidence must however be considered: the enthalpy difference on passing from the liquid to the solid is comparatively low, melting enthalpies of organic compounds being of the order of $10^2 \, J\,g^{-1}$; and even the amazing degree of spatial selectivity exhibited in persistent, symmetric crystalline patterns is achieved at a very low price in energy differences [1]. Thus, crystalline molecular ensembles are soft, energetically speaking malleable; crystal polymorphs are frequently observed as the same organic compound crystallizes in two or more solids with different microscopic structure [2], but differences in properties are often as small as differences in energy. The fundamental fact is that a crystal made of the elements of organic chemistry, carbon and its neighbors plus hydrogen, is as weak as the dearth of covalent bonding dictates. Material strength increases on going from molecular crystals to polymers to diamond.

14.2 What are crystal polymorphs?

14.2.1 *The taxonomy of organic crystals*

The very definition of crystal polymorphism is not without controversy. An attempted systematization follows.

A solid is here defined as a lump of matter that does not flow, has an elastic response to strain (Section 11.5), and preserves its outer shape for a time significantly longer than a human lifetime. Then a proper crystal is a solid having, over substantially large volume domains (say, at least 1,000–10,000 times the size of a single molecule), a set of three-dimensional translationally periodic symmetry operations (a 3D-TPSO solid) forming a closed group, and applying to the entire asymmetric unit. When probed by X-rays at room temperature, a single crystal yields more than $5N$ independent Bragg spots for an N-atom molecule, while a crystalline powder yields a diffraction pattern with sharp peaks. Macroscopic crystals can have different morphologies dictated by surface and growth kinetics while having the same internal structure. It may be said that twinning is a special, extreme form of morphology, because the internal symmetry operations do not form a closed group. A defective crystal is a 3D-TPSO solid which at room temperature yields less than $5N$ independent Bragg spots for an N-atom molecule, or whose powder diffraction pattern has broad peaks, or that has fuzzy features in its diffractogram: diffuse spots, streaks between Bragg peaks, or even low-intensity spots in between the main ones. These features indicate the presence of high thermal motion or even of rotational or translational diffusion, including disordered and plastic crystals, where the TPSO apply only to parts of the molecule or to centers of mass. Liquid crystals are in this classification extremely defective crystals. An amorphous material is a solid without TPSO of any dimensionality. An amorphous material must give no well-resolved Bragg peaks. A glass is an amorphous material that exhibits the sharp singularities in thermodynamic function and correlation properties discussed in Section 13.6. In the above definition, the number 5 is a courageous proposal but may be negotiable within reasonable limits.

Solvates are crystals whose constituent species is $A_X B_Y$ where A is a higher molecular weight organic component of constant chemical composition and connectivity, and B is a lower molecular weight component (solvent). X is a small integer, and Y is any number >0. Salts are crystals whose constituting species is $A_X^{n\pm} B_Y^{m\pm}$, where A is as above but with an integer charge revealed by a hypervalent or hypovalent atom, and B is a counterion, either molecular (like A) or inorganic (chloride, sulphate...).

Polymorphs are a set of crystals: (1) with identical chemical composition; (2) made of molecules with the same molecular connectivity as defined, for example, by a molecular graph in the AIM sense (Section 12.1), but allowing for different rotations about single bonds (torsions, Fig. 2.2), if any are possible; (3) with a different 3D-TPSO set, not just a subgroup or a supergroup due to disorder. The transition from a proper to a defective crystal, as well as tautomeric equilibria in crystals, and other transformations brought about by temperature changes without a change in 3D-TPSO ensemble, are not polymorphic transitions. On the other hand, one can have polymorphs of solvates and salts, but the crystal of a salt or solvate of a substance A can never be a polymorph of a crystal of pure substance A. Note however that having a different 3D-TPSO ensemble or a different conformation in the crystal does not ensure that crystalline properties will be different. Small geometrical alterations can cause a symmetry element to "turn off" or "turn on", with a change in space group perhaps even detectable by X-rays, without a substantial change in crystal structure.

Is polymorphism a frequent occurrence for organic solids – whereby "frequent" means often enough to significantly affect chemical operations both in the academic and industrial milieus? The answer to this question may not come from an analysis of polymorph occurrences in the Cambridge Database (see below), which is to this end neither a sufficiently large nor a sufficiently randomized database. The famous saying by Walter McCrone, that the number of polymorphs for a given substance is proportional to the time and effort spent in their search [3], is taken as an authoritative statement of polymorphism ubiquity, and indeed in pharmaceutical companies newly synthesized chemicals with applicative promise are routinely screened for crystal polymorphism, and ever increasing numbers of polymorphs are found, not without a large investment of time and resources. By varying temperature and stress or deposition conditions, the Innsbruck school [4] was able to prepare polymorphs of nearly anything, at the expense of a large amount of patience and ingenuity. And yet McCrone's words could also be read backwards: if no effort and no money are spent, no polymorphs will be found, meaning that most organic substances do oblige and crystallize nearly always in the same crystal form, when not bothered by the use of exotic solvents or unusual temperature, pressure, and crystallization conditions. This is the case for aspirin, prepared in tons per year over decades always in the same crystal form (Fig. 14.1). Add to this the fact that even when polymorphs are found, many solid-state properties are very nearly the same in the various crystal forms, and the phenomenon of polymorphism can perhaps be reduced to more sizeable and manageable terms [5].

14.2.2 *Phenomenology of crystal polymorphism*

To provide some experimental examples of organic polymorphism, the Cambridge Structural Database was searched for appearance of the "polymorph" designator, yielding 2,999 hits overall. Many entries are thus overlooked because some determinations of crystal structures of polymorphs do not carry the designator, since the authors were unaware of polymorphism, or did not care to point it out in their report. The search was further restricted as follows: (1) the maximum number of carbon atoms is set at 30, to avoid having too complex molecular systems with possible large differences in conformation and shape; (2) chlorine is the heaviest accepted element, but molecules containing phosphorous or silicon, not really "organic" elements, are excluded; (3) when entries correspond to multiple determinations of the same structure, perhaps with different cell settings or at different temperatures, the determination closest to room temperature, or the latest determination in time, are

Fig. 14.1. The concern over drug polymorphism is a blend of legal and scientific issues. *"Actually, most doctors now insist on $P4_12_12$ aspirin"*.

preserved; since this analysis is not always straightforward, a few cases of identical structures may have slipped in; and (4) polymorph pairs where the temperatures of the X-ray crystal structure determinations differ by more than 50 K are excluded, because the focus of the survey is on concomitant polymorphs, i.e. those polymorphs appearing at the same ambient conditions, and not on structure change with temperature. Of course, all structures with unacceptable lattice energies or densities, due to some error in the reported intra- or intermolecular information or to some error in the retrieval procedure (recall what was said in Section 8.3), are excluded; for example, many crystals of alcohols or amines where the positions of the OH or NH hydrogens were poorly determined had to be discarded. This leaves the statistical sample with 815 crystal structures for 391 polymorph groups and 475 polymorph pairs, a group with n polymorphs producing $n(n-1)/2$ pairs; so most of the polymorph groups include only two partners. A subset was created from this data set by considering only smaller molecules and excluding crystal structures with more than one molecule in the asymmetric unit, in order to apply the Pixel method (Section 12.2) in the calculation of the lattice energy. The polymorph database is included in the supplementary material, file *polymc.oeh*.

Figures 14.2 and 14.3 show the main landscape for polymorphic pairs in organic compounds. 30% of polymorphic crystal structure pairs have one centrosymmetric and one non-centrosymmetric partner, and 25% of the cases show one partner with more than one molecule in the asymmetric unit. No correlation appears between crystal density differences, and either centrosymmetricity or difference in the number of molecules in the asymmetric unit. Crystal density differences range from 0 to 10%, and lattice energy differences, as computed in a preliminary way with atom–atom UNI

WHAT ARE CRYSTAL POLYMORPHS?

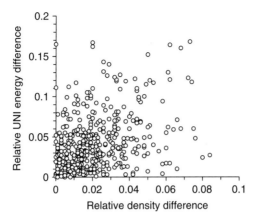

Fig. 14.2. Relative lattice energy differences, $\Delta E/E$, as a function of relative density difference, $\Delta D/D$. 475 polymorph pairs.

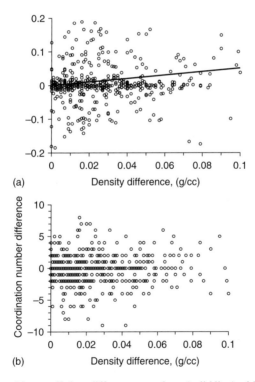

Fig. 14.3. (a) The packing coefficient difference correlates (solid line) with density difference. The variation in self-occupation factor (circles) does not. (b) The difference in number of molecules in the coordination sphere plotted against density differences: complete scatter. 475 polymorph pairs.

potentials (Section 8.8), range between zero and $20\,\text{kJ}\,\text{mol}^{-1}$, never exceeding 15% of the total lattice energy; but for the overwhelming majority of cases (Fig. 14.2) the density difference is just a few hundredths of a g cc^{-1} or less than 5%, and the energy difference is just a handful of $\text{kJ}\,\text{mol}^{-1}$. This is preliminary computational evidence that organic crystal polymorphism is a phenomenon of marginal energetic value, as already mentioned in the opening paragraphs – but a deeper analysis, involving also intramolecular energy differences, will be conducted using the Pixel method and quantum chemical calculations. The UNI lattice energy shows a faint tendency to be more stabilizing for the more dense polymorph, while the point-charge coulombic energy is completely insensitive to density, as expected.

The difference in packing coefficient $C_{\text{pack}} = ZV_{\text{M}}/V_{\text{cell}}$ (equation 8.2) correlates strictly with the density difference (Fig. 14.3(a)): since density is $D_{\text{x}} = ZM_{\text{W}}/V_{\text{cell}}$, this result shows that molecular volume differences between molecules in polymorphic crystals are marginal. On the other hand, differences in self-occupation factors (shape factors, equation 1.11) can be very high, showing that conformational differences are quite common between polymorphs; however, the associated difference in molecular shape has no consistent effect on crystal density, because a higher density may as well go with a more compact or a less compact molecular shape, as demonstrated by the equal distribution of points above and below the zero line in Fig. 14.3(a). The coordination sphere, defined as the number of molecules that contribute more than 3% of the total UNI lattice energy, can also be largely different, without consequences on the resulting crystal density (Fig. 14.3(b)): organic molecules can pack at the same density in spite of large differences in molecular shape/conformation, and with a vast choice between accepting only one, very nearby partner or many, more separated neighbors. All that floppiness is no good news for crystal structure prediction.

All the above results confirm precisely, ten years later and on a database about twice the former size, the results obtained in a previous analysis of crystal polymorphism [2c]. The open question is, how many of the considered crystal pairs are "real" polymorphs, when the same compound crystallizes in two or more significantly different and recognizable crystal structures, and how many are just small variations on the same packing theme, as a consequence of the weakness and scarce directionality of intermolecular forces in organic crystals, or even multiple determinations of the same structure. The indications of the authors of the corresponding papers cannot always be trusted in full because polymorph phases are sometimes overlooked, sometimes designated with different names in different papers. Perusal of cell parameters and space group is often complicated by the fact that crystal cell parameters are never unique, so at first glance it is not easy to see if one is dealing with a different crystal or just with different cell and space group settings for the same crystal. Tables 14.1 and 14.2 show some examples of the difficulties one meets when searching for genuine polymorphism in the Cambridge Structural Database.

For DCLANT (9,10-dichloroanthracene) one sees three determinations in $P\bar{1}$, in all evidence corresponding to the same polymorph, although cell parameters are different because the authors did not care to use the same reduced cell. Practicing crystallographers tend to consider the standardization of cell setting as a minor whim, while it could save a great deal of effort of theoretical chemists and solid state chemists

WHAT ARE CRYSTAL POLYMORPHS?

Table 14.1 Problems encountered in the analysis of polymorphic crystal structures in the Cambridge Database. See text for discussion. Cell parameters in Å and degrees

Refcode	Space group	a	b	c	α	β	γ	T(K)
DCLANT01	$P\bar{1}$	8.582	16.825	3.869	97.9	97.2	76.0	295
DCLANT10	$P2_1/a$	7.041	17.950	8.613	90.0	103.0	90.0	295
DCLANT11	$P\bar{1}$	3.900	8.763	16.508	105.3	90.1	101.0	293
DCLANT12	$P\bar{1}$	3.873	8.585	16.727	102.4	95.3	97.2	295
ANTCEN07	$P 1 1 2_1/b$	11.174	8.554	6.016	90.0	90.0	124.6	295
ANTCEN15	$P2_1/a$	8.542	6.016	11.163	90.0	124.6	90.0	295
ANTCEN17	$P2_1/n$	8.553	6.021	22.333	90.0	124.5	90.0	293
DMANTL01	$P2_12_12_1$	8.942	18.798	4.893	90.0	90.0	90.0	295
DMANTL07	$P2_12_12_1$	8.694	16.902	5.549	90.0	90.0	90.0	295
DMANTL08	$P2_12_12_1$	4.865	8.873	18.739	90.0	90.0	90.0	100
DMANTL09	$P2_12_12_1$	5.538	8.580	16.795	90.0	90.0	90.0	100
DMANTL10	$P2_1$	4.899	18.268	5.043	90.0	118.4	90.0	100
IHEMIR	$P2_1/n$	11.374	3.811	15.721	90.0	110.4	90.0	173
IHEMIR01	$P2_1/n$	3.776	11.171	15.229	90.0	95.0	90.0	174
CAPLEK	$C2/c$	13.835	10.422	9.427	90.0	97.1	90.0	295
CAPLEK01	$A2/a$	9.432	10.432	13.851	90.0	97.1	90.0	295
CBENPH01	$C2/c$	25.161	6.104	7.535	90.0	102.0	90.0	295
CBENPH03	$I2/c$	24.527	6.064	7.457	90.0	100.1	90.0	164
DQUNDS02	$P2_1/c$	13.602	9.540	13.112	90.0	116.4	90.0	295
DQUNDS21	$P 1 1 2_1/b$	14.010	13.108	9.550	90.0	90.0	120.0	295
EPOPDO	$P2_1/c$	17.782	8.221	16.252	90.0	115.1	90.0	295
EPOPDO01	$P2_1/c$	15.191	8.375	16.819	90.0	95.0	90.0	295
FEGWAP	$P2_1$	6.124	8.202	13.818	90.0	99.0	90.0	295
FEGWAP01	$P2_1$	7.469	8.568	10.094	90.0	92.3	90.0	295
QIJTOS	$P2_1$	7.218	14.042	9.160	90.0	110.6	90.0	100
QIJTOS01	$P2_1$	4.510	14.462	13.231	90.0	98.8	90.0	100
QAXMEH	$P2_1/c$	3.945	18.685	16.395	90.0	93.8	90.0	295
QAXMEH01	$P2_1/n$	8.500	16.413	8.537	90.0	91.8	90.0	295
QAXMEH02	$P\bar{1}$	7.492	7.790	11.911	75.5	77.8	63.6	295
QAXMEH03	$P2_1/n$	7.976	13.319	11.676	90.0	104.7	90.0	295
QAXMEH04	$P\bar{1}$	4.592	11.249	12.315	71.2	89.9	88.2	295
QAXMEH05	$Pbca$	13.177	8.021	22.801	90.0	90.0	90.0	295

who are trying to interpret and use the crystallographic information, while being less familiar with crystallographic conventions. On the other hand, the $P2_1/a$ phase appears to be a genuine polymorph. For ANTCEN (anthracene), the use of the nonstandard $P2_1/b$ setting generates confusion [6], while the structure is the same as the $P2_1/a$ structure; one of the cell edges of the the $P2_1/n$ structure is exactly twice one of the cell edges of the $P2_1/a$ phase, suggesting that these two structures are in fact modulations of the same structure with perhaps minor orientation differences, and therefore not significant in the study of real polymorphism. For DMANTL (D-mannitol) the first structure is the same as the third and the second is the same as the fourth, in spite of different settings. For IHEMIR (benzene-1,3-dicarbaldehyde) the

Table 14.2 Some examples of semi-polymorphic crystal structures found in the Cambridge Database. See text for discussion. Cell parameters in Å and degrees

Refcode	Space group	Z	Z'	a	b	c	α	β	γ	T'K
AFLATM	$P2_1$	2	1	7.930	6.210	14.040	95.8	90.0	90.0	295.0
AFLATM01	$P2_12_12_1$	4	1	7.840	6.360	28.350	90.0	90.0	90.0	295.0
KIFHUC	$P112_1/a$	4	1	20.253	10.304	7.132	90.0	90.0	89.4	295.0
KIFHUC01	$P2_1/n$	4	1	11.006	10.172	14.046	90.0	102.8	90.0	295.0
LHISTD04	$P2_1$	2	1	5.166	7.385	9.465	90.0	98.2	90.0	295.0
LHISTD13	$P2_12_12_1$	4	1	5.175	7.315	18.750	90.0	90.0	90.0	295.0
PANQUO	$Pbca$	8	1	6.719	7.168	44.206	90.0	90.0	90.0	295.0
PANQUO01	$Pbca$	8	1	12.783	6.969	23.369	90.0	90.0	90.0	295.0
QAMNAT	$P2_1/c$	4	1	8.061	12.449	11.132	90.0	109.7	90.0	168.0
QAMNAT01	$P2_1/c$	8	2	15.934	12.471	11.436	90.0	110.5	90.0	296.0
BIPHEN04	$P2_1/a$	2	1/2	8.120	5.630	9.510	90.0	95.1	90.0	295.0
BIPHEN06	Pa	4	2	7.770	11.140	9.440	90.0	93.7	90.0	22.0
BIPHEN07	Pa	2	1	7.770	5.570	9.440	90.0	93.7	90.0	20.0
SINZIY	$P2_1$	2	1	5.341	8.342	10.521	90.0	102.8	90.0	213.0
SINZIY01	$P2_1/m$	2	1/2	5.319	8.320	10.433	90.0	103.3	90.0	130.0
ZZZIYE03	Cc	8	2	12.902	10.435	14.464	90.0	96.9	90.0	200.0
ZZZIYE05	$F\bar{1}$	16	2	12.921	10.464	28.805	90.1	97.3	90.3	200.0

two structures may seem largely different because the length of the unique axis, b, is largely different; however, the screw direction (the b axis) of one structure becomes a pure translation (the a axis) in the other structure, and vice versa, while the third cell dimension is identical in the two structures. Here too the suspicion arises that minor deformations lead to apparently large cell differences in two structures that are essentially identical.

CAPLEK and DQUNDS are further examples of identical structures in different settings. The other entries in Table to 14.1 are all dubious cases, although one is never too critical, because crystal structures can be really different even within the same space group and with nearly the same cell parameters, see for example the case of FEGWAP. For contrast, consider the well known case of QAXMEH [7], Scheme 14.2, a conformationally flexible bicyclic molecule, that has six polymorphs at room temperature with different space groups and/or cell dimensions, different molecular conformations, different crystal colors, etc., clearly identifying six different crystalline materials.

A fairly common occurrence, which could be called semi-polymorphism, is that of structures that are identical except for the doubling of one cell parameter, changing from a given space group into a subgroup with loss of one symmetry element (Table 14.2). Another similar circumstance is the formal loss of a symmetry element with a doubling of the number of molecules in the asymmetric unit, which, more often than not, are almost exactly related by that symmetry operation. The diffracted

X-ray intensities are sufficiently accurate to detect the symmetry breaking due to these occurrences, which are however scarcely significant in terms of overall structure and cast little light on the phenomenon of true polymorphism. They should be considered different modulations of the same crystal structure, the origin of the phenomenon being again the general weakness of cohesion forces and the relatively easy deformation of the resulting structures.

14.2.3 *Crystal structure fingerprints: Detecting real polymorphism*

The need is felt for some accurate and objective indicators of crystal identity, which would also solve the problem of comparing different crystal structures for the same compound and of sorting out real crystal polymorphs. One obvious method, based on the human tendency to rely on eyesight, is the comparison of crystal packing diagrams to extract geometrical patterns like ribbons, layers, ring stacks, etc. While sometimes useful, as for instance when a clear difference in hydrogen bonding pattern appears, this method is often inconsistent because packing diagrams in projection are nearly always complex and confusing, even for small molecules, not to mention the hardly interpretable mess that usually results when a moderately large and flexible molecule packs in a high-symmetry space group (as $P2_1/c$ is considered to be for organic compounds). How subjective (and deceptive) all geometrical methods can be is illustrated in Fig. 14.4, which shows that depending on a small change in perspective two aromatic rings can be judged to be fully overlapping or offset.

Objective definitions of crystal structure fingerprints should preferably use numerical indicators deriving from the physics of the intermolecular interaction, rather than from geometry, with the advantage that such an identification also opens the way to the recognition of the properties of different materials. Total lattice energies are unsuitable for this purpose, because the intrinsic uncertainty of the calculation is of the same order of magnitude as the energy differences that the calculation has to deal with (recall Section 8.9). Appropriate indices should therefore describe the mutual arrangement of the molecular objects in the crystal, and must therefore proceed from the calculation of molecule–molecule energies. As already discussed at several places throughout this book, the molecule–molecule view of crystal structure is more appropriate than the atom–atom view, because a molecule is an immediate, well defined chemical entity, while an atom in a molecule requires either more questionable assumptions, or complex calculations for its identification.

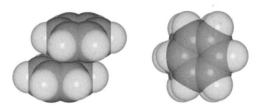

Fig. 14.4. The same dimer (two parallel offset aromatic rings) in two different views.

The construction of an intermolecular energetic fingerprint of an organic crystal then proceeds through the following steps.

1. Collect cell dimensions, space group information and atomic coordinates from an X-ray diffraction experiment; use point-group symmetry, if any, to reconstruct an integer number of entire molecules.
2. Having chosen one reference molecule, use crystal symmetry to generate molecules in its coordination sphere, that is, the twelve molecules whose centres of mass are closest to the center of mass of the reference molecule.
3. Associate with each coordinated molecule the distance between its centre of mass and that of the reference one, R_i; the label of the symmetry operator connecting the two molecules, O_i (T, I, S, G, translation, inversion centre, screw axis, glide plane, respectively, or no operation for different molecules in the asymmetric unit); and the molecule–molecule interaction energy, E_i. For planar molecules, also the dihedral angle between the two average molecular planes can be relevant.

A set of R_i, O_i, and E_i may be called a structure determinant, and a set of structure determinants is an unequivocal, inexpensive, and accurate fingerprint of a crystal structure. A consideration of those molecules whose E_i is greater than 3–5% of the total lattice energy is more than adequate, and more often than not, it turns out that less than twelve molecules do interact significantly with the reference molecule. The chances that these fingerprints may be identical for different structures are very low, and the chances that identical structures may give different energetic fingerprints are null. There is little need for high accuracy in the calculation of the E_is, since even an approximate (but not unreasonable) energy scheme like the atom–atom formulation is more than enough for the purpose. Certainly, better energies are welcome, and for many organic molecules the calculation of these dimer energies by accurate quantum chemical methods is feasible. Also relevant and informative would be the subdivision into coulombic and dispersive terms obtained by the Pixel method, providing more insight into the detail of similarities and differences in the packing of molecules in crystal polymorphs.

Table 14.3 shows some examples, demonstrating how the identity of a crystal structure and its relationship to a prospective polymorphic partner can be established through the structural fingerprint. The two structures of DQUNDS are identical, as the use of standard settings would have revealed at once. The structures of the two polymorphs for AFLATM and LHISTD appear almost identical, in spite of the doubling of the length of one cell edge and the change in formal space group (Table 14.2). Packing differences are relegated to minor changes in the more remote structure determinants. The two putative polymorphs of SINZIY are in fact identical: the product of inversion center and mirror is screw, and hence the interchange between I and S in the structure determinants can be understood. The change in space group depends only on the neglect of the point-group mirror plane symmetry operation in $P2_1$, and its consideration in $P2_1/m$, an example of sometimes optional procedures in the

WHAT ARE CRYSTAL POLYMORPHS?

Table 14.3 The energetic fingerprints of some polymorphic crystal structures. Each entry has the distance between centres of mass (Å), the molecule–molecule energy (kJ mol^{-1}) and the symmetry operator symbol

AFLATM	AFLATM01	
6.21 −52.4 T	6.36 −50.8 T	
7.33 −30.5 S	7.28 −30.7 S	
9.84 −22.5 S	9.88 −22.4 S	
7.93 −17.7 T	7.84 −18.5 T	
9.82 −17.6 S	8.72 −15.8 S	
LHISTD04	**LHISTD13**	
5.17 −30.7 T	5.18 −30.4 T	
6.48 −25.1 S	6.43 −25.1 S	
9.47 −21.8 T	9.38 −21.8 S	
6.10 −18.5 S	6.41 −18.6 S	
6.49 −13.4 S	6.38 −14.2 S	
7.16 −14.1 S	6.72 −14.1 S	
SINZIY	**SINZIY01**	
7.40 −38.7 S	7.37 −38.1 I	
5.34 −37.4 T	5.32 −37.0 T	
5.36 −27.9 S	5.32 −28.7 I	
DMANTL07	**DMANTL08**	**DMANTL10**
5.55 −69.3 T	4.87 −72.7 T	4.90 −68.7 T
5.48 −63.4 S	5.73 −55.9 S	5.09 −41.4 T
7.89 −27.7 S	8.20 −28.6 S	9.52 −18.7 S
6.89 −12.9 S	7.21 −12.8 S	5.04 −39.2 T
IHEMIR	**IHEMIR01**	
3.81 −28.6 T	3.78 −29.0 T	
5.06 −17.4 I	5.10 −17.4 I	
6.13 −9.2 I	6.14 −11.4 I	
7.11 −8.7 S	7.58 −6.1 S	
6.76 −8.2 I	7.23 −6.6 I	
8.24 −5.3 G	7.77 −8.2 G	
8.09 −5.2 G	8.09 −6.6 G	
EPOPDO[a]	**EPOPDO01**	
6.28 −29.0 2 2 I	6.18 −29.4 2 2 I	
6.39 −29.4 1 1 I	6.15 −32.5 1 1 I	
6.64 −27.4 2 1 G	6.54 −29.6 1 2 G	
6.64 −23.1 1 2 T	6.81 −22.4 1 2 T	
6.48 −23.5 2 2 S	6.58 −21.6 2 2 S	
FEGWAP	**FEGWAP01**	
8.06 −36.9 S	7.01 −49.0 S	
8.20 −25.2 T	6.78 −22.0 S	
8.35 −15.4 S	7.47 −21.9 S	
QAXMEH	**QAXMEH01**	
3.95 −75.4 T	5.80 −58.0 I	
8.41 −16.0 I	7.23 −38.8 G	
7.07 −25.2 G		
QAXMEH02	**QAXMEH03**	
6.35 −49.7 I	6.28 −46.9 G	
6.63 −41.2 I	6.74 −44.4 I	
7.10 −39.3 I	7.94 −26.2 G	
6.11 −34.7 I		
QAXMEH04	**QAXMEH05**	
4.59 −63.3 T	5.22 −45.4 G	
6.43 −28.3 I	6.48 −43.6 I	
5.68 −26.5 I	7.79 −27.1 S	
7.95 −18.2 I		

[a] The integers 1 or 2 label the two different molecules in the asymmetric unit.

refinement of X-ray crystal structures (Section 5.4). So the analysis of structure fingerprints also helps in clarifying the relationships between point-group and space-group symmetry.

The IHEMIR case, mentioned above, is borderline: the two structures agree for the first two determinants, but then differ in the arrangement of the further members of the coordination sphere, as if the molecule were happy to satisfy just a couple of packing requirements, and then didn't care much about the arrangement of less close partners. For EPOPDO, the coordination spheres of the two polymorphs, in the same space group and both with two molecules in the asymmetric unit, are very similar, although not identical. In any case, when structural fingerprints are very similar, one at least knows in advance that no large differences in cohesion energies, attachment energies, morphology, solubility, etc., are to be expected. On the other hand, the fingerprint analysis convincingly confirms that the two structures for DMANTL and for FEGWAP are real polymorphs; in particular, if the first determinant is different, the crystal structures are different and the analysis of further terms of the coordination sphere is unnecessary. A glance at the structure fingerprints for the six QAXMEH polymorphs reveals the amazing polymorphic versatility of this compound; it can pack successfully in a compact coordination sphere (QAXMEH), where the first pair provides most of the crystal cohesive energy, and other molecules in the coordination sphere contribute very little; or in a diffuse coordination sphere (QAXMEH02), with four molecular pairs with almost equal contributions. The molecular constitution of this compound can comply with molecular recognition requirements over any symmetry element, with the possible exception of the screw axis, which appears less frequently among the closest partners.

The cases discussed so far show that there is great flexibility in molecular packing in organic crystals. Small, and sometimes not so small, structural changes may come and go at a minor energetic expense. X-ray diffraction on single crystals is a highly sensitive analytical technique, and detects these changes through variations in the number and relative intensity of Bragg peaks, but the energy fingerprint analysis shows that on energetic ground these "crystallographic polymorphs" are in fact, to a large extent, the same crystal structure, barring a few minor variants on the periphery of the coordination spheres. They are good cases for academic dispute but are of scarce interest to crystal chemistry or materials science.

Another way of comparing crystal structures is through the comparison of simulated powder patterns, which plot diffraction intensities as a function of θ angle (see note [5], Chapter 5) and hence are insensitive to different choices in the settings of cell and space group (see Section 5.5 and 5.8). Figure 14.5 shows that the simulated powder patterns for the DQUNDS pair confirm the complete coincidence of the two crystal structures, and that the patterns for the two partners in the AFLATM pair are different. The sensitivity of powder profiles to minor changes in cell parameters and molecular position can also be a disadvantage, because the comparison may return a misleading impression of wide disparity for crystal structures which instead have an essentially identical energetic texture.

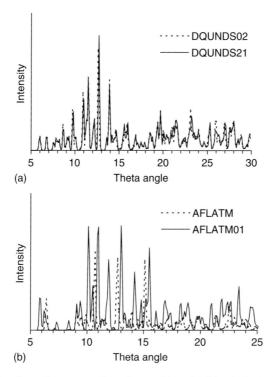

Fig. 14.5. Simulated powder patterns for polymorphic pairs: identical (a), different (b). Standard parameters (see Section 5.8, equation 5.37; copper wavelength, constant thermal factor of 4 Å^2).

14.2.4 *Analysis of crystal polymorphism by Pixel and quantum chemical calculations*

The Pixel intermolecular interaction energy decomposition allows a better understanding of the forces that are at the root of the phenomenon of polymorphism. Moreover, the same quantum chemical calculation (MP2/6-31G**) that provides the electron density for the Pixel calculation also provides total electronic energies and so, in principle, a way of also estimating intramolecular energy differences between polymorphs.

Things, however, are not as straightforward as they may seem. The quantum chemical calculation is extremely sensitive to minor variations in molecular geometry that may accumulate over different temperatures of the X-ray determination, over different accuracies of the overall X-ray work (as reflected in the *R*-factor), over minor undetected disorder, and even due to plain typos. For hydrogen-bonding substances, especially for alcohols and primary amines, where there is no simple geometrical rule for normalization of hydrogen positions, the hydrogen-bonding protons are sometimes placed by X-ray investigators in approximate positions, because they are anyway scarcely relevant to structure factor calculations. In carboxylic acids, quite often the two C—O distances in the carboxyl group are almost identical, because of partial ionization to a COO$^-$ group, or just because of orientational disorder. The position of the acid proton is consequently poorly determined. These uncertainties are unavoidable "geometrical noise" which, however minor, may easily result in random energy fluctuations of tens of kJ mol^{-1} in quantum chemical calculations of energy differences, if one recalls that these differences are very small numbers resulting from differences between the very large numbers of electronic energies. The task of double-checking and correcting all structures, in databases that may contain hundreds or thousands, is clearly beyond fair duty, and most problems cannot be solved without introducing some other approximation. There is quite often no other choice than to discard the most offending cases and hope for the best for the remaining ones; when dealing with large databases, one must live with a certain amount of computational noise.

The disturbing aspect of this disparity of accuracy in structure determinations is that its effects are amplified when more accurate computational methods are used. Molecular geometry optimization should be carried out, but this would destroy the conformational information from the X-ray structure determination, and moreover, it is out of the question when considering statistical samples of hundreds of large molecules. Comparisons should be made over fully energy optimized crystal structures, but that also would somewhat distort the experimental structural information, because of the neglect of thermal energy in the temperature-less potential energy calculations. Once again one sees here a combination of adverse factors which concur in making the uncertainty of any lattice energy calculation of the same order of magnitude as the energy differences of interest.

A detailed analysis of some exemplary cases will further clarify these points. Scheme 14.1 shows a few cases where loss of an intermolecular hydrogen bond is compensated by gain of an intramolecular hydrogen bond (ACBNZA, CIMETD). The corresponding energies calculated by Pixel and by ab initio are of the correct order of magnitude, in further confirmation of the good performance of the Pixel method. The OCHTET case, however, reveals the mix of difficulties one meets in such calculations: the three polymorphs correspond to two different conformations of the main ring, but the large intramolecular destabilization of OCHTET03 is an artifact due to the scarce geometrical accuracy of the determination (R factor of 11.5%). The coulombic energy from the Pixel calculation is scarcely accurate for the two polar space groups, where the molecules pack with their large dipoles all parallel to the polar axis, in the worst possible scenario for a coulombic calculation.

In the case of BIYSEH, the large intramolecular energy difference is due to a different conformation at the ring, but is presumably enhanced by the low accuracy of both determinations (R factor around 9%). This is even more evident for TMETTS, where one of the structure determinations has an R factor as high as 20%, and for DMURAC and CABTUV, where the structure with the higher R factor has a much less stabilizing intramolecular energy. Apparently, such an explanation does not apply to the case of BEBMAX, where both crystal structures have a low R factor, the molecular

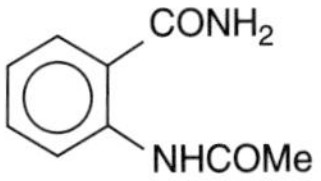

WHAT ARE CRYSTAL POLYMORPHS?

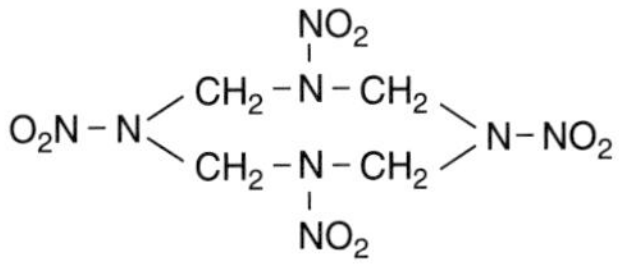

ACBNZA: benzene ring with CONH$_2$ and NHCOMe substituents (ortho)

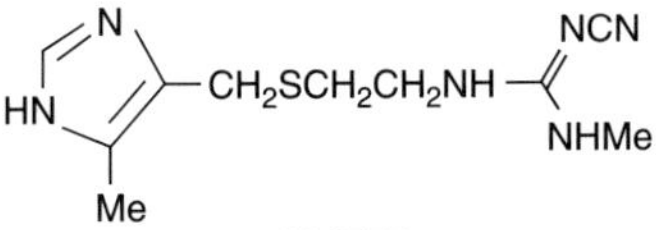

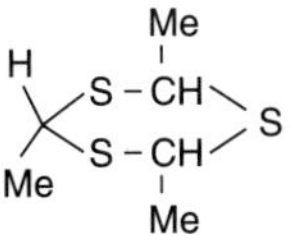

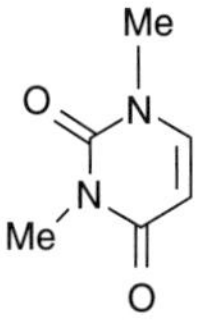

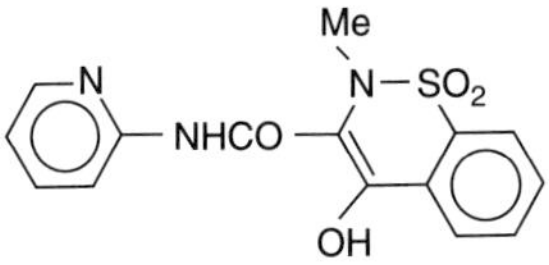

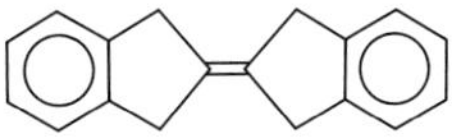

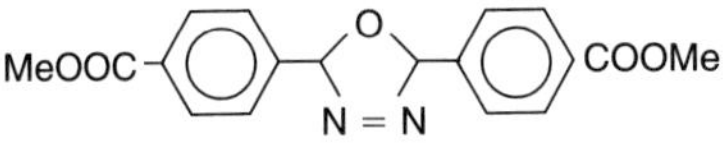

CIMETD: imidazole (HN, N, Me) —CH$_2$SCH$_2$CH$_2$NH—C(NCN)(NHMe)

OCHTET: O$_2$N—N(CH$_2$—N(NO$_2$)—CH$_2$)(CH$_2$—N(NO$_2$)—CH$_2$)N—NO$_2$

TMETTS: C(H)(Me)(S—CH(Me))(S—CH(Me))S

DMURAC: N,N'-dimethyluracil (Me on both N; two C=O)

BIYSEH: pyridine-2-NHCO—C(=N(Me)-SO$_2$-aryl)—C(OH)=...

CABTUV: bis-indane alkene

BEBMAX: MeOOC—C$_6$H$_4$—CH(—O—)—C$_6$H$_4$—COOMe with N=N in dioxolane

compound refcode	R factor	difference in E_{INTRA}	difference in E_{INTER}	structural difference
ACBNZA	–	0	0	intramolecular H-bond
	–	+42	–34	no intramolecular H-bond
CIMETD	–	0	0	intramolecular H-bond
	–	+34	–16	no intramolecular H-bond
OCHTET	3.5	0	–	boat ring conformation
	5.9	–14	–	chair ring conformation
	11.5	+44	–	boat ring conformation
BIYSEH	8.8	0	–	–
	9.6	–33	–	–
TMETTS	4.0	0	0	equatorial Me group
	5.7	–23	–7	axial Me group
	20	+39	–3	equatorial Me group
DMURAC	5.6	0	0	–
	10.5	+22	+15	–
CABTUV	3.8	0	0	–
	5.6	+43	+10	–
BEBMAX	3.4	0	0	–
	5.0	–28	–27	–

Scheme 14.1.

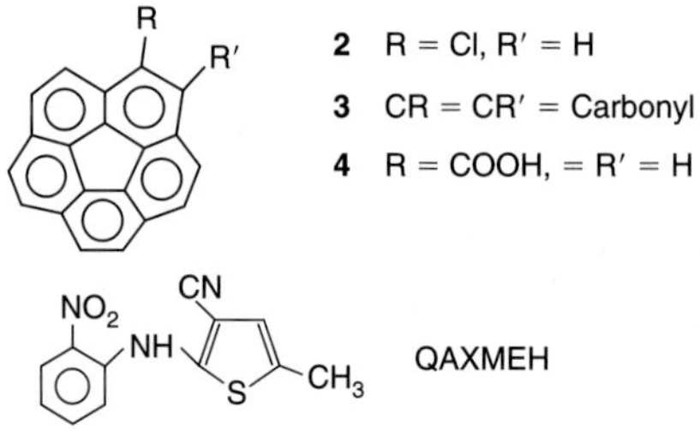

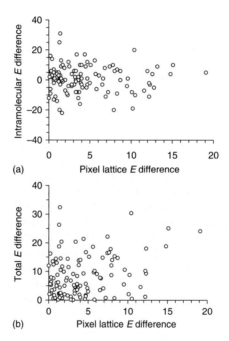

Scheme 14.2.

Fig. 14.6. (a) A scatterplot of differences in intramolecular energy (MP2/6-31G** calculation on the unoptimized molecular structure extracted from the crystal) versus intermolecular energy differences (calculated by Pixel) for 116 polymorphs pairs. Differences cancel for points below the zero line and add for point above the zero line. (b) Total energy differences (intramolecular + intermolecular) as a function of intermolecular energy differences. Units kJ mol^{-1}.

conformation is identical in the two structures, but the calculated energy differences are too large to be realistic. There is no evident explanation for this result.

In view of the previous analysis, the systematic study of energy differences between polymorphs must be taken with a pinch of salt; nevertheless, some undisputable

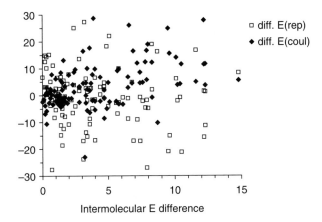

Fig. 14.7. Same sample as in Fig. 14.6: relative importance of coulombic and repulsion energy in total intermolecular energy differences (Pixel calculations) between polymorph pairs (kJ mol^{-1}). Small total energy differences result from cancellation of large partial energy differences.

if broad trends are observable. Figure 14.6(a) shows a scatterplot of intra- versus intermolecular energy differences between crystal polymorph pairs. Except for the cases discussed above, there is no clear prevalence of compensation over concurring destabilization. Figure 14.6(b) shows the same fact on a different perspective, and also shows that total energy differences between polymorphs are usually below 10 kJ mol^{-1} and very seldom exceed 15 kJ mol^{-1}. Figure 14.7 shows that more stabilizing lattice energies usually go with higher coulombic energy and higher repulsion energy (the corresponding differences have the same sign), as expected since crystal structures reach a higher stabilization by establishing closer contacts, which also lead to increased repulsion. Differences in dispersion energy are less important, while the coulombic term is the most significant one for the larger differences in intermolecular energy (> 5 kJ mol^{-1}), confirming the structurally selective role of coulombic energy.

Overall, the above analysis shows that polymorph pairs where the difference in lattice energy exceeds 10 kJ mol^{-1} or the total energy difference exceeds 15 kJ mol^{-1} are a minority. Thus, these results support the view that polymorphism in organic crystals involves subtle energetic modulations, rather than radical packing energy differences. On a gross scale, these numbers are comparable to the expected activation energies for nucleation, as discussed in Chapter 13.

14.3 The construction of crystal structures by computer

The procedure for determining the structure of a real crystal goes through chemical synthesis of the constituting compound, followed by purification and recrystallization to obtain adequate single crystals, and X-ray diffraction analysis. Each of these steps may succeed or fail, and even if they succeed, they do have drawbacks and costs. The theoretical chemist is more fortunate, because the construction of crystal structures in the computer is much easier, much less expensive, and always successful. Of course, there is no guarantee that computational crystals will ever turn into real ones.

Assuming a known and fixed molecular conformation – either the molecule is rigid, or structure optimization has been carried out by some computational method – and that there is no more than one molecule in the asymmetric unit, the variables describing the crystal structure are three coordinates of the center of mass, three orientation angles, and the cell parameters, which range in number from one for a hypothetical cubic crystal to six for a triclinic crystal. These variables define the multidimensional space that must be scanned by the computational procedure. It is a relatively easy matter to instruct a computer to search this space and to provide computational polymorphs for a given molecular structure.

14.3.1 *"Brute force" approach*

The easiest way of obtaining computational crystal structures might be called the 'brute force' approach, consisting of the following steps: (1) consider space groups in turn, starting with the most common ones (check with Table 8.2); (2) select a range of variation for the cell parameters, taking into account the limitations due to crystal system (e.g. all angles 90° in orthorhombic); (3) set a range of variation for the coordinates of the center of mass and for the orientation angles, taking into account only the independent part of cell space; (4) build a crystal model for each point through the ranges (2) and (3) using the space group symmetry; (5) accept raw structures that comply with the density and lattice energy prescription sketched in Section 8.11 or even with less restrictive criteria; (6) for each accepted raw structure, carry out a lattice energy optimization to reach close-packed, well behaved crystal structures. The big advantages of the brute force approach are that very little or no knowledge of geometrical crystallography is needed, all space groups being equally easy to access, so that writing a small computer program to perform the above described routines is no demanding task; and that the method, when used in full deployment, ensures a thorough sampling of parameter space. The disadvantage is that even if only ten steps are explored for each free parameter (a very conservative estimate) one easily runs into unacceptable numbers of cases to be sieved, something like 10^{12} for triclinic crystals, reducing (!) to 10^{10} for monoclinic crystals. Nevertheless, with modern computers that can handle a complete lattice energy calculation with atom–atom potentials in milliseconds, the brute force approach has its appeal, for example in some space groups or when some of the positional parameters may be restricted, as happens for molecules in special positions.

14.3.2 The "Prom" sequential approach

A number of procedures for a more ingenious exploration of molecular coordination geometries in crystals have been devised for use in computational crystal structure generation [8,9]. A brief description of the procedure, called the Prom approach, which led to the first attempt at systematic crystal structure generation by computer [1] will now be given. For convenience, the description will proceed along the sections into which the corresponding computer software is organized (see supplementary material, Prom module of the OPiX package). The method exploits a view of crystal packing that proceeds from the molecule through the symmetry operations to reach the space groups, the same view at the basis of the presentation of crystal symmetry in Section 5.1.

The opening stage of the procedure reads in atomic coordinates of the molecular model, estimates an expected lattice energy through the options described in Section 8.11, and establishes the intermolecular force field. Needless to say, the force field must be a simple one, because of computing time requirements; very advisable choices are, for example, the standard UNI or Williams atom–atom formulations (see Section 8.8). The next stage combines two or more molecules into clusters over a first symmetry operator; the cohesion energy of these primitive crystal nuclei are calculated, and they are accepted or discarded according to preset energy windows. These primitive clusters can be combined into secondary clusters by adding a second symmetry operator, or further by adding a third symmetry operator. These more complex aggregates are also sorted by their cohesive energy. A final stage adds the missing cell translations and performs lattice energy minimizations. Starting from one independent molecule, eventually space groups with up to eight equivalent positions can be generated: $P1, P\bar{1}, P2, Pm, P2_1, Pc, P2/m, P2/c, P2_1/c, P2_1/m, P2_12_12_1, P2_12_12, Pca2_1, Pna2_1, C2, Cm, Cc, C2/c, Pbca, Pnma$. All the most populated space groups for organic compounds are present in this list.

14.3.3 Molecular clusters with one symmetry operator

Table 14.4 Summarizes the the possible cases. The simplest nucleus is just a single molecule. There is no cluster generation, and pure translation then leads to space group $P1$ (see Fig. 5.1), or $P\bar{1}, Z = 1$ if the molecule itself is centrosymmetric. Figure 14.8 shows the various setups for the construction of other clusters. A twofold rotation axis (symbol A) or a mirror-plane (symbol M) produce a nucleus made of two molecules, whose cohesive energy is calculated as a function of the molecular orientation and of the distance between the molecular center of mass and the twofold axis or the mirror plane. For an I-nucleus, the inversion center is placed at each point of a grid around the starting molecule, and the interaction energy between the original and the inversion-related molecule is computed.

Translation of an I nucleus leads to space group $P\bar{1}, Z = 2$, (see Fig. 5.2). For an S_y cluster (a screw axis with translation along y), the variables are the distance between the center of mass and the screw axis, the screw pitch, and three molecular

Table 14.4 A summary of procedures for the search molecular clusters with one symmetry operator; see also Fig. 14.8

Symmetry	Variable orientation	Variable coordinates	Resulting translation elements	Resulting structure
I	No	x,y,z	none	$P\bar{1}$ nucleus
S_y	Yes	y,z	$b = 2y$	$P2_1$ ribbon
G_{xz}	Yes	y,z	$c = 2z$	Pc ribbon
A_y	Yes	z	None	$P2$ nucleus
M_y	Yes	y	None	Pm nucleus
C	Yes	x,y	$a = 2x, b = 2y$	C layer

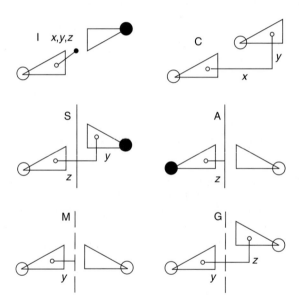

Fig. 14.8. A summary of the variable parameters in the search for cohesive nuclei over a single symmetry operator (I inversion, S screw axis, G glide plane, A twofold axis, M mirror plane, C centering). Broken lines denote traces of planes. The black circle denotes "up", the white circle "down".

orientation angles (compare Figs 5.3 and 14.8). A ribbon of molecules along y is thus generated. The b cell axis of any crystal structure generated from the ribbon must be twice the chosen screw pitch, and translation of this ribbon along x and z leads to space group $P2_1$ (recall Fig. 5.4). For a G_{xz} ribbon (a glide plane parallel to the xz plane, with glide translation along z) the variables are the distance (along y) between the molecular center of mass and the glide plane, the glide translation, and three molecular orientation angles. The c cell axis must be equal to twice the chosen glide translation, and translation of the ribbon leads to space group Pc.

To introduce centering, a search is conducted on two possible orthogonal translations (x,y) of the original molecular object, the other three variables being, as usual, the molecular orientation angles; a layer is obtained. In any crystal structure resulting from these layers, the a cell parameter must be equal to twice the x translation, and b to twice the y translation.

The only way of dealing with multimolecular asymmetric units in this approach is to provide an optimized structure of the symmetry-less dimer, and to use this dimer as the repetition unit throughout the procedure. The total lattice energy is then the sum of the lattice energy of the dimer and of the dimer cohesion energy (Section 8.7).

14.3.4 Combination of two or three symmetry operators

Any molecular nucleus, ribbon or layer obtained by using a first symmetry operator can be exposed to a second operator, thus yielding larger clusters and product symmetries. The reciprocal orientation of the operators should be properly chosen to produce space groups according to conventions in the International Tables of Crystallography. Whenever the combination leads to fixing some translational periodicities, the nuclei can be expanded into higher-order clusters, e.g. dimers or strings into layers, and layers into three-dimensional crystal structures. These second-order aggregation units are also accepted or rejected according to some energy prescription. A few examples are now described.

1. S_y cluster, then I operator (Fig. 14.9): the I operator, located at a distance $z°$ along the z axis perpendicular to the screw direction, generates centrosymmetric pairs of screw ribbons. The new structure can be optimized by rotation of the S ribbon as a whole around its own propagation axis. The c cell axis must be $c = 4z°$; since the b cell axis was fixed during the S_y search, the new cluster is

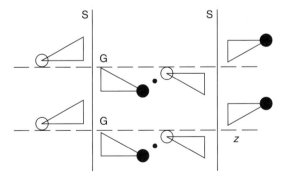

Fig. 14.9. Sample combination of two operators: a screw ribbon is duplicated by an inversion operation. Same symbols as in Fig. 14.8. Recall Fig. 5.5.

equivalent to a crystallographic layer: for example, translation of the layer along a third direction (allowing for the monoclinic angle) leads to space group $P2_1/c$. A similar procedure involves using an I operator first, and then a screw operator, but the results may be very different, depending on the energetic screening. Crystals from the first procedure will have compact screw ribbons, crystals from the second procedure will contain compact centrosymmetric dimers. Similar reasonings apply to all binary combinations of an inversion center with a screw or glide operator.

2. S_y cluster, then S_z operator: the variables in the S_z search are the position in x and y of the second screw operator, and the screw translation along z (refer to Fig. 5.6) The b cell axis is already fixed as twice the pitch of the first S ribbon, c must be twice the pitch of the second ribbon, and, according to the standard crystallographic settings, a must be four times the distance between the two screws along x. In this case, the final "cluster" defines in fact a full crystal structure, in space group $P2_12_12_1$, as no further translational freedom exists. A moderately experienced crystallographer should easily realize that if $z = 0$ a $P2_12_12$ layer results instead.

Unfortunately, there is no hope of systematizing these operations for all space groups and orientations of symmetry operators, so the computer code must contain a special section for each combination and desired space group. The most common and useful combinations of two symmetry operators are summarized in Table 14.5. A reader who cares to follow in detail all the manipulations shown in that table learns all there is to learn about the geometrical crystallography of organic compounds [10].

Table 14.5 Combinations of symmetry operations in organic crystals: A, twofold rotation; M, mirror reflection; G, glide reflection; S, twofold screw rotation; I, inversion through a point; Ct, centering translation. The labels preceding each space group symbol are as follows: C, cluster, R, row, L, layer and 3D, full three-dimensional structure. When several possibilities are given for an arrangement, they depend on the relative orientation of the symmetry operations

	A	M	G		S	
M	C $P2/m$					
G	R $P2/c$	L Cm	L L 3D	Cc $Pca2_1$ $Pna2_1$		
S	L $C2$ L $P2_12_12$	R $P2_1/m$	L L 3D	$P2_1/c$ $Pca2_1$ $Pna2_1$	L 3D	$P2_12_12$ $P2_12_12_1$
I	C $P2/m$ R $P2/c$	C $P2/m$ R $P2_1/m$	R L	$P2/c$ $P2_1/c$	R L	$P2_1/m$ $P2_1/c$
Ct	L $C2$	L Cm	L L	Cm Cc	L	$C2$

THE CONSTRUCTION OF CRYSTAL STRUCTURES BY COMPUTER   389

14.3.5 *The translation search: Sorting and ranking*

The most promising clusters (hence the name of the procedure) found at each combination stage are in turn and sequence passed on to further stages, but when the operator combination stage leads only to partial structures like ribbons or layers, some cell translations remain to be determined, as discussed in the examples above. A final stage in the Prom procedure takes care of finalizing the determination of the unit cell by finding the missing translations. When the search is complete, full crystal structures are determined, but they satisfy only the preliminary close packing and lattice energy requirements. Typically this output contains hundreds or, for the most popular space groups, thousands of crystal structures, many of which are very similar because the search is largely redundant. The next stage is clustering to eliminate duplicate structures, accomplished by comparing reduced cell parameters [11], cell volumes, and lattice energies, to obtain a sorted set of a few hundred crystal structures. Energy optimization is then carried out using the Symplex method (ref. [9], Chapter 2) and the optimized structures are sorted again, so that structures that have fallen into the same energy minimum are discarded. Combinations of sorting and optimization cycles quickly lead to the truly independent crystal structures, typically 10–50 within an energy range of $10\,kJ\,mol^{-1}$.

14.3.6 *The Prom algorithm: Pros and cons*

The duration of the crystal structure search and the quality of the results depend on the ambitions and skills of the operator. An experienced crystallographer may wish to bias the search by restricting the distance and orientation of molecules with respect to crystal symmetry elements; during the promising nuclei search stage, all chemical or structural information available on the compound or on a series of similar compounds can and should be exploited. For example, it is known that condensed aromatics will form herringbone or stack patterns, and in the Prom formalism these structural features can easily be imposed by an appropriate limitation of the ranges of the molecular orientation angles, thus predetermining to some extent the structure of the clusters. When generating possible crystal structures for carboxylic acids, amides, and the like, it may be advisable to start with an inversion center to generate a hydrogen-bonded dimer; hydrogen bonding requirements should be fulfilled as soon as possible anyway. A chemist with a different training and no knowledge of crystallography may prefer a totally unbiased search, using the procedure as a black box. In this case, appropriate ranges for all the variable parameters in the search can be set automatically by using the available information on the size and shape of the constituting molecule. In fact there are two separate versions of the computer software: a full hands-on version whose user must somewhat scratch his head, and an automatic version where all that is required from the user is to type in the desired space group (respectively, Promany and Prom modules in the Supplementary material).

A less attractive side of the successive cluster building procedure is that the choice of the first nucleus biases all the downstream procedure. For example, when a very cohesive centrosymmetric dimer is chosen, then all resulting crystal structures will contain that dimer, and if the most stable structure does not, that structure will never be reached. To obviate this difficulty, several different starting points can be taken; for example, a thorough search for space group $P2_1/c$ will have to consider all starting points using the I, S, and G clusters. Prom does not handle translation as the first operator, so crystal structures where pure translation is the largely dominating structure determinant are difficult to reach. For the same reasons, the stable cluster bias causes more difficulty in reaching crystal structures with scattered coordination spheres (that is, where a molecule is surrounded by many weakly interacting neighbors). To extend the survey to such cases, it may be advisable to use energy windows that also include less strictly cohesive nuclei, to avoid biasing all ensuing crystal structures towards a compact coordination sphere. In general, the setting of the thresholds and windows for acceptance of clusters is a critical point in the whole procedure, since accepting too few will cause one to overlook the relevant structures, while accepting too many may lead to inordinately large numbers of duplicate structures.

Computing times may be surprisingly small. In most space groups, the first acceptable crystal structures spring out of the computer in a matter of minutes or even seconds. A search over 4 I nuclei in the first stage, and 100 S clusters in the second (a total of 400 trial clusters) plus the translational search producing a few hundred structures in $P2_1/c$ may run in something like an hour for a 50-atom molecule on any of the most common desktop PCs, and in a matter of a few minutes on high-performance computing devices. A complete search of all possible nuclei in the same space group may be run overnight. The subsequent sorting and optimization stages may then require the best part of an afternoon. Generally speaking, even for a rather large molecule, on average one space group can be explored fully in one working (human time) day. Computer generation of crystal structures is orders of magnitude faster than its experimental counterpart ...

14.3.7 Some examples of crystal structure generation

The Prom procedure is encoded in a mature software that has been used over the years with considerable success [1,9,12]. As an example of the general landscape that a crystal structure predictor must face, an exercise in crystal structure generation by Prom will now be discussed.

The first test is chlorocorannulene, **2**, Scheme 14.2, a bowl-shaped molecule without strong attachment points for intermolecular recognition, for which cohesion is expected to depend mainly on dispersion energy, enhanced by close packing [13]. The space groups accessible to Prom were all explored, either in the automatic version of the cluster search for the most common space group, or in the "brute force" approach to finding crystal structures in rare space groups. A total of more than 1,000 different minima in the potential energy hypersurface (barring a few mistakes, fluctuations or duplications here and there) were found (Fig. 14.10): one is impressed by the continuous population of possible crystal structures encompassing an energy range of about 30 kJ mol^{-1}. Table 14.6 shows that although $P2_1/c$ (Fig. 14.11) is the

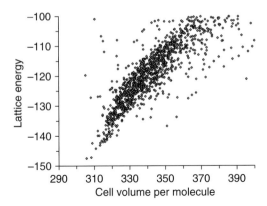

Fig. 14.10. Lattice energies of computational crystal structures for chlorocorannulene. Each point in the graph corresponds to a separately optimized crystal structure. Volumes in Å^3 and energies in kJ mol^{-1}.

favourite, as expected, almost any space group possessing at least one close-packing symmetry operator (screw, glide, inversion, but also pure translation or centering) provides a reasonable packing niche for this molecule. In fact, if one remembers that packing energies are only approximate, that fluctuations of up to a few kJ mol^{-1} cannot be ruled out, and that the search protocols for the different space groups may have been more or less efficient, there are almost fifteen different space groups that could provide reasonable polymorphs, with packing energy within 90% of the best one (the 148–130 kJ mol^{-1} range). Some of this almost incredible versatility may be due to the regular molecular shape, only modestly perturbed by the protruding chlorine substituent; the main intermolecular recognition requirement here is the formation of columns of stacked bowls, while the way in which columns are juxtaposed is less important. This also explains the unusual stability of structures in systems higher than orthorhombic (e.g. $P4_1$ or $P3_1$ structures).

The stabilization energy decreases with decreasing packing coefficient, and, roughly, with the increasing number of non-close packing operators; serious trouble in finding a convenient packing for this compound begins only in frankly non-close packing space groups, especially those including mirror planes, Pm, Cm, $P2_1/m$. In spite of a great deal of effort, and of the presence of the center of symmetry, not even a raw, tentatively cohesive crystal structure could be found in $P2/m$, a space group that contains two anti-close packing operators. A computational tour de force led to the preparation of the cubic crystal structure with 48 molecules in the unit cell (Supplementary material, file *chlorcor_pn3b.oeh;* see the packing in Fig. 14.12), which, with a packing coefficient of only 0.38, is perhaps more suitable for printing good Christmas cards than for a scientific document. Interestingly, the $P\bar{1}$ crystal packing with two molecules per asymmetric unit includes a bump-in-hollow asymmetric dimer, Fig. 14.13, very similar to the one adopted by corannulene itself in its

Table 14.6 Crystal structure generation by the Prom procedure for chlorocorannulene. Raw structures: output of the first search; first sorting: clustered structures from raw structures (no optimization); final sorting: number of optimized crystal structures; best energy (kJ mol^{-1}): best lattice energy of final sorting. For some space groups, results for the dicarbonyl derivative are also reported. Crystal data are collected in the Supplementary material, file *chlorcor.oeh* and *co2cor.oeh*

Space group[a]	No.of raw structures	First sorting	Final sorting	Best energy[b]	Packing coeff.
$P2_1/c$					
Cl derivative	22504	1734	249	−148	0.772
C=O derivative	26973	1308	249	−164	0.776
$P2_1$	—	172	26	−144	0.759
$P\bar{1}$					
Cl derivative	—	810	109	−142	0.751
C=O derivative	12603	933	178	−153	0.762
Pbcn	—	—	2	−141	0.744
$P2_12_12_1$	3230	206	40	−141	0.748
$Pca2_1$	—	54	20	−139	0.747
Cc	906	430	69	−140	0.744
$Pna2_1$	—	70	34	−139	0.748
*P*1	—	386	96	−135	0.724
*C*2/*c*	161	102	35	−135	0.733
*C*2	746	488	105	−134	0.739
Pnma	—	—	14	−134	0.672
Pbca	90	58	23	−133	0.722
$P4_1$					
Cl derivative	—	—	8	−133	0.750
C=O derivative			27	−133	0.696
$P3_1$	—	—	18	−132	0.747
$P2_12_12$	62	26	4	−131	0.689
$P6_1$	—	—	25	−130	0.748
$P\bar{1}, Z = 4$	10022	2078	181	−129	0.727
$P2/c$	31	21	11	−127	0.711
$P4_12_12$	—	—	38	−125	0.737
*Fdd*2	—	—	13	−122	0.678
$P2_1/m$	836	55	27	−118	0.670
Pc	66	14	4	−115	0.661
Cm	552	152	17	−108	0.617
*P*2	2010	282	68	−104	0.617
Pm	136	14	4	−75	0.542
$Pn\bar{3}n$	—	—	1	−58	0.382

[a] The number of molecules per cell is equivalent to the number of symmetry operations ($Z' = 1$) unless otherwise stated.

[b] UNI atom–atom energy plus point-charge coulombic energy.

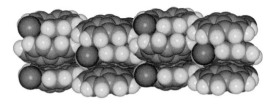

Fig. 14.11. The best $P2_1/c$ crystal structure for chlorocorannulene. The molecule has the shape of a baseball glove: "up" and "down" columns alternate with partial stacking of the calottes. Chlorine is the large grey sphere.

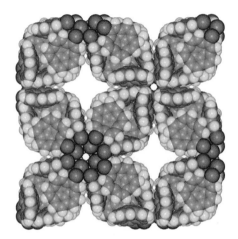

Fig. 14.12. A molecular decoration: a computer-generated pseudo-crystal structure for chlorocorannulene in space group $Pn\bar{3}n$, with 48 molecules in the unit cell.

Fig. 14.13. The asymmetric dimer in the $P\bar{1}$, $Z = 4$ computer-generated crystal structure for chlorocorannulene.

Fig. 14.14. The best $P2_1/c$ crystal structure for dicarbonyl-corannulene, **3**. In spite of the slightly different viewing, the structure is similar to that of the chloroderivative in Fig. 14.11, with alternating up and down molecular columns. Inter-column contacts involve favorable coulombic interactions between oxygens and hydrogens (dark and white calottes, respectively).

crystal structure, Fig. 8.3. Indeed, the data in Fig. 14.10 and Table 14.6 are a computer confirmation of Kitaigorodski's close-packing principle.

The second test molecule is a dicarbonyl derivative of corannulene, **3**, whose molecular structure was constructed from standard bond distances and angles. Atomic point charges were as for chlorocorannulene, except for a +0.30 and −0.35 electron charge on carbonyl carbon and oxygen, respectively. The results of a search in $P2_1/c$ were quite similar to those for the chloro derivative, and the crystal packing is essentially the same (compare Fig. 14.11 with Fig. 14.14); the molecules arrange themselves in columns, as before, but the polar carbonyl substituted molecule draws a little extra crystal stability by contacts between positively charged hydrogen regions and negatively charged carbonyl regions. There is little doubt that crystal structures similar to those of the chloro-derivative could be found in all space groups for this compound too, perhaps even more easily thanks to the small but significant cohesion enhancement due to larger coulombic terms.

The third test molecule is the carboxylic acid of corannulene, **4**. The–COOH group has standard geometric dimensions, but has no point charges, because the coulombic energy over the hydrogen bond is incorporated in the parameterization of the UNI force field and adding a separate coulombic term results in large overestimations of the lattice energy. Figure 14.15 shows the crystal energy landscape for this hydrogen-bonding molecule; the density of points is lower than for the chloroderivative, Fig. 14.10, as a consequence of the higher structural selectivity due to the presence of the hydrogen bond. The detailed results of the structure search in some common space groups are shown in Table 14.7.

The dispersion-enhanced cohesive force due to stacking of aromatic rings competes with hydrogen bonding: in fact the upper cluster in Fig. 14.15 corresponds to crystal structures that do not form a hydrogen bond. The lattice energies are in this case systematically lower by about 20 kJ mol^{-1} than lattice energies of hydrogen-bonded structures. Figure 14.16 shows that an optimal packing is reached when stacking goes hand in hand with hydrogen bonding: all structures shown in this figure form a cyclic dimer hydrogen bond, but the $P\bar{1}$ structure has little stacking; the $P2_1$ structure has ring stacking with all ring columns equioriented; the $P2_12_12_1$ structure has ring

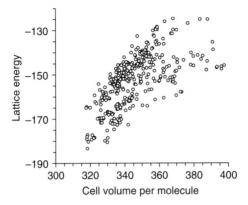

Fig. 14.15. Lattice energies of computational crystal structures for corannulene carboxylic acid. Each point in the graph corresponds to a separately optimized crystal structure. Volumes in Å^3 and energies in kJ mol^{-1}. Structures in the upper cluster with energies above -165 kJ mol^{-1} are closely packed with aromatic ring stacking but are without hydrogen bonds.

Table 14.7 Crystal structure generation by the Prom procedure for corannulene carboxylic acid. See Table 14.6 for the meaning of the headers. Complete crystal data are collected in the Supplementary material, file *coohcor.oeh*

Space group[a]	No. of raw structures	First sorting	Final sorting	Best energy[b]	Packing coeff.
$P2_1/c$	7686	413	133	-183.3	0.788
$P2_12_12_1$	9985	379	34	-180.3	0.786
$P2_1$	3106	464	34	-178.4	0.778
$P\bar{1}$	2700	1055	151	-171.7	0.769

stacking with columns of alternating molecular concavity; finally, the $P2_1/c$ structure has the best opportunities for the accommodation of neighbor columns with alternating concavity and slight rotation of the columns along their axis, favoring further interdigitation and close packing. The ordering of lattice energies and packing coefficients for the best structure in each space group reflects these structural effects. The energy range is anyway only about 5% of the total lattice energy, quite acceptable for coexisting polymorphs, recall Fig. 14.2.

14.4 Crystal structure prediction by computer

It has been shown in the preceding section that it is very easy indeed to generate many, a great many crystal structures for a given organic compound. The Prom procedure is

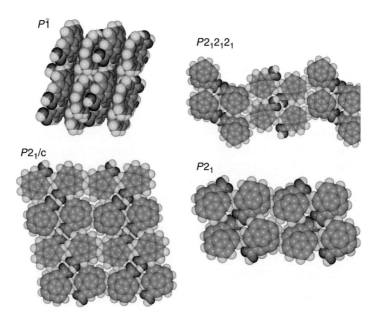

Fig. 14.16. Packing diagrams for computational crystal structures of corannulene carboxylic acid, **4**. In the $P\bar{1}$ structure the hydrogen bond cannot be seen; in the other diagrams, oxygen atoms are darker spheres.

but one among many, and may not be the most efficient one [9]. These computational polymorphs are appealing in all respects, with quite reasonable packing energies, packing coefficients and densities, often if not always forming packing patterns that show a tempting similarity to the ones found in experimental crystal structures of related compounds. Human desire is difficult to quench: the hope arose that crystal structure generation could turn into crystal structure prediction, and that the list of cell parameters, atomic coordinates, and space group for the structure with the most stabilizing lattice energy among those generated by computer could help crystal chemists to dispense with the much more expensive synthetic and X-ray diffraction experiments. It is not known whether executives of companies that produce X-ray diffraction equipment ever worried, but this author has a personal reminiscence of a Coeditor of *Acta Crystallographica* worrying, half seriously, that these methods could be used to forge or at least to doctor up some experimental X-ray diffraction data.

Consider the computational crystal structures generated for chlorocorannulene, from the most stable $P2_1/c$ one, fully abiding by the rules of close packing and having the expected good manners for an organic crystal structure, down to the most unorthodox $Pn\bar{3}n$ structure, with its ridiculous packing coefficient of about 0.3. Do these crystal structures exist? If by "existing" one means that one can hold the actual crystal in one's hand in ordinary laboratory conditions, then the answer ranges from "yes, with small probability" for the most stable structures down to an unequivocal "no" for the unstable ones. And yet each and every one of these structures has its place in phase space, and its internal energy is the configurational average over that space, while its entropy is unequivocally connected with the size of phase space that can be swept at a certain temperature. The phase space basins could be well separated, or some structures could "spill over" into different basins on changing temperature, pressure, and heaven knows what other experimental conditions. If one had the perfect quantum mechanical molecular potential and a computer so powerful that one could play the dynamics of each structure for an infinite time, the free energy of each structure could be calculated, and the structure would, thermodynamically speaking, "exist" in the sense that it could find its place in the phase diagram of the compound and wait there for an experimenter ingenious enough to reach it.

Put this way, the problem of crystal structure prediction is poorly defined – in the worst sense, because any computer-generated crystal structure would be a legitimate successful prediction. The case must be reduced to a subset that, while preserving the integrity of first principles, may be meaningful and tractable in the real world and produce some results that can be comfortably checked against experiment [14]. The desirable features of a computational machine for crystal structure prediction will be now analyzed, with their thermodynamic underpinnings and practical aspects.

14.4.1 *The aims*

They are here formulated in decreasing order of feasibility.

1. Reproduce the complete phase diagram of a substance and its phase equilibria as a function of temperature and pressure, allowing full prediction and control of melting, mesogenicity, and polymorphism.
2. Reproduce the most stable part of the phase diagram and some phase equilibria at room temperature and pressure.
3. Reliably calculate some isolated points in the phase diagram, to predict all observable polymorphs at room temperature and pressure and from all crystallization conditions (by sublimation, from melt, and from solution with different solvents), ranking them in order of stability and chance of appearance.
4. Predict the polymorph that is most likely to appear, and usually appears at the most common conditions of temperature, pressure, and crystallization.
5. Suggest a range of 1 to 5 possible polymorph structures without rank of stability or likelihood.
6. Sample the crystal structure and energy landscape, to establish the basic recognition modes of the compound, to acquire an overall impression of its packing abilities, and to estimate its crystal density and lattice energy.

The community of computational crystallographers and theoretical chemists has fully accomplished task (6), is well on the way to a sensible answer to the query of task (5) at least for rigid molecules, and has had only very limited success for task (4). Tasks from (4) upwards are still uncharted land for real organic molecules, although there have been explorations in the simulation of phase diagrams using simple monoatomic fluids, or n-alkanes represented by just n interaction sites, and similar small pseudomolecules (Section 13.10).

14.4.2 *The tools*

A structure prediction procedure should:

1. include a robust routine for surveying as much as possible of structural space, including at least all intramolecular degrees of freedom, the twenty most populated space groups (consideration of all 230 space groups is almost certainly a waste of time), the possible interactions between point-group and space-group symmetry ($Z' < 1$) for mirror planes, twofold axes and center of symmetry, and possible plurimolecular asymmetric units ($Z > 1$); a dynamic or Monte Carlo routine for enhancing the surveying capabilities and for cross-checking the possible connections among generated structures would be welcome;
2. use an intermolecular potential with the highest possible accuracy for the lowest price: in the search stage, the potential energy scheme should require less than 5 seconds for the calculation of one complete lattice energy, including optimization, on a standard present day PC, while in later stages the computing times could go up to perhaps as much as 5 minutes per optimization or even to 30 minutes for a single lattice energy calculation in the very final ranking stages (these time estimates can be updated in the future by scaling with new computer power enhancement factors); at present, atom–atom potentials qualify for all tasks, the Pixel method qualifies for the latter, and periodic orbital quantum chemical methods are inapplicable, unless some parametric contribution is added to represent correlation effects (Section 6.2);
3. apply a decision protocol for ranking the relative chances of appearance of the generated polymorphs; this may include a consideration of lattice energies, lattice vibration frequencies, zero point energies, vibrational entropies, elastic moduli, morphology, growth rate, the presence or absence of cleavage planes, and other criteria, either of general applicability or peculiar to the case at hand.

14.4.3 *Are crystal structures predictable?*

The perspective of crystal structure generation and prediction by computer has been extensively reviewed over the years [15]. Needless to say, the selection criteria at point (3) above are the real stumbling block of the whole crystal structure prediction business. Consider Table 14.8: the order of relative energy easily changes on changing the method for the calculation of the potential. The ease with which so many crystal structures are generated for any given molecule and the impossibility of crystal structure prediction based on total lattice energies alone are different faces of the same physical fact, namely, the small differences in energy between polymorphs.

Table 14.8 UNI force field, Williams force field, and Pixel lattice energies for computational crystal structures

Space group	Best energy and rank (UNI)		Best energy and rank (Williams)		Best energy rank (Pixel)	
chlorocorannulene						
$P2_1/c$	−148	1	−136.9	3	−145	1
$P2_1$	−144	2	−131.2	6	−141	3
$P\bar{1}$	−142	3	−137.0	2	−143	2
$Pbcn$	−141	4	−137.1	1	−137	6
$P2_12_12_1$	−141	5	−131.8	5	−138	5
$Pca2_1$	−139	6	−127.1	10	−140	4
Cc	−140	7	−128.7	8	−136	7
$Pna2_1$	−139	8	−121.9	16	−135	8
$P1$	−135	9	−133.4	4	−129	11
$C2/c$	−135	10	−124.0	12	−127	15
$C2$	−134	11	−124.5	11	−125	16
$Pnma$	−134	12	−128.3	9	−127	14
$Pbca$	−133	13	−123.7	13	−131	10
$P4_1$	−133	14	−122.7	14	−132	9
$P3_1$	−132	15	−120.8	17	−128	12
$P2_12_12$	−131	16	−129.2	7	−128	13
$P6_1$	−130	17	−122.6	15	−123	17
corannulene carboxylic acid						
$P2_1/c$	−183	1	—		−177	1
$P2_12_12_1$	−180	2	—		−167	2
$P2_1$	−178	3	—		−164	3
$P\bar{1}$	−172	4	—		−154	4

In a landscape with many nearby valleys of very similar depth, kinetics is likely to play a dominant role. No method so far has even dared approach the kinetic problem, or the other lurking issues of equilibrium defects, crystal domains, mosaicity, and internal texture, all factors that are likely to generate enthalpic and entropic fluctuations that may largely outweigh those arising from differences in overall crystal structures or in lattice vibrations. Differences in lattice-vibrational entropy (Section 6.3) are likely to be small, and $T\Delta S$ terms are of the same absolute magnitude as the ΔH terms. Besides, enthalpy–entropy compensation is at work, because less dense crystals have less stabilizing cohesion enthalpies but also softer intermolecular vibrations, hence more available low-energy vibration modes, more energy dispersion, and more stabilizing entropic terms, so that the ΔGs are small or even negligible. The usual thermodynamic prescription of minimum free energy can hardly be applied to predict polymorph appearance.

As discussed in several places in this chapter, when predicting the crystal structures of flexible molecules, intra- and intermolecular energies must be optimized together, an added difficulty because the size of structural space to be searched increases by orders of magnitude, and because it is very difficult to develop suitably parameterized force fields, accounting at the same time for the strong valence bonding energies and for the weak intermolecular interaction energies. Restricting the prediction exercise to rigid molecules cuts out 90% of the real chemical world.

The influence of the solvent is also crucial in crystallization from solution, the most common crystallization technique in practice, but, as discussed in Section 13.10, there are practically no instances in the literature where this problem has appropriately been dealt with using dynamic simulation on realistic systems – not just the Lennard-Jones fluid. It is probably appropriate to say that if real progress in predicting the appearance of crystal polymorphs is desired, the current effort in improving intermolecular potentials and search algorithms for comparing static, final crystal structures should be redirected to the study of pre-crystallization states, and to the improvement of dynamic methods for the simulation of the kinetics of crystal nucleation and growth.

A related question is, does the chemical community really need to predict crystal structures in complete and detailed fashion, including space group, exact cell dimensions, and the final positional parameters of all atoms? This is certainly a formidable intellectual challenge, one that has a special fascination in fundamental research, but in a realistic perspective one must be satisfied that although some partial success can be obtained for some special, favourable cases, a general answer for all molecules of real chemical interest is not within reach and will not be in the near future. But consider points (5) and (6) of the goal list in the former section; they are easily within reach, and there is a wealth of chemical information on the crystal energy and structure landscape that is infallibly and cheaply generated by the computer, if one cares to look, and if one is prepared to abandon the almost obsessive target of competing with X-ray diffraction [16]. Lattice energy and crystal density are predicted to great accuracy because the values for the eventually observed crystal structure cannot be too different from those of the computational polymorphs. Approximate bulk moduli and other elastic properties can be estimated from potential energy calculations (see Section 11.5). A fair estimate of attachment energies (Section 13.9.1), even if the crystal structure is wrong, can be obtained to simulate dissolution behavior. And even if the complete detailed crystal structure is not known, a study of all the generated structures shows the basic packing modes for the molecule under study: in our present example, the corannulene core is definitely seen to require stacking in columns, witness the almost ubiquitous presence of 3.8 or 7.6 Å cell axes; columns prefer an up–down alternation, and benefit also from rotational adjustment to enhance interlocking. One also knows what will not appear in the crystal: in this example, T-shaped molecular dimers with aromatic rings at high interplanar angles appear only in structures with two molecules in the asymmetric unit. The largest contribution to the stabilizing cohesive energy of the crystal comes from dispersion due to the rich coating of polarizable electrons in the large aromatic core, and coulombic-polarization terms play a minor role in the bulk energy but possibly a relevant role in adjusting the inter-column arrangement. The introduction of a hydrogen-bonding group does not destroy this tendency to dispersion-dominated stacking (Table 14.9), because, due to

Table 14.9 The partition of Pixel lattice energies (kJ mol^{-1}) for the $P2_1/c$ crystal structures of the three substituted corannules

	E_{coul}	E_{pol}	E_{disp}	E_{rep}	E_{tot}
chlorocorannulene	−38	−22	−217	131	−145
corannulene biscarbonyl	−48	−25	−191	120	−144
corannulene acid	−98	−56	−199	199	−177

the favorable flat and circular shape of the molecule, the stacking requirement and the hydrogen bonding requirement can be satisfied at the same time.

Crystal structures can indeed be predicted, if the word is taken in a sensible meaning. Compare this situation with astrophysics: we know mass, age, distribution, speed, and spectral properties of galaxies, although we do not (and we do not need to) know the absolute position of each star in a galaxy. In a similar fashion, computational crystallography can easily predict the essential parameters of the bulk texture of organic matter in the solid state, although it cannot tell the exact position of each atom in a crystal structure.

References and Notes to Chapter 14

[1] The fact that the lattice energies of crystal polymorphs of rigid molecules are very similar was established by computation (Gavezzotti, A. Generation of Possible Crystal Structures from the Molecular Structure for Low-Polarity Organic Compounds, *J. Am. Chem. Soc.*1991, **113**, 4622–4629).

[2] (a) Bernstein, J. *Polymorphism in Molecular Crystals*, 2002, Oxford University Press, Oxford. (b) Threlfall, T. L. Analysis of organic polymorphs, a review, *Analyst*, 1995, **120**, 2435–2460. (c) Gavezzotti, A.; Filippini, G. Polymorphic forms of organic crystals at room conditions: thermodynamic and structural implications, *J. Am. Chem. Soc.* 1995, **117**, 12299–12305.

[3] See *passim* ref. [2a], especially in the introductory chapter.

[4] The famous organic crystal polymorph school at the University of Innsbruck started with L. and A. Kofler, continued with M.Kuhnert-Brandstaetter and A. Burger, and is now led by Ulrich Griesser.

[5] There is presently great concern about polymorph patents (see ref. [2a]). To this author, blissfully unaware of legal intricacies, trying to patent a different crystal polymorph for the same chemical compound appears to be like trying to patent a different package for the same drug. True, this author is also to a large extent unaware of technical details of organic synthesis, purification and final testing and production details.

[6] The most popular space group for organics, $P2_1/c$, unfortunately lends itself to numerous different settings. In $P2_1/c$, $P2_1/a$ and $P2_1/n$ the unique (screw direction) axis is b, but in the first setting the glide plane direction is along c, in

the second it is along *a*, and in the third it is along the *ac* diagonal. Sometimes the choice is dictated by the need to keep the β angle as close to 90° as possible. When other settings are used (unfortunately, they are or have been used) it is necessary to specify in detail which axis is the unique one with a complicated conventional notation (e.g. $P112_1/b$ means that the screw axis is along *c* and is perpendicular to the glide direction along *b*).

[7] Yu, L.; Stephenson, G. A.; Mitchell, C. A.; Bunnell, C. A.; Snorek, S. V.; Bowyer, J. J.; Borchardt, T. R.; Stowell, J. G.; Byrn, S. R. *J. Am. Chem. Soc.* 2000, **122**, 585. The case of six concomitant polymorphs is quite uncommon, if it makes a hit paper in a major journal. On the other hand, structural variation by formation of binary crystals is far more common: Bingham, A. L.; Hughes, D. S.; Hursthouse, M. B.; Lancaster, R. W.; Tavener, S.; Threlfall, T. L. Over one hundred solvates of sulfathiazole, *Chem. Comm.* 2001, 603–604.

[8] The fathers of *de novo* crystal structure generation and prediction were Leiserowitz and Hagler (Leiserowitz, L.; Hagler, A. T. The generation of crystal structures of primary amides, *Proc. Roy. Soc. London* 1983, **A388**, 133–175). They used no computer, relying only on their excellent knowledge of crystallography. Their conceptual procedure is in some places similar to the Prom sequential structure construction.

[9] For a complete overview of methods see the three papers resulting from a blindfold contest on crystal structure prediction organized by the Cambridge Crystallographic Data Center: Lommerse, J. P. M.; Motherwell, W. D. S.; Ammon, H. L.; Dunitz, J. D.; Gavezzotti, A.; Hofmann, D. W. M.; Leusen, F. J. J.; Mooij, W. T. M.; Price, S. L.; Schweizer B.; Schmidt, M. U.; van Eijck, B. P.; Verwer, P.; Williams, D. E. *Acta Crystallogr.* 2000, **B57**, 697; Motherwell, W. D. S.; Ammon, H. L.; Dunitz, J. D.; Dzyabchenko, A.; Erk , P.; Gavezzotti, A.; Hofmann, D. W. M.; Leusen, F. J. J.; Lommerse, J. P. M.; Mooij, W. T. M.; Price, S. L.; Scheraga, H.; Schweizer B.; Schmidt, M. U.; van Eijck, B. P.; Verwer, P.; Williams, D. E. *Acta Crystallogr.* 2002, **B58**, 647; Day, G. M.; Motherwell, W. D. S.; Ammon, H.; Boerrigter, S. X. M.; Della Valle, R. G.; Venuti, E.; Dzyabchenko, A.; Dunitz, J. D.; Schweizer, B.; van Eijck, B. P.; Erk, P.; Facelli, J. C.; Bazterra, V. E.; Ferraro, M. B.; Hofmann, D. W. M.; Leusen, F. J. J.; Liang, C.; Pantelides, C. C.; Karamertzanis, P. G.; Price, S. L.; Lewis, T. C.; Nowell, H.; Torrisi, A.; Scheraga, H. A.; Arnautova, Y. A.; Schmidt, M. U.; Verwer, P. *Acta Crystallogr.*, 2005, **B61**, 511.

[10] A technical, but by no means trivial operation that has to be carried out at each stage is moving the origin of coordinates from a local molecular or cluster reference frame to the appropriate location for compliance with crystallographic standards for space group construction; for example, in centrosymmetric space groups the origin must invariably be moved to the center of symmetry. For full detail on the possible combinations of two or three operators, the reader is directed to the Supplementary Material, Section "Documentation", document on the Promany module.

[11] Program PARST, written by M.Nardelli (University of Parma) is used.

[12] Gavezzotti, A.; Filippini, G. Computer prediction of organic crystal structures using partial X-ray diffraction data, *J. Am. Chem. Soc.* 1996, **118**, 7153–7157; Gavezzotti, A; Filippini, G. Crystal packings and lattice energies of polythienyls: calculations and predictions. *Synthetic Metals* 1991, **40**, 257–266; Braga, D.; Grepioni, F.; Sabatino, P.; Gavezzotti, A. Molecular organization in crystalline [$CO_2(CO)_8$] and [$Fe_2(CO)_9$] and a search for alternative packings for [$CO_2(CO)_8$], *J. Chem. Soc. Dalton Trans.* 1992, 1185–1191; Filippini G; Gavezzotti A. The crystal structure of 1,3,5-triamino-2,4,6-trinitrobenzene. Centrosymmetric or non-centrosymmetric? *Chem. Phys. Letters* 1994, **231**, 86–92; Gavezzotti, A. Polymorphism of 7-dimethylaminocyclopenta(c)coumarin: packing analysis and generation of trial structures *Acta Cryst.* 1996, **B52**, 201–208; Gavezzotti, A.; Filippini G.; Kroon, J.; van Eijck B. P.; Klewinghaus, P. The crystal polimorphism of tetrolic acid: molecular dynamics study of precursors in solution, and a crystal structure generation, *Chem. Eur. J.* 1997, **3**, 893–899; Gavezzotti, A. Computer simulations of organic solids and their liquid state precursors, *J. Chem. Soc. Faraday Disc.*, 1997, **106**, 63–77; Filippini, G; Gavezzotti, A; and Novoa J. Modeling the crystal structure of 2-hydro nitronyl nitroxide radical (HNN): observed and computer-generated polymorphs, *Acta Cryst.* 1999, **B55**, 543–553; Dunitz, J. D.; Filippini, G.; and Gavezzotti, A. Molecular shape and crystal packing: a study of $C_{12}H_{12}$ isomers, real and imaginary, *Helv. Chim. Acta* 2000, **83**, 2317–2335; Boese, R.; Kirchner, M. T.; Dunitz, J. D.; Filippini, G.; Gavezzotti, A. Solid-state behavior of the dichlorobenzenes: actual, semi-virtual and virtual crystallography, *Helv. Chim. Acta* 2001, **84**, 1561–1577; Destri, S.; Pasini, M.; Porzio, W.; Gavezzotti, A.; Filippini, G. X-ray diffraction studies and computer simulations of the crystal and molecular structure of 2,5, di-(9,9-dimethylfluoren-2-yl)-3,4-dih*e*xyl-thiophene-1,1-dioxide, a photoluminescent material, *Cryst. Growth Des.* 2003, **3**, 257–262.

[13] A molecular model was constructed from the parent hydrocarbon, substituting one hydrogen atom with chlorine. The lattice energies were calculated by the UNI atom–atom potentials; to compensate for the minor part of coulombic energy which is missing in that formulation, in addition, explicit coulombic terms were calculated using approximate point charge parameters assigned to nuclear positions, with the usual C–H charge separation of about 0.10 electrons, a small positive charge on carbon atoms not carrying hydrogen, and a negative charge of −0.10 electrons on the chlorine atom.

[14] In a famous comment line, the editor of *Nature*, J. Maddox, said it is a "continuing scandal" that "it remains in general impossible to predict the structure of even the simplest solids from a knowledge of their chemical composition" (*Nature*, 1988, **335**, 201). Characteristically, for the editor of such a flamboyant journal, he did not care to distinguish between metallic, inorganic, and organic materials, a crucial divide in real science. Maddox's comment became a citation classic and generated a further, somewhat confusing debate (Cohen, M. L. Novel materials from theory, *Nature* 1989, **338**, 291–292; Hawthorne, F. C. Crystals from first principles, *Nature* 1990, **345**, 297; Ball, P. Scandal of crystal design ..., *Nature*

1996, **381**, 648–650; de Rosier, D. J. Who needs crystals anyway? *Nature* 1997, **386**, 26–27). Characteristically, for *Nature*, science is mixed with tinges of journalist hype. When the results of the careful and systematic studies of ref. [9] were sent to *Nature*, they were considered too down to earth and boring for the readership of the journal, and rejected without even editorial consideration.

[15] Gavezzotti, A. Are crystal structures predictable? *Acc. Chem. Res.* 1994, **27**, 309–314; Dunitz, J. D. Are crystal structure predictable? *Chem. Comm. Focus Article* 2003, 545–548; Verwer, P.; Leusen, F. J. J. Computer simulation to predict possible crystal polymorphs, in *Reviews in Computational Chemistry*, 1998, edited by Lipkovitz, K. B. and Boyd, D. B., Wiley, New York, Vol. 12, Chapter 7; Gavezzotti, A. Methods and current trends in the simulation and prediction of organic crystal structures, in Can crystal structures be predicted? *Nova Acta Leopoldina*, NF 1999, **79**, 33–46; Gavezzotti, A.; Filippini, G. Self-organization of small organic molecules in liquids, solutions and crystals: static and dynamic calculations, *Chem. Comm.* Feature Article 1998, 287–294; Beyer, T.; Lewis, T.; Price, S. L. Which organic crystal structures are predictable by lattice energy minimisation? *CrystEngComm.* 2001, **44**, 1–35; Gavezzotti, A. Ten years of experience in polymorph prediction: what next? *Cryst. Eng. Comm.* 2002, **4**, 343–347; Day, G. M.; Chisholm, J.; Shan, N.; Motherwell, W. D. S.; Jones, W. Cryst. An assessment of lattice energy minimization for the prediction of molecular organic crystal structures, *Cryst. Growth Des.* 2004, **4**, 1327–1340; Dunitz, J. D.; Scheraga, H. A. Exercises in prognostication: crystal structures and protein folding, *Proc. Natl. Acad. Sci.* 2004, **101**, 14309–14311; McArdle,P.; Gilligan, K.; Cunningham, D.; Drak, R.; Mahon, M. A method for the prediction of the crystal structure of ionic organic compounds, *CrystEngComm.* 2004, **6**, 303–309. For accessory criteria in judging computer-generated crystal structures see for example Coombes, D. S.; Catlow, C. R. A.; Gale, J. D.; Rohl, A. L.; Price, S. L. Calculation of attachment energies and relative volume growth rates as an aid to polymorph prediction, *Cryst. Growth Des.* 2005, **5**, 879-885. The discussions in the papers mentioned in ref. [9] are also helpful and illuminating.

[16] The bottom line from the present status of the blind tests conducted on crystal structure prediction [9] is that there are now a few positive results for rigid molecules, as compared with none only ten years ago; but also that the results have sometimes a scientifically unpleasant taste of haphazardness. Hits and misses come and go on differences of fractions of a $kJ\,mol^{-1}$; certainly, as the result of chance rather than of accuracy.

15

Epilogue: A theory of crystallization?

– We had difficulties in simulating these molecular systems.
– I have a suggestion for a way out of these difficulties…
– Yes? What is it?
– Stop studying molecular systems.
Dialog at question time in a computational crystallography meeting.

15.1 Laws and theories

At the beginning of the twenty-first century, chemistry is a well developed and powerful science. Organic chemistry proceeds by well established practical rules with solid, predictable, and reproducible results, supplying the world with a myriad of new chemicals. Instrumental analytical chemistry can spot a single molecule in a crowd as large as the population of Italy, so that cyclists and runners can no longer conceal the use of illegal drugs, and customers are unnecessarily bothered because nanodoses of pollutants in mineral water, harmless as they may be, can now be detected. Biological chemistry can track the fate of millions of bioactive compounds, and medicinal chemistry can, sometimes with success, interfere with those fates, sometimes even for the benefit of patients. Polymer chemistry can synthesize just anything and is almost anywhere, flooding the planet with insane gadgets, but it is also well on its way to replacing biologically made, vital parts of the human body.

A law is a form of knowledge that predicts future facts on the basis of experience gained on similar facts gleaned from the past. Laws are applied using mathematical or experimental methods. A theory is an independent conception of the human spirit that proceeds from first principles and produces master equations to regulate the facts. No wonder, there have been many laws but very few theories in the history of science. Chemical thermodynamics is a collection of laws, gravitation and quantum mechanics are theories. In common parlance the two terms are often interchanged, and "theory" is taken to mean more or less anything that does not have to do with experiment; one may speak for example of a theory of chemical reactivity, instead of, more properly, of the laws of chemical reactivity within the framework of the quantum chemical theory; or even of the "Pixel theory" (Section 12.2.1), although one should more properly speak of the Pixel method. The only truly independent theory of chemistry is quantum mechanics, which is very seldom (if ever) called into action, because of its lack of practical applicability. Most of the time, we write the Hamiltonian and the Schrödinger equations for the system under study just to state that the former cannot be used in practice and the latter cannot be solved. Even theoretical chemistry is in fact a dilute version of quantum mechanics, simmered in molecular orbital or density functional approximations. Methods and techniques in chemistry proceed not from theory, but from laws derived from experience and found to apply to very wide classes of facts and phenomena.

Even within the less ambitious category of laws, there is a distinction between cognitive and predictive aspects: the former are mere categorizations of facts, the latter are what gives value to a law, allowing the prediction or control of the outcome of future experiments. A first-rate law has a high degree of cognitive and predictive power, suffering few or no exceptions; an example is the first law of thermodynamics. A second-rate law has good categorization power, but little predictive ability; an example is the close-packing principle in crystals, a necessary but not sufficient condition for the appearance of a given crystal structure. A third-rate law has little cognitive and little predictive power; an example is Ostwald's rule of stages, a virtual law more often violated than observed. At present, in molecular crystallization there are no first-rate laws, very few second-rate laws, and many third-rate laws. The first-rate laws of thermodynamics become in such a context third-rate laws because the concept of phase is ill-defined in most if not all of the transformations involved in the evolution from a disperse molecular system to a molecular aggregate. There is a wide gap between the ever increasing ease with which the aggregation and crystallization phenomenon can be studied, thanks to calorimetry, X-ray diffraction, NMR, atomic force microscopies, molecular simulation (Chapter 13), and the degree of understanding and control that may be gained from these experiments. If even laws are weak, the way to a theory seems even more problematic. The title of this chapter, including a question mark, and its very shortness, testify to this.

15.2 Aggregation stages

In the dynamic part of the aggregation process, a number of molecules proceed from a disperse structure in which the time- or space-average distance between molecular centers of mass is large with respect to the molecular dimension, and forces between molecules are very weak or null, to a cohesive structure in which molecular centers of mass are in as close a contact as is compatible with the molecular size and shape, and the forces between molecules are strong enough to prevent them from going back to the disperse structure. This happens in solidification from vapor, or in solute nucleation and growth from solution, but not in crystal nucleation from the melt. The following discussion will deal almost exclusively with nucleation and growth from solution, and to ordinary laboratory conditions for typical non-ionic organic molecules where the intermolecular forces, coulombic, dispersive, or hydrogen-bonding, produce crystals whose melting temperature is of the order of 70–150°. Reference is made to real chemical objects, like molecules with a mass of 200 amu and two or three conformational degrees of freedom, as opposed to model entities like Ar atoms or the Lennard-Jones fluid. These are by far the most common working conditions in ordinary organic chemistry.

It is convenient to distinguish some categories, depending on the number of molecules, N_{agg}, that compose one aggregate structure.

15.2.1 Oligomers

The first class is that of oligomers, aggregates in which N_{agg} is of the order of 2 to 10^2. These oligomers must by definition be without crystalline structural periodicity. There may be some partial ordering, as there is for example in fatty acid micelles, but there is no geometrical capacity for the construction of a real crystal precursor, because of the sheer size factor. These entities have no bulk structure because they have no bulk, and must be fluxional, or, if one wants, liquid-like, although they need not even show all the features of the equilibrium structure of the bulk liquid. In the limit of very small N_{agg}, of the order of a few units, one recognizes what has been called a growth unit: "the essential building block that transfers structural information from the solution to the crystal surface" [1]. This concept holds better for rather strong intermolecular bonding, and a particularly important case is the formation of hydrogen-bonded units with $N_{agg} = 2$, where the dimerization energies are of the order of 20–60 kJ mol^{-1} (Table 12.11). The preceding chapters have shown examples where these aggregates have been seen through the eyes of spectroscopic or computational methods.

Oligomers of sizeable organic molecules have no distinct shape or surface of their own, because the continuing rearrangements and exchanges of molecules with the surroundings put the intrinsic molecular shape in competition with the aggregate shape. In the literature, these species are sometimes called fractal nuclei, or just fractals [2]. Their energy is strongly influenced by the surrounding medium, especially when it consists of other molecules, as in a solution, but also when it is in a vacuum. When speaking of the surface free energy for such molecules and in such a size regime, one is probably alluding to some empirical quantity that may roughly describe any deviation from ideal, homogeneous thermodynamic behavior. For the same reasons, any phenomenological model based on the shape or size of the particles is at best parametric.

15.2.2 Nanoparticles

A second class may be called nanoparticles, with N_{agg} of the order of 10^2 to 10^4. Their size is sub-microscopic; Fig. 15.1 shows that a particle with 10^4 medium-size organic molecules is still an order of magnitude too small to be seen, even assuming a generous microscope resolution of 0.5 μm. In nanoparticles, the distinction between bulk and surface components becomes significant. Consider as a very simple model a spherical aggregate in which N_{agg} spherical molecules are uniformly distributed between a spherical core and surface layer of thickness of 1.1 times the average molecular diameter. Assuming a packing coefficient of 0.6, intermediate between

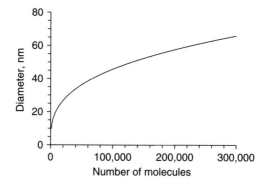

Fig. 15.1. The diameter of a particle composed of organic molecules of medium size (molecular volume 300 Å^3) as a function of the number of molecules. The diameter has been estimated as the diameter of a sphere of volume equal to molecular volume times number of particles, allowing for an estimated packing factor of 0.6.

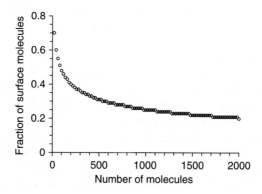

Fig. 15.2. Fraction of surface molecules in a spherical cluster, estimated as described in the text and in the caption to Fig. 15.1.

solid and liquid, the volumes of the two zones and their population can be calculated; the results are shown in Fig. 15.2. The core of a molecular aggregate contains the majority of members only when there are at least 500 molecules in the aggregate. Most likely, there is some difference between the molecular organization at the surface and in the core of the nanoparticle.

Nanoparticles can conceivably develop in different ways: (1) simultaneous coalescence of all molecules into the aggregate, a very unlikely process that would dispense with oligomers; (2) growth from oligomers by molecular additions; and (3) coalescence of oligomers. In addition, each of the aggregation entities present in solution may interact with the solvent molecules, S. Consider the corresponding expressions for the formal equilibrium constants between a solute species A and a solvent species S:

$$\text{Simultaneous} \quad nA \leftrightarrow A_n \quad K = [A_n \text{ clusters}]/[A]^n \quad (15.1)$$

$$\text{Addition} \quad A_n + A \leftrightarrow A_{n+1} \quad K = [A_{n+1} \text{ clusters}]/[A_n \text{ clusters}][A] \quad (15.2)$$

$$\text{Coalescence} \quad A_n + A_m \leftrightarrow A_{n+m} \quad K = [A_{n+m} \text{ clusters}]/$$
$$[A_n \text{ clusters}][A_m \text{ clusters}] \quad (15.3)$$

$$\text{Solvation} \quad A + S \leftrightarrow AS \quad K = [AS]/[A][S] \quad (15.4)$$

$$A_n + S \leftrightarrow A_n S \quad K = [A_n S]/[A_n \text{ clusters}][S] \quad (15.5)$$

For $n = 2$, equation 15.1 describes the fundamental association equilibrium that can be reliably studied by spectroscopic methods in the gas phase or in solution, as well as by quantum mechanical or approximate simulation methods as described in Chapter 12, and provides a basic understanding of the strength of the intermolecular association. For $2 < n < 5$, the equilibrium described by equation 15.1 is well documented, as has been shown in Chapter 13, but that equation is clearly unrealistic for n-values in excess of a few units, and such equilibria cannot even be imagined for n of the order of 10^2–10^3. Equations 15.2 and 15.3, assuming that the symbol [ ] represents a "concentration" or number of clusters per unit volume, describe conceivable equilibrium situations even for very large n, at least at low supersaturations. In particular, equation 15.3 with n large and $m = 2$ is a very likely occurrence for those molecular species that form stable dimers in solution, as has been demonstrated, for example, for glycine [1]. Note that for large n, equations 15.2, 15.3, and 15.5 correspond to the interaction of single molecules or of oligomers with the surface layer of the nanoparticle, almost certainly a more isotropic, less specific, and faster process than addition of a molecule to a crystalline surface template.

The core of larger molecular aggregates may well have (but has no thermodynamic or kinetic obligation to have) a crystalline structure. Most likely, the very first time period of the newborn nucleus is spent in a liquid-like form, which then evolves into the crystalline periodic symmetrical structure by some process that must be similar to nucleation from the melt. The actual lifetime of these liquid-like states may be short on a human timescale, say 10^{-3} seconds, which is, however, eternity on a molecular timescale, allowing for millions of molecular moves in a viscous regime, a vigorous scan of phase space. Whether those liquid-like aggregates "exist" is, again, a question with a large dialectic component; there is no technique that allows their experimental documentation, and the proposition must rely upon the results of molecular dynamics simulations, which invariably point to the almost inevitable conclusion that the time spent by the aggregate in a liquid-like structure may be short but cannot be zero. Otherwise, one must assume that all molecules are already pre-oriented for optimum docking onto the forming substrate – in violation of the laws of molecular dynamics and of common sense.

If the above scheme indeed holds in crystal nucleation, the conclusion is that a mother solution does not contain crystalline nuclei of all possible crystal polymorphs, but contains the mother of all polymorphs, a fluxional droplet ready to evolve in several different crystalline directions. The mechanism of this process could be a mirror image of surface melting; while in surface melting the isotropic surface layer rushes in toward complete melting, in crystallization within the nanoparticle the molecular ordering of the inner core successively templates the growing outer layer.

15.2.3 *Mesoparticles*

A further class of aggregation comprises what may be called mesoparticles, with $10^4 < N_{agg} < 10^7$. At the upper limit of this size domain, the volume of such particles is approximately 0.2 μm, and they become almost visible, but even crystalline ones may not yet have a well developed morphology. This size domain could well be the turning point: either crystallization is bound to occur, in which case it does; or crystallization will never occur, in which case the particle stays liquid-like and grows into an oily or waxy product, as many practicing organic chemists know all too well. Very little is known about this stage, which is crucial for the prediction of crystal structure, and nothing certain is known about the reasons why, for the same molecular size, some organic compounds crystallize readily and others do not, or at least they require long induction times or a substantial increase in supersaturation (recall that supersaturation is defined as the amount of concentration in excess of equilibrium solubility). For those particles that do crystallize, evolution to one of the different possible polymorphs is in part determined by intrinsic molecular recognition factors, that is, average predominance of certain pairing modes within the pure solute part of the system; but also by a host of local, transient, and scarcely predictable, let alone controllable, conditions: particle size, supersaturation, temperature, impurities, total volume of the vessel, interaction with the solvent at the particle surface, or even long-range polarity of the solvent. Another likely turning point along the path to crystallization is a rapid change from monodisperse to a frankly heterogeneous character of the suspension: larger particles have a higher chance of attracting new molecules because of their larger surface, so the accretion process becomes autocatalytic, with larger clusters growing at the expense of smaller ones:

$$A_n + A_m \leftrightarrow A_n + A_{m-1} + A \leftrightarrow A_{n+1} + A_{m-1} \quad \text{for } n \gg m \qquad (15.6)$$

15.3 Macroscopic crystals

Growth units larger than mesoparticles certainly must be able to develop translational correlation, that is, to grow into macroscopic crystals with periodic symmetry. At this point the surface of the aggregate becomes ordered very much as the inner part is, specific crystal faces develop anisotropically, and the crystal grows by attachment of new molecules to the template provided by crystal surfaces. Owing to the conceptual problems associated with the definition of crystal faces (recall Fig. 13.10) for large organic molecules with complex shapes the traditional distinction between "flat" and "rough" surfaces becomes equivocal because the shape of the constituent molecular object itself can be to some extent "rough", i.e. rich in cavities and kink attachment points.

The growth of a macroscopic crystal conceivably proceeds via the traditional stages of (1) adsorption, that is attachment of an incoming molecule to the crystal surface in a random position, (2) surface migration, that is a sort of tumbling diffusional motion of the adsorbed molecule on the surface, and (3) final docking, or the arrival of the newly crystallized molecule at its proper place in the periodically symmetric substrate. In this picture, the cohesive energy is by definition less stabilizing in stages (1) and (2) than in stage (3). However, if the cohesive power at a defective site is smaller than at the proper place but is still competitive with it, one may have growth "mistakes" with a number of possible consequences, including local defects (the least damage), or more extensive misfits like stacking faults, dislocations, or twinning. In a very alarming scenario, at least as far as crystal structure prediction and control are concerned, those defective structures are enthalpy – or entropy stabilizing for the macroscopic aggregate, and thus play a key role in the determination of the final crystal form. If that is the case, as it certainly is, for example, in intermetallic compounds and alloys, then any simulation model based exclusively on long-range order is of limited use or may even be misleading. Fortunately, the complex shape of organic molecules may also be a bonus here because its directing power is sometimes superior to that of more spherical and isotropic objects.

The crystal form that finally appears may well win its race at growth stage, since it is conceivable that the final crystal crop may proceed at the expense of smaller, more sluggish crystallites of other polymorphic forms.

15.4 Thermodynamics, kinetics, and symmetry

Chapter 13 in this book has a survey of current methods in the study of crystal nucleation and growth. Section 13.4, in particular, illustrates some of the concepts and results of classical nucleation theory. As already discussed there, equilibrium thermodynamics has little to say for the crucial stages of aggregation of complex organic molecules. Consider for example equations 15.1 – 15.5; are these real equilibrium conditions, in the sense of classical thermodynamics? As discussed in Section 7.5, writing an equilibrium constant implies a definition of activity of each of the components in the mixture of the evolving system, with respect to the activity of the pure component. This is relatively easy when one is dealing, for example, with well-specified molecular species in a homogeneous reacting medium; but one wonders what is really meant by the symbol "[A_n clusters]" in equations 15.1–15.5, and how one can define the necessary activities and reference states for evanescent nuclei composed of a fluxional number of particles. If the definition of equilibrium constants is problematic, so is the definition of the corresponding free energy differences. Expressions like equation 13.2 and the ones that stem from it are thus bound to be phenomenological approaches in which the thermodynamic functions are in fact disguised parameters [3]. The concepts of stability and metastability also lose much of their definition, and might well be substituted by expressions like "not prone to evolution" or "ready to evolve".

If thermodynamics is inaccessible, one might think of using pure phenomenological kinetics, by writing and solving rate equations on the forward and backward reactions implied in equations 15.1 – 15.6 [4], possibly also including the influence of inhibitors, heterogeneous components, etc., on the way to a complete kinetic account of the "soup" sketched in Fig. 13.11. Kinetic equations would by necessity be of a very simple form, including velocity terms with Arrhenius-type rate constants and first-order dependence on pseudo-concentrations of the participating species. It is obviously impossible to take into account each species appearing for each value of n, so these models require the introduction of a rather small number of more or less arbitrary mesoscopic species. As could easily have been predicted on considering the complexity of the system, one ends up with extremely complicated differential equations whose solution is impossible without drastic approximations. Besides, even these hardly gained solutions cast little light on the molecular aspects of the phenomenon and, more important, depend in a critical way on the assumptions made for the distribution of cluster sizes. The predictions of such kinetic treatments can hardly be verified experimentally, at least in the case of the very small and transient nuclei formed during organic crystallization.

The basic difficulties in dealing with a system like a molecular solution on the verge of crystallization are: (1) the events at each microscopic site within the system are random, in the sense that they depend on a large number of unpredictable local boundary conditions, and are thus not correlated, either in time or in space, with the events at other microscopic sites; (2) at some stage, some of these events trigger an autocatalytic evolution, when the right path to crystallization has been entered; and (3) the critical entities, the nuclei, are short-lived and very small on a human size and time scale, so that our present experimental techniques are hardly applicable. On the theoretical side, the key to the difficulty in describing (and even worse, predicting) the recognition modes to be adopted in the eventual crystallization step lies in the numbers given in Table 12.11: organic molecules all have a number of possibilities in the gray zone of recognition modes between 5 and 20 $kJ\,mol^{-1}$, allowing the crystallizing molecules to more or less freely roam the recognition energy landscape for a small fee in enthalpy. Besides, if the final process is autocatalytic, then all models and predictions based on enthalpy and entropy may well be overthrown. Even rather unstable (enthalpy) or highly improbable (entropy) structures may be propagated, being the species that, once formed even in very small proportions, validate and sustain their own existence.

In this respect, an issue strictly connected with the possible autocatalytic behavior is symmetry with translational periodicity. It is at first sight amazing that an ultra-fast

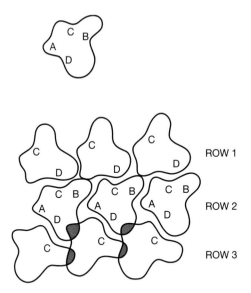

Fig. 15.3. A molecular object with prospective interaction pairs A...B and C...D. Row 2: satisfies A...B match and sets the seed of translational symmetry. Row 1 to row 2: satisfies C...D match with a rotation of the molecules. Row 3: tries its best to satisfy again C...D match, with another molecular rotation, at the expense of some frustration (overlapping, repulsive areas, shaded).

stochastic process like crystallization may produce a fully symmetric periodic structure, hence the awe inspired by the observation that the large majority of solids are crystalline. On closer analysis, however, this occurrence is less mysterious or phenomenal than it might seem. Consider Fig. 15.3. As temperature decreases, a molecule finds itself more and more short of kinetic energy and its spectrum of aggregation choices becomes restricted. Let A ... B or C ... D be pairs of molecular sites with complementary potential for intermolecular recognition (either complementary features on the molecular electric potential, or favorable alignment for charge transfer, or just lock-and-key steric compatibility, etc.) When a molecule finds a locking position by matching its A-site with the B-site of another molecule, there are always the B-site of the first molecule and the A-site of the second molecule available for repeating the same exercise, thus propagating the structural motif (row 2, Fig. 15.3); the intramolecular conformation also adjusts to these translational requirements. This mechanism is at the root of translational symmetry.

In a further step, molecules build a second motif in another direction in space, by matching sites C with D sites (rows 1–2, Fig. 15.3). This may require an arbitrary rotation of the whole molecule and/or further conformational adjustment. But the recognition motifs are always to some extent malleable, recall the estimated vibrational amplitudes in Table 12.11, and the whole construction adjusts itself so as to reach the best compromise between the requirements of the A–B and of the C–D matches. The C...D match may be satisfied in different ways, for example as shown in the contact between row 2 and row 3 of Fig. 15.3, and then the same matching process may be repeated in other directions. Aggregates like the one shown in Fig. 15.3 are thus in a partially structured, pre-symmetric state, ready for transition into a fully symmetric periodic structure by small collective adjustments. When this happens along one of the successful routes, the further evolution of the crystalline nucleus becomes autocatalytic. The adoption of long-range overall symmetry is at this stage indispensable, because it helps to optimize the propagation ability of all matches, and it is a powerful enthalpy-advantageous process that favors close packing at almost no further expense in entropy.

Matching modes along A...B, C...D, etc., have different interaction energies. The molecular pairs joined by stronger attraction will of course be the first candidates for matching, but frustration may occur at any point along this process because the energy gain from a given match may be overruled by a number of concomitant, opposing factors. The energy loss due to repulsion in other parts of the molecules (recall the destabilizing repulsive contacts between O...O, O...N moieties, etc., in Table 12.11) must be avoided. Some A...B matches may be discouraged by loss of close packing. A less obvious source of frustration is that some matches can be too stabilizing, that is, they exhaust all the potential for intermolecular "valence" of the molecule and the dimer formed by that match cannot propagate into a periodic structure. Otherwise, in some cases there may be many matching points of very nearly the same strength, molecular shape and electrostatic potential being flat and featureless. This is frequently the case with organic molecules.

A crystallizing system is then an ensemble of molecules in a high-density liquid state (in the case of crystallization from solution, the inner part of a nano- or mesoparticle) in which a number of instantaneous multiple recognitions occur, until one of the combinations picks up enough evolutionary momentum to become autocatalytic, that is, to propagate itself into a macroscopic crystal. The system is thus funneled through several evolution traps in a sequence of moves constituting a highly guided pathway through phase space. The mechanism is an inextricable mixture of kinetics and energetics; the crucial autocatalytic step can be moderately downhill or moderately uphill in energy – moderately, because a number of prescriptions cannot be violated, the most obvious ones being the avoidance of repulsion and the necessity of close packing. At the final stages, the number of choices is relatively small, because the combination of productive modes becomes more and more restricted; this number is small on a molecular scale, if one considers the full extension of phase space, but may be still too large for a consistent crystal structure prediction, as computational experiments show (Section 14.4).

Depending on molecular features and on local factors, the mechanism described above opens the way to a range of events:

1. compounds that always crystallize in the same polymorph: one propagation mechanism predominates;

2. compounds with temperature- or pressure-dependent polymorphism: temperature and pressure influence the relative importance of different matching modes and the final propagation mechanism;
3. compounds with concomitant polymorphism: several matching and propagation mechanisms coexist at the same temperature and pressure, and the system has a potential for allowing different decisional events by different local conditions within its boundaries;
4. compounds that do not crystallize: either no productive matching and propagation mechanism exists, or too many mechanisms are competing and the system oscillates between many possible paths without developing enough decisional momentum;
5. supercooled liquids: some matching modes are so strong that they cannot be readily overcome by the available kinetic energy, and the autocatalytic relaxation to a symmetric structure is difficult;
6. crystals with two molecules in the asymmetric unit: either the asymmetric match between the two molecules is very strong, so that the aggregation unit must be the asymmetric dimer, or the match is not particularly strong, but the asymmetry helps in adjusting the final structure when the autocatalytic periodicization occurs;

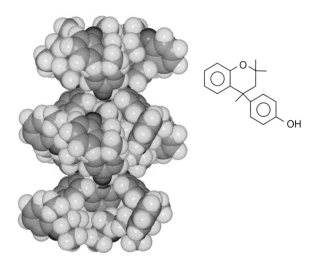

Fig. 15.4. A striking molecular liaison motif in the crystal structure of Dianin's compound (CSD refcode QQQESP01, see ref. [5]). In the upper row, three molecules stick upwards and are bound by O–H···O hydrogen bonds to three inversion-center related molecules sticking downwards, forming an hourglass-shaped structure held together by a cycle of six hydrogen bonds. The motif is repeated by translation downwards, so that the upward-pointing molecules of the lower hourglass interdigitate with the downward-pointing molecules of the upper hourglass, forming a cage that may include several guest molecules. The threefold axis is vertical in this picture.

7. solvates or clathrates: the system cannot proceed to the primary phase separation process that leads to aggregates of pure solute molecules, and the process of matching, selection and autocatalytic evolution must involve matches between the solute and the solvent molecule; the considerations at point (6) above also apply to the heteromolecular solute–solvent dimer.

Remarkably, the above mechanisms are not incompatible with the bewildering experimental observation that one and the same polymorph is invariably obtained for a simple nearly spherical molecule like benzene and that a consistent aggregation obtains for such a complicated molecule as Dianin's compound, whose crystal structure (Fig. 15.4) is a delicate Venetian lace involving a cyclic sextet of O–H$\cdots$O hydrogen bonds and a hourglass-shaped cage, with 18 symmetry-related molecules in the unit cell [5]. The structure of the benzene crystal is not the outcome of the most favorable molecule–molecule contact, but of the one and only favorable path toward symmetry periodicization. The complexity of the shape and potential of the Dianin molecule must be a source of structural information rather than of structural randomization.

15.5 The language of the theory

The mechanisms sketched out in the preceding section are little more than a collection of rough qualitative working hypotheses, the clay in which a scientific law or theory is molded. The analysis may however be a worthy one if it helps to sort out a number of turning points that deserve further study, or if it can discourage some conceptual approaches that are not likely to lead to better theorization.

The analysis conducted in this chapter, along with the perspective offered in Chapter 14, strongly discourages the attempt to develop a predictive theory of crystallization of organic molecules solely on the basis of classical nucleation theory or of phenomenological reaction kinetics. Atomic level consideration is necessary, but, given the extreme variability of shapes in organic molecules, a theory based on a few general principles seems too ambitious a goal. More realistically, the aim should be a collection of laws and methods that include a large amount of structural analysis and computer simulation.

There can be little arguing about the fact that the primary event in crystal formation for an organic molecule of even a moderately complex shape and electrostatic potential is the binary matching of molecular recognition sites. In this respect, one might think of going back to Table 12.11, or perhaps to an even richer collection of dimerization energies, and to rank the possible preconceived bimolecular recognition modes for predicting the final crystal structure(s). Some of these "synthons" have also been postulated on geometrical grounds without regard to quantitative energetic considerations [6]. The obvious objection is that even if the prediction of strong interaction between pairs of molecules can be successful (this is the case in practice only for the two or three strongest hydrogen bonds worth more than 30 kJ mol^{-1}, see Table 12.11), one still has a long way to go to predict the further aggregation and propagation modes of these minuscule aggregates, modes that depend on weaker interactions in the 10–20 kJ mol^{-1} aggregation pool – the previously mentioned "gray zone" of molecular recognition, clearly borne out also by the analysis of structure determinants (Section 14.2.3). If it is sometimes helpful when considering closely related series of compounds, or when used by experienced hands with due consideration of boundary conditions, the general use of the synthon approach is problematic, because too many synthons are competitive, and because the overall stability of the system depends on extensive compromise among three-dimensional matches and frustration of periodicization.

The static approach to computer crystal structure prediction, in which many candidate structures are generated and compared with respect to lattice energy or other properties of the fully developed crystal (Section 14.4) is a more worthy one, but suffers from the lack of kinetic insight. Clearly, in the final crystal structure no trace is left of the preceding autocatalytic step, the one that actually traced the path to that particular structure, irrespective of small lattice energy fluctuations. Thus, the highest prediction stage that is accessible to static approaches is the production of a number of plausible crystal structures out of which the final kinetic choice will be made. Correspondingly, even experimental methods that are able to accurately measure free energy differences among observed crystal structures are not predictive, because the crystal structure with the lowest free energy is the most stable one in a platonic sense and is by no means bound to appear or to predominate in actual crystallization experiments.

Real organic molecules are conformationally flexible, and this fact does make a difference with inorganic covalent or ionic crystals. Any theory of organic nucleation and growth must come to grips with the fact that the aggregating unit changes its shape and electrostatic potential while crystallizing, further stressing the need for a dynamic treatment. It can be safely stated that no simulation method will even come close to the solution of the crystal structure prediction problem (point (3) in Section 14.4.1) without including some dynamics, but it is also clear that straightforward molecular dynamics simulation is not the proper answer, at least until a million-fold increase in computing power is available (see Section 13.10). Smart sweeping of phase space is a promising approach [7], but the most valuable addition to the present arsenal of tools for a viable approach to robust crystal structure prediction would be some biased, "smart move" dynamic or evolutionary simulation method that might take appropriately into account the autocatalytic nature of the growing molecular clusters.

References and Notes to Chapter 15

[1] Davey, R. J.; Allen, K.; Blagden, N.; Cross, W. I.; Lieberman, H. F.; Quayle, M. J.; Righini, S.; Seton, L.; Tiddy, G. J. T. Crystal engineering – nucleation, the key step, *CrystEngComm* 2002, **4**, 257–264.

[2] "If interactions between monomers are nonspecific…long times will be required for proper alignment of the molecules to ordered structures. Therefore, random

aggregation (fractal or amorphous precipitate formation) will be energetically favored more than nucleation via coalescence. Nuclei may then form via aimed restructuring of the fractal clusters." Umbach, P.; Georgalis, Y.; Saenger, W. Time-resolved small-angle static light scattering on lysozyme during nucleation and growth, *J. Am. Chem. Soc.* 1998, **120**, 2382–2390. "A line of recent theories and simulations have suggested that the nucleation of protein crystals might, under certain conditions, proceed in two steps: the formation of a droplet of a dense liquid, metastable with respect to the crystalline state, followed by ordering within this droplet to produce a crystal". Vekilov, P. G. Dense liquid precursor for the nucleation of ordered solid phases from solution, *Cryst. Growth Des.* 2004, **4**, 671–685. These quotations, however, belong in protein crystallization, where molecular mechanisms may be different from those at work in "small" molecule aggregation.

[3] "We will restrict ourselves to the limiting case where an equilibrium state exists on the surface between the monomers and the g-mers. Note that this state does not actually exist in any real process, but this approximation allows us to employ the tools of thermodynamics in our analysis." Liu, X. Y. Generic progressive heterogeneous processes in nucleation, *Langmuir* 2000, **16**, 7337–7345.

[4] See, e.g., the formulation and the references in Wattis, J. A. D.; Coveney, P. V. Mesoscopic models of nucleation and growth processes: a challenge to experiment, *Phys. Chem. Chem. Phys.* 1999, **1**, 2163–2176. This paper draws conclusions from comparisons with experiments for barium sulfate. For an example with organic materials see Burnham, A. K.; Weese, R. K.; Weeks, B. L. A distributed activation energy model of thermodynamically inhibited nucleation and growth reactions and its application to the β–δ phase transition of HMX, *J. Phys. Chem. B* 2004, **108**, 19432–19441.

[5] Imashiro, F.; Yoshimura, M.; Fujiwara, T. Guest-free Dianin's compound, *Acta Cryst.* 1998, **C54**, 1357–1360. Thanks are due to C. Eckhardt for pointing out this structure.

[6] Desiraju, G. R. Supramolecular synthons in crystal engineering – a new organic synthesis, *Angew. Chem. Int. Ed. Engl.* 1995, **34**, 2311–2327.

[7] Raiteri, P.; Martonak, R.; Parrinello, M. Exploring polymorphism: the case of benzene, *Angew. Chem. Int. Ed. Engl.* 2005, **44**, 3769–3773.

Index

List of abbreviations AIM = Atoms in Molecules theory, DFT = Density Functional Theory, EHT = Extended Hückel Theory, MD = molecular dynamics, QM = quantum mechanics, MC = Monte Carlo.

activity coefficients 184–5, 411
additives, designed for crystal growth 292
amorphous 368
anharmonic behavior 35
antisymmetrization 66, 89, 101
arene-prefluoroarene 323
aspirin "polymorphs" 274, 369–70
asymmetric unit 122, 153, 211, 387
atom-atom model 106–10
   potential 108–9
atomic basins, in AIM 73, 297
   displacement parameters (ADP) 137, 170, 196, 275
   force microscopy (AFM) 355
   orbitals 61, 68
   prejudice, in structural chemistry 139
   radii, see non-bonding radii
Atoms In Molecules (AIM or QTAIM) theory 72, 102, 297–8
aufbau 66
autocatalytic accretion/nucleation 410–14

Bader, R. 72
band gap 158, 285
   structure 158
Bartell, L. S. 39, 47, 83
BASIC language 264
basis-set superposition error (BSSE) 81–2, 158
benzene dimer 310–11
benzoic acid dimer 312
Berkovitch-Yellin, Z. 110
binary notation 254
birefringence 289
bits (computer) 254
Bloch functions 155, 157, 160 (Figure)
   theorem 163, 168
BLYP functional, in DFT 79
Bohr, N. 53
   magneton 64
Boltzmann, L. 174, 181–2, 233
   distribution 174, 177
bond length tables, 8
   critical point, in AIM 73, 297–8
Born-Oppenheimer assumption 68
Boyd, R. H. 39
Bragg, L. 137
   law 132
   "reflections" 138
Bravais lattices 120
Brillouin zone 155, 164
   scattering 284
bugs (computer) 260
bulk modulus 278–81, 282 (Table), 400
Bürgi, H.-B. 198

CADD (computer aided drug design) 41
Cambridge Crystallographic Data Centre (CCDC) 197
   Structural Database (CSD) 196–8, 220, 276, 286, 299, 351, 369, 372
Carnot, S. 181
Casalone, G. 145
CECAM (Centre Europeen de Calcul Atomique et Moleculaire) 47
central processing unit (CPU) 256
charge, atomic 38, 92
   AIM 93
   ESP 43, 93
   Gasteiger-Marsili 49
   Hirshfeld 93
   Mulliken 74, 93
   overview 114
   rescaled EHT 94, 216
   distribution 90
charge-transfer complex 285
chemical bond 4–5, 296
   in AIM 73
   bonding 5, 102
   equilibrium 184
   potential 183–5
chirality 9, 292
chlorobenzene dimer 311–12
chromophore 285
C-H...X interactions 318–19
classical nucleation theory 338, 416
Clausius, R. 179, 181
Clausius-Clapeyron equation 190–1
clearing temperature 346
Clementi, E. 75
closed shell 66
close packing 204–5, 335, 406, 414

INDEX

CNDO (Complete Neglect of Differential Overlap) 75
Cohen, M. 290
collision avoidance procedure, in Pixel 307
color 284–8, 394
compilers (computer) 258
compliance matrix 279
compressibility 279–80
computer program checking 260
   publication and reproducibility 261
condensation level, in Pixel 305, 310–13
conduction band 161, 292
configuration interaction (CI) 77, 158
convergence problems, in lattice sums 212, 314
coordinates
   cartesian 7, 31
   internal 7, 31
   normal 31, 44
      crystal vibrational 164
   polar 56
   symmetry 31
   transformation 43–5
correlation energy 115
   matrix 223
   functions, in MD 241–2, 336–7
counterpoise method 82
critical nucleus 339, 341
crystal
   coordination sphere 371–2, 376–8, 390
   defective/disordered 368
   density, estimated 397, 400
   dissolution 273
   Einstein 340
   growth 347, 351, 354, 411
      by sublimation 341
      MD simulation 357
      mistakes 411
      unit 407
   lattice 122
   morphology 292, 351–4
   nucleation 347
      from the melt 338, 409
      frustration 414
      homogeneous 339
      kinetics 412–16
      NMR study 350
      of ice 341
      rate 339, 349, 357
      X-ray scattering and diffraction 350
   orbital method 157
   packing and color 287
   polymorph 191, 196, 274, 349
      computational 384, 396, 399 (Table)
      concomitant 370
      coordination sphere 376–8
      crystallite competition 411
      "crystallographic" 378
      definition 367–9
      density difference 372
      energy differences 211, 382–3
      Innsbruck School 369
      nucleation 415
      packing coefficients 372
      patents 401
      real 372, 375
      semi 374
      transitions 368
   plastic 333, 368
   proper 368
   solution interface 353 (Figure)
   structure
      acceptable 225
      determinants 376–7
      fingerprints 375
      prediction 211, 395–401
         blind tests 404
         decision protocol 398
         static approach 417
   surface 348, 352 (Figure)
   systems 121–6
Curtin, D. Y. 290

DeBroglie, L. 53
   relationship 56, 171
Debye equation 135, 141
Debye-Waller factor 137
decimal representation 255
deformation density 93–4
density matrix 71
   functional theory (DFT) 78–81, 105
   of states, electronic 162
      vibrational 167
Destro, R. 145
Dewar, M. J. S. 75
dielectric properties, estimation, in MD 247
diffraction intensity 137
dipole moment 91, 222
dipole-dipole energy 345
Dirac, P. A. M. 63
direct methods 137, 148
dispersion 89, 99–101
distance distribution function (DDF) 299–304
distributed multipole analysis (DMA) 92
donor-acceptor 99, 288 (Figure)
Drude model 115
Dunitz, J. 198
dynamic variables 54

eigenfunction/eigenvalue 55
Einstein, A. 53
electric conductivity 284, 292
   field 98, 210
      in crystals 307
electron correlation 68, 99, 304

INDEX

density, in QM 71
  X-ray 139
  micrography 355
  spectroscopy 163
electrostatic potential 72, 90, 98
energy
  activation, viscous flow 280
  attachment 213, 274, 354, 400
  Coulomb 38, 92–6, 305
    of hexachlorobenzene crystal 313
  dispersion 100, 307
  electronic (atom) 62
  equipartition 178, 210, 233
  from QM exchange 101, 115
  expectation value, in QM 57
  intermolecular, landscape 326, 394, 397, 400
  internal 173, 178, 233
  kinetic 232, 333–4
    operator, in QM 55
  lattice 207–13, 217, 221 (Figure), 313, 325
    acceptable 225
    accuracy 217, 380
    comparison in different force fields 399 (Table)
    estimated 397, 400
  LCAO-MO 71
  minimization methods 38
  molecule-molecule 325, 375
  packing (PE) 209
  packing potential (PPE) 208
  polarization 98, 103, 209–10, 306
  repulsion 101–3, 308
  rotational, in QM 58
  strain 36
  translational, in QM 57
  units 105, 116
  vibrational, in QM 60
  vibrational potential 33, 35, 231
  zero-point 100, 210
enthalpy 9, 331
  difference between polymorphs 211
  of dissolution 274
  of melting 188 Fig, 190 (Table), 333, 367
  of sublimation 109, 192 (Table), 193 (Figure), 201–2, 217, 221 (Table), 325, 332
  of vaporization 243, 245
enthalpy-entropy compensation 399
entropy 9, 173, 180–3, 331
  of melting 188 (Figure), 190 (Table), 332–3
  of sublimation 213–14
equations of motion, in MD 232
equilibrium constant 185–6
  of association 409
  solid-liquid 332
  thermodynamics and nucleation 411
  vapor-liquid 341
  vapor-solid 341
equipartition principle 178

equivalent positions 125, 153
exchange integral 71
  correlation functional, in DFT 79
expectation value, in QM 56
Extended Hückel Theory (EHT) 75, 286

Fermi, E. 330
Fermi level 158
Filippini, G. 145
fluxional droplet, in nucleation 410
Fock matrix 71
Fock-Roothaan equations 67–71
  periodic 157
force 232
  constants 33–5, 34 (Table)
    intermolecular 326 (Table)
  Ehrenfest 297
  Feynman 297
  field 37–43
    compilation 43
    for crystals 214
    GROMOS 42
    OPLS 43, 244–5
    UFF 42, 50
    UNI 214, 215 (Table), 325, 385, 394
    Urey-Bradley 37
    water 250
    Williams 216 (Table), 385
Fortran 258, 265
Fourier synthesis 139, 149
free energy
  Gibbs 184
  Helmholtz 184, 247
  of activation, in nucleation 332, 339
  of dissolution 274
  perturbation method 248
  simulations, liquid-crystal equilibrium 340
  surface 332, 339, 407
  thermodynamic integration method 248, 340
frequency doubling 289
Fukui, K. 76

Galileo, 27
gaussian basis function 67
GAUSSIAN package 76, 309
Gay-Berne potential 345–6
glasses 342, 368
G-matrix method 44
Gramaccioli, C. M. 145
Gulliver, L. 327

Hamiltonian 54
  operator, in QM 54, 67, 406
hardware (computer) 256
harmonic approximation 33
  oscillator, in QM 60

potential 34
Hartree-Fock limit 77
  method 67–71, 314
heat
  capacity 173, 179–80, 186–8, 277, 331
    crystal vibrational 167
    in glass transition 344 (Figure)
    MD estimate 244
  of formation 39,191
  of combustion 39
  of sublimation, *see* enthalpy of sublimation
Heisenberg, W. 53
Hellman-Feynman theorem 4, 102
Hillary, Sir E.
Hohenberg-Kohn theorem 78
Hoffman, R. 76, 82
Hooke's potential 43, 279
HOMO-LUMO 73, 285–8
host-guest complex 212
hydrogen atom positions, renormalization 199
hydrogen bonding 105–6, 312, 319, 321–3, 389, 394

IBM 258
  1620 computer 45, 145, 255 (Figure)
  7040 computer 147
ideal gas 6, 340
improper dihedrals 40, 232
impurities, in solution 353
  tailor-made 364
inelastic neutron scattering 168
inertial axes 11
"in silico" 273
insulator 161
integrals, in QM 71
interface tension 347
interference 132
intermolecular bond/bonding 299, 302–3, 326
  radii, *see* non-bonding radii
Intermolecular Perturbation Theory (IMPT) 116, 309, 315
International Tables for X-ray Crystallography 125, 387
irreversible process 180
isothermal compressibility 244

jellium 168
Jorgensen, W. L. 43, 244
Joule, J. P. 179, 182, 233

Karplus, M. 41
Kitaigorodski, A. I. 14, 107, 205–6, 394
Körner, W. 362
Kohn, W. 78
Kohn-Sham assumption 78

Kollman, P. 42
Koopman's theorem 73
Kuchitsu, K. 39

lattice dynamics 163, 257, 277
lattice-vibrational frequencies 110, 167
LCAO-MO equations 69, 71 (Table)
Leiserowitz, L. 110, 202
Lennard-Jones fluid 339, 400, 407
Lewis, G. N. 5
lifetime, of liquid-like particles 409
Lindemann criterion 337
liquid crystals 345, 368
Lifson, S. 39
liquids 230
local density approximation (LDA) 79
LUMO 73

machine language (computer) 258
Maddox, J. 403
magnetic properties, 293
Mariani, C. 145
Marsh, R. 145
McCrone, W. 369
mean square displacements, in MD 242
melting temperature 23, 186, 189, 332
  and lattice vacancies 335
  simulation 333
  volume change 24 (Table)
mesoparticles 410
metal, electronic structure 161
Metropolis algorithm 237
Miller indices 137
Møller-Plesset method 77
molecular
  configuration/conformation/constitution 7
  connectivity 4
  dipole 291
  dynamics (MD) 230–6
  mechanics 35–43
  models 30
  orbitals 69
  packing 10
  reorientation 23–4
  self-density 16
  shape 11–12, 20–24, 333
  simulation types 271
  size 11–18
  surface 14–18
  vibrations 30–35
  volume 14–20, 203
moments of inertia 11, 12 (Table), 176
momentum, linear and angular, in QM 54–7
  space 171
monoclinic system 124
Monte Carlo (MC) 236–7

aggregation-volume-bias 357
  Gibbs ensemble 249
  kinetic 356
  smart move 357
  time definition 356
Morokuma analysis 104, 315
Moser, C. 47
mother solution 410
Mugnoli, A. 145, 226
Mulliken, R. S. 73
multipoles, central and distributed 90–2

nanoparticles 407–8
Nardelli, M. 402
neutron scattering, liquids 241
  glasses 342
NMR, experiments 242
  in crystal nucleation 350
  quantum basis 65
  relaxation 336
nodes, of wavefunctions 57
non-bonding radii 13, 14 (Table), 299
non-bonded interactions 37, 232
non-linear susceptibility 289
normal modes 31, 176
  of liquids 294
normalization, of wavefunction 54
Noyes-Nernst dissolution model 213
Noyes-Whitney equation 274
NPT-NVT, in MD 234
nucleation, see crystal nucleation
numerical integration, in MD 232
  in Pixel 315

"oiling out" 363
oligomers 407
open shell, in QM 66
operating system (computer) 257
operator, in QM 54, 55 (Table)
optical modes 165, 170
  properties 289
orbital, see atomic orbital or molecular orbital
order parameter 9, 346
orthogonalization matrix 150, 309
orthorhombic system 126
Ostwald's rule 406
overlap integral 71
  over densities 20, 103, 288, 308
  population 74
  volume 20

$P\bar{1}$ space group 122, 385
P2$_1$/c space group 125, 128, 198, 200, 375, 387, 390
  different settings 401

$P2_12_12_1$ space group 126, 388
packing coefficient 203, 205 Fig, 206 (Table), 371, 391
  acceptable 225
  diagrams 129, 375
  energy, see energy, packing
PAILRED diffractometer 146
pairwise additivity 100, 103, 209
parallelization (computer) 259
parameter fitting, in force fields 40
PARST computer program 402
particle size in crystal nucleation 349, 408 (Figure)
partition function 174–7
Pasteur, L. 269
Paul, I. C. 290
Pauli, W. 63
Pauli exclusion principle 66, 68–9
  "forces" 66
Pauling, L. 5, 148
Peierls distortion 161
periodic bond chain 354
periodic boundary conditions in QM 156, in MD 234, 357
perturbation theory, in QM 77
phonon 170
  dispersion 165
phase
  change/transition 331, 333, 336, 338
  diagram 397
  difference 132
  equilibria 331, 348, 397
  of structure factor 136
  problem in X-ray crystallography 137
  space 231, 271–2, 332, 397, 414
pigments 285
π-interactions 323–4
Pixel (SCDS) method 112, 304–9, 398
  parameters 308
polar axis 289–91
polarizability 97, 46
  atomic 97 (Table), 306 (Table)
  molecular 222
polarization
  electrical 89, 96–7, 246, 289
  light 289
Pollini, M. 330
POLYATOM package 83
polymorphism, see crystal polymorph
Pople, J. 77
population analysis 73, 92
potentials, see force field
powder pattern 141, 378–9
precession camera 144
pre-melting trajectory 335
pressure 173, 233
principal component analysis 223
"Prom" approach, crystal construction 385–90

promolecule 93
propagation velocity, mechanical 280
proximity coefficients, of atoms in crystals 310 (Table)
pseudopotential MO 86

QSAR (quantitative structure-activity relationships) 41
QSPR (quantitative structure-property relationships) 41
quantization, source 57–8
    vectorial 59
Quantum Chemistry Program Exchange (QCPE) 82, 86, 265

radial distribution functions 143, 238–40, 300, 336
    functions, hydrogen-like, in QM 62 (Table)
random numbers 25
Raphson-Newton method 49
real space 135
reciprocal centimeters 48
    space 135, 155
refraction index 289
rescaling, of P and T, in MD 235
reversible process 181
R-factor 137, 197–8, 218, 379
rigid-body approximation 170
rigid molecules 400
    spheres model 91
roughening transition 354

Sagan, C. 251
salts (in crystals) 368
Sayre, D. 148, 264
scanning tunneling microscopy (STM) 356
scattering factor, atomic 133, 134 (Figure), 151 (Table)
    point charges 131
SCDS method, *see* Pixel-SCDS method
SCF approach, in LCAO-MO 70
Scheraga, H. 39
Schmidt, G. M. J. 290
Schrödinger, E. 53
Schrödinger equation 54, 65, 176, 406
    periodic 156
second-harmonic generation 289
secular equation 44, 71
self-diffusion coefficient 242, 280, 339
self-occupation coefficient 21, 205–6 (Table), 371
serendipity 263
SHAKE algorithm 236
shear modulus 279
Simonetta, M. 82, 144, 148
Slater determinant 68

Slodowska, M. 53
soft modes 338
software (computer) 256
    "black-box" 262
solid 368
solubility 213
solvates (in crystals) 368, 416
source code (computer program) 258
space-group frequency 207 (Table)
    matrix notation 129
    multiplication table 129
    polar 213
    symmetry 127–9
Spackman, M. A. 110
spectroscopy, in intermolecular association 349
spheres and caps method 16
spherical harmonics 59 (Table)
spin 63–5
spin-orbital 65
state functions 179
stationary state, in QM 55
steepest-descent method 49
steric interactions 37
    repulsion 102, 298
stiffness coefficients 282 (Table)
    matrix 279
Stone, A. J. 92
stress and strain 278
structure 5
    -activity correlation 270, 330, 367
    correlation 198
    descriptors 6
    factor, X-ray 134–6
Sun, H. 43
supercooling 332, 342, 415
supersaturation 410
supplementary material 199, 201, 294, 370, 385, 389, 391–2, 395, 402
surface melting 337
    flat and rough 348, 411
    molecules, number of 408 (Figure)
    tension energy 281–3
Swift, Jonathan 296
symmetry 8
    group 8, 128–9
    number 175, 333
    operations/operators 121–3, 153, 309, 386
        close-packing 391
        combination 388 (Table)
    translational, origin of 413 (Figure)
symplex method 49, 389
synthons 416–17
systematic absences (extinction) 153

temperature 173, 193, 232, 237
thermal conductivity 277

INDEX

expansion coefficient 182, 244–5, 275, 276 (Figure), 277 (Table)
thermodynamics first principle 179
  second principle 181
  third law 183
thermogram 342, 343 (Figure)
timescale of diffusion 230
  problem, in MD 357
torsion angle 7, 36, 232
trajectory, in MD 232
transferability, of potential parameters 112
triclinic system 122
Trueblood, K. 148

umbrella sampling 341, 358
uncertainty principle 57, 60, 100
unit cell 121, 153
united atom approach, 42, 245

valence band 161
van der Waals radii, *see* non-bonded radii
van Gunsteren, W. 42

van't Hoff equation 190
vapor pressure, solids 213
variational principle 69
vibrational spectroscopy 33
virial 234
virus (computer) 258
viscosity 280
von Laue, M. 130

Walden's rule 333
water 249–50, 253
  freezing simulation 341
wavefunction 54
wave vector 155
Weissenberg camera 144
Williams, D. E. 107, 225

X-ray crystal structure analysis 137

Young modulus 278

DATE DUE

MAY 1 5 2010		
DEC 2 8 2010		

GAYLORD　　　　　　　　PRINTED IN U.S.A.

SCI QD 921 .G38 2007

Gavezzotti, Angelo.

Molecular aggregation